The Handbook of Environmental Chemistry

Founded by Otto Hutzinger

Editors-in-Chief: Damià Barceló • Andrey G. Kostianoy

Volume 21

The Handbook of Environmental Chemistry

Recently Published and Forthcoming Volumes

The Llobregat: The Story of a Polluted Mediterranean River
Volume Editors: S. Sabater, A. Ginebreda, and D. Barceló
Vol. 21, 2012

Emerging Organic Contaminants and Human Health
Volume Editor: D. Barceló
Vol. 20, 2012

Emerging and Priority Pollutants in Rivers: Bringing Science into River Management Plans
Volume Editors: H. Guasch, A. Ginebreda, and A. Geiszinger
Vol. 19, 2012

Global Risk-Based Management of Chemical Additives I: Production, Usage and Environmental Occurrence
Volume Editors: B. Bilitewski, R.M. Darbra, and D. Barceló
Vol. 18, 2012

Polyfluorinated Chemicals and Transformation Products
Volume Editors: T.P. Knepper and F.T. Lange
Vol. 17, 2012

Brominated Flame Retardants
Volume Editors: E. Eljarrat and D. Barceló
Vol. 16, 2011

Effect-Directed Analysis of Complex Environmental Contamination
Volume Editor: W. Brack
Vol. 15, 2011

Waste Water Treatment and Reuse in the Mediterranean Region
Volume Editors: D. Barceló and M. Petrovic
Vol. 14, 2011

The Ebro River Basin
Volume Editors: D. Barceló and M. Petrovic
Vol. 13, 2011

Polymers – Opportunities and Risks II: Sustainability, Product Design and Processing
Volume Editors: P. Eyerer, M. Weller, and C. Hübner
Vol. 12, 2010

Polymers – Opportunities and Risks I: General and Environmental Aspects
Volume Editor: P. Eyerer
Vol. 11, 2010

Chlorinated Paraffins
Volume Editor: J. de Boer
Vol. 10, 2010

Biodegradation of Azo Dyes
Volume Editor: H. Atacag Erkurt
Vol. 9, 2010

Water Scarcity in the Mediterranean: Perspectives Under Global Change
Volume Editors: S. Sabater and D. Barceló
Vol. 8, 2010

The Aral Sea Environment
Volume Editors: A.G. Kostianoy and A.N. Kosarev
Vol. 7, 2010

Alpine Waters
Volume Editor: U. Bundi
Vol. 6, 2010

Transformation Products of Synthetic Chemicals in the Environment
Volume Editor: A.B.A. Boxall
Vol. 2/P, 2009

Contaminated Sediments
Volume Editors: T.A. Kassim and D. Barceló
Vol. 5/T, 2009

The Llobregat

The Story of a Polluted Mediterranean River

Volume Editors: Sergi Sabater · Antoni Ginebreda ·
Damià Barceló

With contributions by

V. Acuña · R. Aguilera · M. Agulló-Barceló · J.-L. Armenter · C. Barata, · D. Barceló · J. Blanco · B. Catllar · E. Custodio · M.S. Díaz-Cruz · E. Eljarrat · M. Farré · D. Fernàndez-Garcia · E. García-Berthou · M.J. García-Galán · A. Ginebreda · D. Guillén · J. Honey-Rosés · J. Jofre · M. Llorca · M. López de Alda · R. López-Serna · F. Lucena · A. Manzano · L. Matia · R. Marcé · J. Martín-Alonso · N. Mastroianni · L. Moragas · A. Munné · I. Muñoz · F. Ortuño · V. Osorio · J.J. Pastor · M. Petrovic · F. Pérez · S. Pérez · Y. Picó · C. Postigo · N. Prat · E. Queralt · M. Rieradevall · M.-E. Rubiano · S. Sabater · X. Sanchez-Vila · C. Solà · M. Terrado · L. Tirapu

Springer

Editors
Prof. Dr. Sergi Sabater
Institute of Aquatic Ecology
University of Girona
and
Catalan Institute of Water Research (ICRA)
Scientific and Technological Park of the University of Girona
Girona, Spain

Prof. Dr. Damià Barceló
Department of Environmental Chemistry
IDAEA-CSIC
Barcelona, Spain
and
Catalan Institute for Water Research (ICRA)
Scientific and Technological Park of the University of Girona
Girona, Spain

Dr. Antoni Ginebreda
Department of Environmental Chemistry
IDAEA-CSIC
Barcelona, Spain

The Handbook of Environmental Chemistry
ISSN 1867-979X ISSN 1616-864X (electronic)
ISBN 978-3-642-30938-0 ISBN 978-3-642-30939-7 (eBook)
DOI 10.1007/978-3-642-30939-7
Springer Heidelberg New York Dordrecht London

Library of Congress Control Number: 2012939136

Printed on acid-free paper

Springer is part of Springer Science+Business Media (www.springer.com)

The Handbook of Environmental Chemistry Also Available Electronically

The Handbook of Environmental Chemistry is included in Springer's eBook package *Earth and Environmental Science*. If a library does not opt for the whole package, the book series may be bought on a subscription basis.

For all customers who have a standing order to the print version of *The Handbook of Environmental Chemistry*, we offer free access to the electronic volumes of the Series published in the current year via SpringerLink. If you do not have access, you can still view the table of contents of each volume and the abstract of each article on SpringerLink (www.springerlink.com/content/110354/).

You will find information about the

- Editorial Board
- Aims and Scope
- Instructions for Authors
- Sample Contribution

at springer.com (www.springer.com/series/698).

All figures submitted in color are published in full color in the electronic version on SpringerLink.

Aims and Scope

Since 1980, *The Handbook of Environmental Chemistry* has provided sound and solid knowledge about environmental topics from a chemical perspective. Presenting a wide spectrum of viewpoints and approaches, the series now covers topics such as local and global changes of natural environment and climate; anthropogenic impact on the environment; water, air and soil pollution; remediation and waste characterization; environmental contaminants; biogeochemistry; geoecology; chemical reactions and processes; chemical and biological transformations as well as physical transport of chemicals in the environment; or environmental modeling. A particular focus of the series lies on methodological advances in environmental analytical chemistry.

Series Preface

With remarkable vision, Prof. Otto Hutzinger initiated *The Handbook of Environmental Chemistry* in 1980 and became the founding Editor-in-Chief. At that time, environmental chemistry was an emerging field, aiming at a complete description of the Earth's environment, encompassing the physical, chemical, biological, and geological transformations of chemical substances occurring on a local as well as a global scale. Environmental chemistry was intended to provide an account of the impact of man's activities on the natural environment by describing observed changes.

While a considerable amount of knowledge has been accumulated over the last three decades, as reflected in the more than 70 volumes of *The Handbook of Environmental Chemistry,* there are still many scientific and policy challenges ahead due to the complexity and interdisciplinary nature of the field. The series will therefore continue to provide compilations of current knowledge. Contributions are written by leading experts with practical experience in their fields. *The Handbook of Environmental Chemistry* grows with the increases in our scientific understanding, and provides a valuable source not only for scientists but also for environmental managers and decision-makers. Today, the series covers a broad range of environmental topics from a chemical perspective, including methodological advances in environmental analytical chemistry.

In recent years, there has been a growing tendency to include subject matter of societal relevance in the broad view of environmental chemistry. Topics include life cycle analysis, environmental management, sustainable development, and socio-economic, legal and even political problems, among others. While these topics are of great importance for the development and acceptance of *The Handbook of Environmental Chemistry,* the publisher and Editors-in-Chief have decided to keep the handbook essentially a source of information on "hard sciences" with a particular emphasis on chemistry, but also covering biology, geology, hydrology and engineering as applied to environmental sciences.

The volumes of the series are written at an advanced level, addressing the needs of both researchers and graduate students, as well as of people outside the field of "pure" chemistry, including those in industry, business, government, research establishments, and public interest groups. It would be very satisfying to see these volumes used as a basis for graduate courses in environmental chemistry. With its high standards of scientific quality and clarity, *The Handbook of*

Environmental Chemistry provides a solid basis from which scientists can share their knowledge on the different aspects of environmental problems, presenting a wide spectrum of viewpoints and approaches.

The Handbook of Environmental Chemistry is available both in print and online via www.springerlink.com/content/110354/. Articles are published online as soon as they have been approved for publication. Authors, Volume Editors and Editors-in-Chief are rewarded by the broad acceptance of *The Handbook of Environmental Chemistry* by the scientific community, from whom suggestions for new topics to the Editors-in-Chief are always very welcome.

Damià Barceló
Andrey G. Kostianoy
Editors-in-Chief

Volume Preface

The Llobregat basin has received the effect of human presence since historical times. Barcelona, one of the most populated cities in the north bank of the Mediterranean, has historically developed using – and avoiding – the river. The Llobregat has provided its water resources and has been at the base of the industrial development of the city and of the whole region. Therefore if something characterizes the story of the Llobregat is the human pressure.

As such, the story of the Llobregat is not different of some others in humanized rivers elsewhere. However, the pressures on the whole watershed are overwhelming. Hydraulic exploitation, water abstraction, channelization, and damming have added to mining exploitation, eutrophication, pollution, and the arrival of invasive species in successive steps of increasing pressures on the ecosystem. These man-made disturbances add to those naturally derived from its Mediterranean character, that is, the irregular and unreliable provision of water resources.

The wide array of pressures, together with the necessity of preserving the river resources, provides the framework to justify the Llobregat as one of the best studied and monitored river basins in Europe. Several groups from different Universities and Research Institutions, both Spanish and European, have placed their efforts to understand the hazards and resilience of such a river system. The basic and applied research in the Llobregat has been immense, and most of the times has been triggered by the water administration remarkably in the context of the WFD development and its subsequent deployment. The knowledge gathered on the Llobregat is a paradigm of collaboration between scientists and managers. Such collaboration historically characterized the management of the most problematic water issues in Catalonia.

As a whole, the proper management of the water cycle – both natural and anthropogenic – in the Llobregat River basin is a cumbersome (but challenging) task, requiring a lot of "fine-tuning" effort. Its successful achievement will be only feasible with the active and collaborative involvement of science, authorities, and stakeholders participation.

Girona and Barcelona, Spain

Sergi Sabater
Antoni Ginebreda
Damià Barceló

Contents

The Llobregat River Basin: A Paradigm of Impaired Rivers Under Climate Change Threats

Rafael Marcé, Jordi Honey-Rosés, Andreu Manzano, Lucas Moragas, Bernardette Catllar, and Sergi Sabater

Abstract The Llobregat River represents a paradigmatic example of an impaired river subject to emerging global change impacts. This chapter provides an introduction to the main geomorphological, geological, climatic, and biological features of the river basin, as well as an overview on the hydrological alterations and the intense management of water resources in the basin. The Llobregat hydrology has experienced a significant runoff reduction during the last decades. This decrease is related to climatic drift but also to the increasing forest land cover that has promoted a 25% reduction of the streamflow. The chapter also describes the human uses of the Llobregat River waters from a historical perspective, with particular emphasis on the difficulties that an intrinsically unpredictable river like Llobregat posed to the different human uses along history. The historical development makes emphasis on the industrial activities affecting water resources during the twentieth century. The chapter includes a detailed analysis of the present situation on water extractions, discharges, and diversions that define the Llobregat River Basin as a deeply impaired ecosystem, especially in the most downstream reaches.

Keywords Climate, Geology, History, Land use change, Water diversions, Water flow trend, Water treatment

R. Marcé (✉) and J. Honey-Rosés
Catalan Institute for Water Research (ICRA), Scientific and Technological Park of the University of Girona, Emili Grahit 101, 17003 Girona, Spain
e-mail: rmarce@icra.cat

A. Manzano, L. Moragas, and B. Catllar
Catalan Water Agency (ACA), Provença 204-208, 08036 Barcelona, Spain

S. Sabater
Catalan Institute for Water Research (ICRA), Scientific and Technological Park of the University of Girona, Emili Grahit 101, 17003 Girona, Spain

Faculty of Sciences, Institute of Aquatic Ecology, University of Girona, Campus Montivili, 17071 Girona, Spain

S. Sabater et al. (eds.), *The Llobregat: The Story of a Polluted Mediterranean River*, Hdb Env Chem (2012) 21: 1–26, DOI 10.1007/698_2012_152,

Contents

1 The Llobregat River Basin

The Llobregat River is located in NE Spain, running for 165 km to drain an area of 4,948 km^2 (Fig. 1). Long-term mean discharge at the mouth is 603 hm^3 $year^{-1}$ or 19 m^3 s^{-1}. It has two main tributaries, the Cardener and Anoia Rivers (Fig. 1). The headwaters of the Llobregat and Cardener Rivers lie in the rather pristine Eastern Pyrenees, but soon downstream they are already impacted by sewage treatment plants and industrial effluents (estimated at 137 hm^3 $year^{-1}$ or 4.3 m^3 s^{-1}), as well as by potash-mining activities in the Cardener reaches just upstream Manresa. The lower course flows through one of the most densely populated areas of the Mediterranean region (Barcelona Metropolitan Area, over 3 million people), and the waters receive large inputs from industry and urban origin. The case of the Anoia River is somehow different: its headwaters lie in an agricultural area, downstream it flows through an industrialized zone (paper mills, tannery, and textile industries), and near the confluence with the Llobregat River the main land use is vineyard fields (Fig. 2).

Water resources from the Llobregat basin are fundamental for the water supply scheme of the Barcelona Metropolitan Area. In fact, due to the intrinsic variability of the Mediterranean climate, three large dams were built in upstream sections of the Cardener and Llobregat Rivers to ensure water supply during low flow periods (Fig. 1). Presently as much as 30% of the long-term mean annual flow is eventually diverted for human uses, with intakes concentrated in the downstream section of the basin. Aquifer withdrawals are also numerous in the deltaic region and play an important role when water quality or quantity of the Llobregat River is insufficient. A new seawater desalination plant with a capacity of 60 hm^3 $year^{-1}$ started operation recently to help overcoming the threats posed by abnormal, high-frequency water scarcity episodes during the last decade.

Due to the ever increasing human pressure by the end of the 1980s the Llobregat River was one of the most polluted and degraded in Western Europe, and the

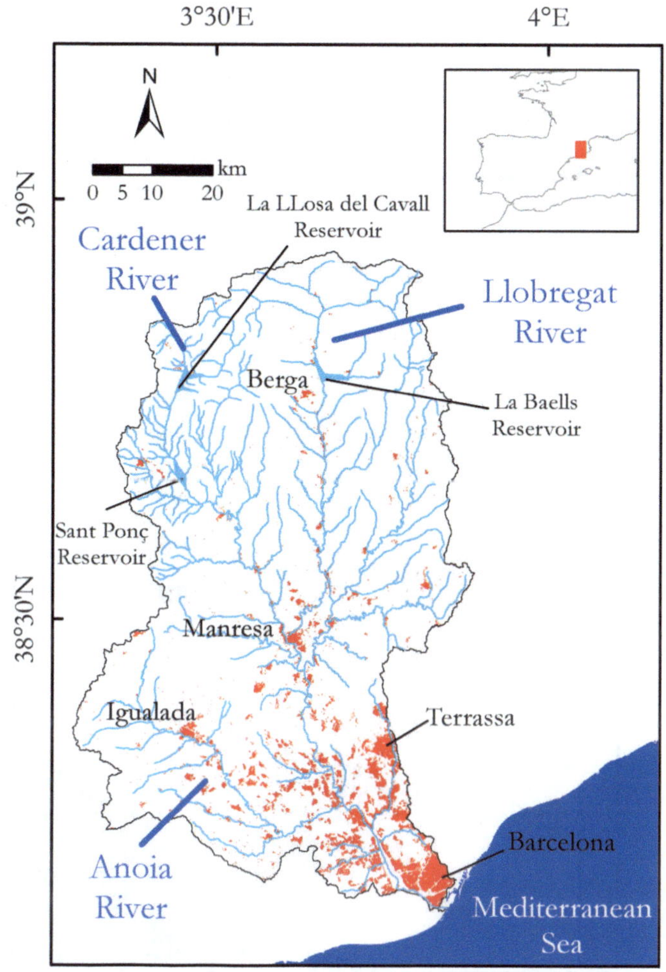

Fig. 1 The Llobregat basin with indication of the main rivercourse and its tributaries, and the location of the three large reservoirs. Urban settlements are also depicted as red areas; only the largest ones are named

overexploitation of the underground water led to salinization of the deltaic aquifer, rendering 30% unusable. Since then, wastewater treatment plants with tertiary facilities have been built along the basin, and the situation has improved dramatically. Nowadays, infrastructures prevent excessive pollution of the river by intercepting specific effluents, such as the channels receiving treated urban wastewater from the Riera de Rubí (near Terrassa) and those collecting brine from the salt-mining sites. Apart from these, there is a major irrigation channel on the right side of the river, the Canal de la Dreta, which provides water for horticulture in the Delta. On the left side of the downstream reaches, the Canal de la Infanta was also built for irrigation purposes, but now its main role is to prevent wastewater from

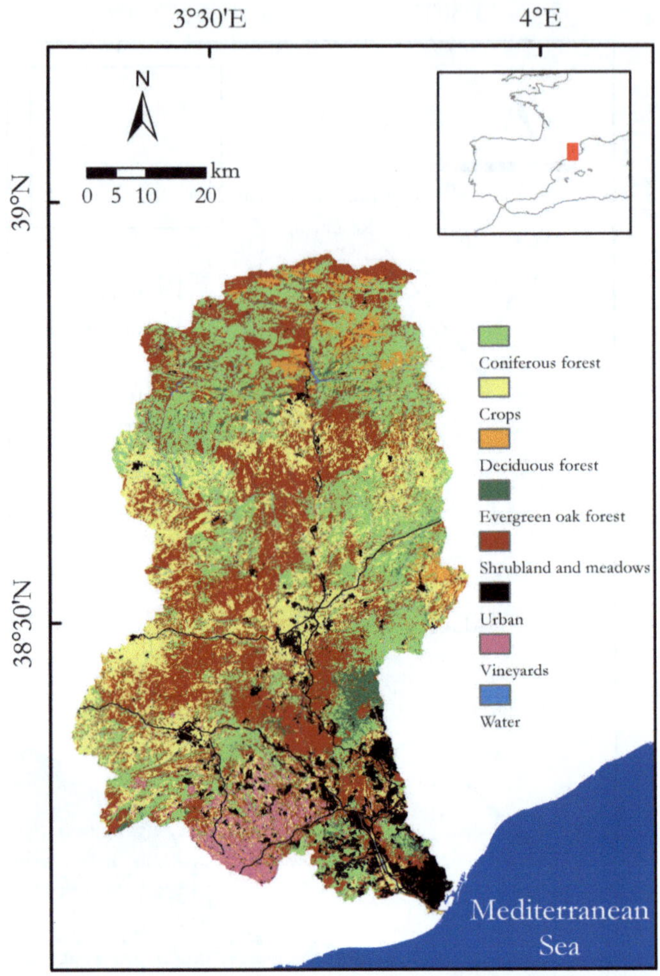

Fig. 2 Land uses distribution in the Llobregat basin. Based on the information gathered by the Thematic Mapper sensor of the Landsat satellite during the year 2002 (30 m grid resolution). Source: Institut Cartogràfic de Catalunya (ICC)

industries and towns reaching the river upstream from main water supply intakes. See Sect. 3 for a comprehensive description of those infrastructures.

The delta of the Llobregat River lies to the south of Barcelona and covers about 100 km^2 and constitutes a valuable natural habitat close to a big metropolitan area. Its wetlands are part of the RAMSAR network and play a critical role in the wintering of many migratory birds. The delta aquifer is an important freshwater resource for the metropolitan region, with a groundwater capacity around 100 hm^3, used by numerous industries, agriculture, and cities. The fertile delta farmland supports intensive agriculture supplying the local market. Since the 1960s, the delta's land has been under constant pressure from Barcelona's urban and industrial expansion. Most important logistics and transportation facilities in the region (port, airport, motorway network, and railways) were built in this area taking advantage of the flat landscape. The recent port extension even forced a southward movement of the river mouth. Less than 5% of the original wetlands in the area now remain, and in some municipalities half of agricultural land has been lost in the last decade [1].

1.1 Geology

The Llobregat watershed is the product of tectonic movements during the late Mesozoic over 29 million years ago. Upon the rupture of Pangea, the geological plate carrying Iberia drifted separately from Eurasia and Gondwana—prehistoric Africa and South America. The rotation of Gondwana then pushed the Iberian plate against Eurasia causing a collision that raised the Pyrenees Mountains and erected the first contours of the Llobregat watershed. The convergence of the Iberian and Eurasian plates culminated during the Quaternary period over 2 million years ago [2]. The collision of Iberia with Eurasia initially created a bay, and then a salt water lake. Over time the water evaporated, leaving behind layers of salt and other minerals that were pushed up into the Pyrenees Mountains as the two geological plates converged.

The Llobregat River crosses several present geological units (Fig. 3). Headwaters are situated in the Eastern Pyrenees, in an area composed of Upper Paleozoic to Cenozoic sedimentary (mainly limestones and sandstones) and volcanic rocks [3]. In its central part the Llobregat River flows through a Cenozoic Basin composed of a regressive (marine to continental) Eocene-Paleocene age sedimentary sequence. The transition evaporites (gypsumanhydrite, halite, and potash facies) have an Upper Eocene age [3] and there has been salt-mining activity since Neolithic times in the area. Close to the Mediterranean Sea, the Llobregat River crosscuts the Catalonian Coastal Ranges, made up of Hercynian basement and Mesozoic to Cenozoic cover. The Hercynian basement consists of Paleozoic metasedimentary rocks and late Hercynian granites, upon which the Mesozoic sedimentary sequences have been deposited. The delta, consisting of Quaternary sediments, is composed of conglomerates, sands, and marine marls.

Previous work in this area [3] has shown that the chemistry of most of the stream waters in the Llobregat basin is mainly controlled by the weathering of the Tertiary sediments, mainly limestone and marls deposited in upstream regions and middle reaches (Fig. 3). Despite its small extent, the salt deposits have a dramatic effect in the lower Cardener and Llobregat, enhanced by the mining activities. The signature of water is derived from weathering of limestone and marls (Ca^{2+}, Mg^{2+}, HCO_3^-, Sr^{2+}, and Ba^{2+}), gypsum (Ca^{2+}, SO_4^{2-}, Sr^{2+}, and Ba^{2+}), and halite and sylvite (Na^+, K^+, Cl^-, Mg^{2+}, Ca^{2+}, and SO_4^{2-}).

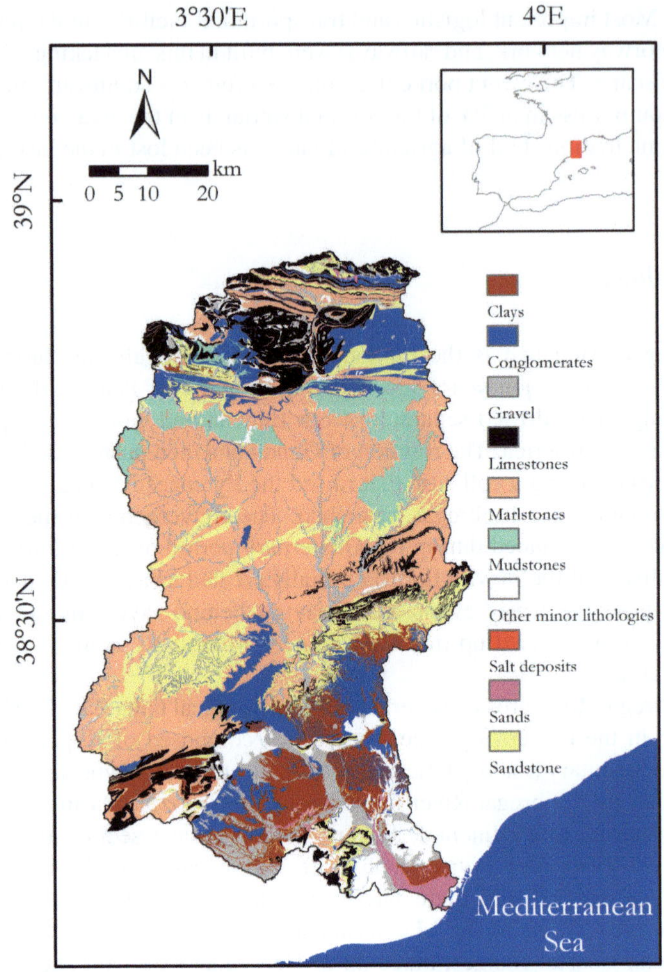

Fig. 3 Main lithology in the Llobregat basin. Source: Institut Cartogràfic de Catalunya (ICC)

1.2 Vegetation

Ice covered the Llobregat watershed during five glacial periods between the Pliocene and Pleistocene. In 8000 BC ice sheets still dominated continental Europe, but with the gradual rise in global temperatures, the ice retreated and temperate forests migrated north. By 2000 BC the composition of tree and shrub species in the Iberian Peninsula began to resemble the native vegetation cover we are familiar with today [4]. Before the massive clearing for agriculture, the Llobregat was covered in oak forests (*Quercus humilis*, *Quercus ilex*) mixed in with Mediterranean white pine (*Pinus halepensis*) and shrubs (*Rhamnus alaternus*, *Phyllirea latifolia*, and *Smilax aspera*).

The lithological variety, climate, and topography of the Llobregat basin encompass the great variety of vegetation formations (Fig. 2). The agricultural areas predominate in the plain areas around Manresa and upstream regions of the Anoia and Cardener Rivers, while native forest patches are abundant in the headwaters of the Llobregat and Cardener. Vineyards dominate the landscape in the southernmost region of the Anoia basin. In general terms, the Llobregat basin is highly influenced by both Mediterranean and Boreoalpine forest species.

The vegetation characteristic of the Boreoalpine region appears in the mountainous Pyrenees, particularly well-developed in the northern slopes. The region has no continuity beyond the Pyrenees, as it develops above 1,600 m. Coniferous forest (*Pinus sylvestris*, *Pinus nigra*) dominate between 1,600 and 2,300 m a.s.l., with some Atlantic deciduous species in the most humid places. Subalpine scrubs and alpine meadows are restricted to the highest ranges.

The Mediterranean region encompasses most of the basin. It is constituted by xerophytes formations typical of Mediterranean climate, highly degraded by centuries of human exploitation and forest fires. In fact, human management switched the dominant vegetation toward white pine forest (*Pinus halepensis*). Holm oak forest (*Quercus ilex*) is the theoretical climax in the area, now limited to small patches and areas of the Catalonian Coastal Ranges (Fig. 2). Human management and fires also favored the extension of woody, thorny, and aromatic, evergreen species of shrubs that occupy vast extensions (among others juniper, thyme, and rockrose).

Riparian forest is in general highly degraded along river corridors. One can now find white poplars (*Populus alba*), black poplars (*Populus nigra*), narrow leafed ash (*Fraxinus angustifolia*), or white willows (*Salix alba*) [5]. In the case of main channels of the three main tributaries, only reaches crossing canyon valleys with stepped topography preserve well-developed riparian areas, even in the Pyrenees. Low-order tributaries maintain most of the good shaped riparian strips. Nonetheless, the most downstream reaches of the Llobregat River are highly degraded, frequently channelized (Fig. 4).

1.3 Climate

The Llobregat River spans from the Pyrenees to the Mediterranean Sea, including an altitudinal gradient with profound impact on the basin climate. The Llobregat River headwaters, well rooted in the Pyrenees, show a perhumid climate (following the classification by Thornwhaite), which rapidly transforms in a dry subhumid climate as we approach the middle reaches (Fig. 5). This shift roughly coincides with the distribution of mean annual temperature and precipitation. The amplitude in mean annual temperature in the basin reaches 11°C, with a remarkable increase as we approach the sea, and the difference between maximum and minimum annual precipitation is as high as ~550 mm (Fig. 5). Most precipitation occurs far from the regions were water is most intensively used, and the highly heterogeneous

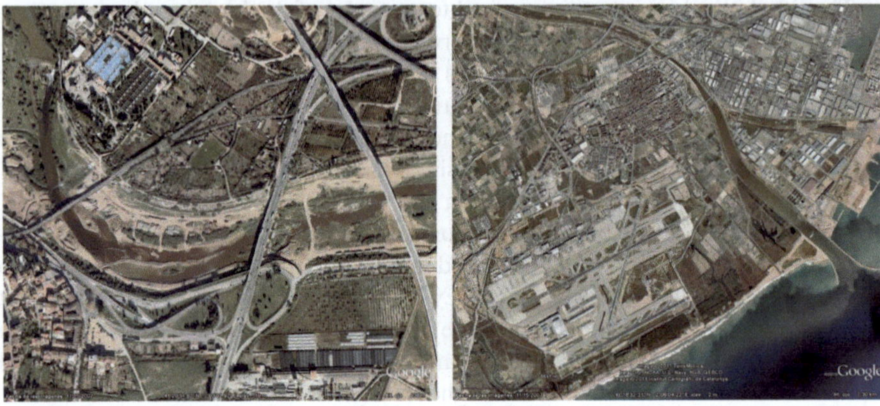

Fig. 4 Two aerial pictures of the Llobregat River near the mouth. On the left, the main channel after the water supply intake at Sant Joan Despí (squared blue patch and upper left-hand side). Note the highly degraded channel, the brown color of the water, and the total absence of riparian strip. On the right, the present mouth of the river, pointing southeast. Note the ancient mouth pointing toward the East, now desiccated. The huge infrastructure occupying most of the Deltaic region and coastal wetlands is El Prat International Airport. The reach on the left is just upstream from that on the right picture

distribution of precipitation and temperature explains the location of the three main reservoirs in the basin. In general vegetation growth is more difficult in the most downstream sections and some central reaches in the middle Llobregat (see hydric deficit; Fig. 5).

However, spatial climate variability hides a strong seasonal variability, especially in the dry areas. Here, there is a strong arid period during summer (Fig. 6), when vegetal communities face a long period of hydric stress that lasts the whole summer. The dominance of xeric Mediterranean species in these areas is an outcome of this climatic challenge. However, the situation in the headwaters is much milder (Fig. 6), with a gentle deficit period in summer and a long lasting saturation period during winter and spring. In fact, deciduous species belonging to the Atlantic climatic region are restricted to areas with hydric deficit below 100 mm (Figs. 2 and 5). This seasonal variability is reflected in the long-term behavior of the Llobregat River at the mouth (Fig. 7). The median seasonal flow shows a marked seasonal pattern, with a clear peak during May, most probably reflecting runoff generated in the Pyrenees by snow melting. It is remarkable that the typical Mediterranean autumn storms do not have a translation in the long-term hydrograph, probably reflecting the erratic nature of these events, and also probably the increased storage capacity of upstream reservoirs after the summer. Note that minimum flows (Fig. 7) are quite similar along the year, showing that drought periods may extend during the whole year. The temporal evolution of the daily streamflow shows the erratic water flow in a Mediterranean river like the Llobregat, defining very long periods below 5 $m^3 s^{-1}$, and abrupt transitions to wetter periods that can increase the water flow by a factor of 50 (~300 $m^3 s^{-1}$) in a single day (Fig. 8).

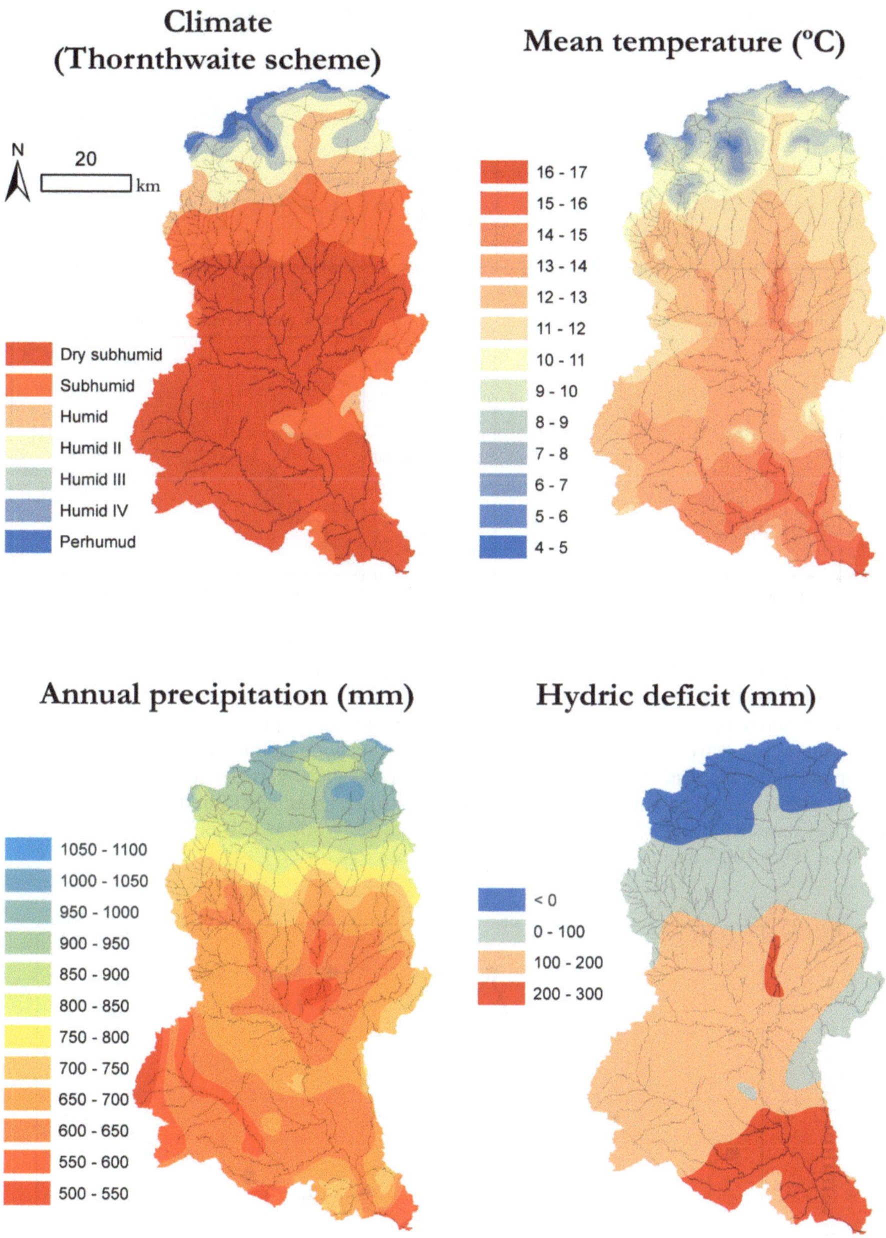

Fig. 5 Spatial distribution of climate descriptors of the Llobregat basin

Interannual variability is also a characteristic trait of the climate in the Llobregat basin, best reflected in the interannual variability of runoff. Figure 9 shows mean annual streamflow for the period 1945–1998, which shows a strong interannual

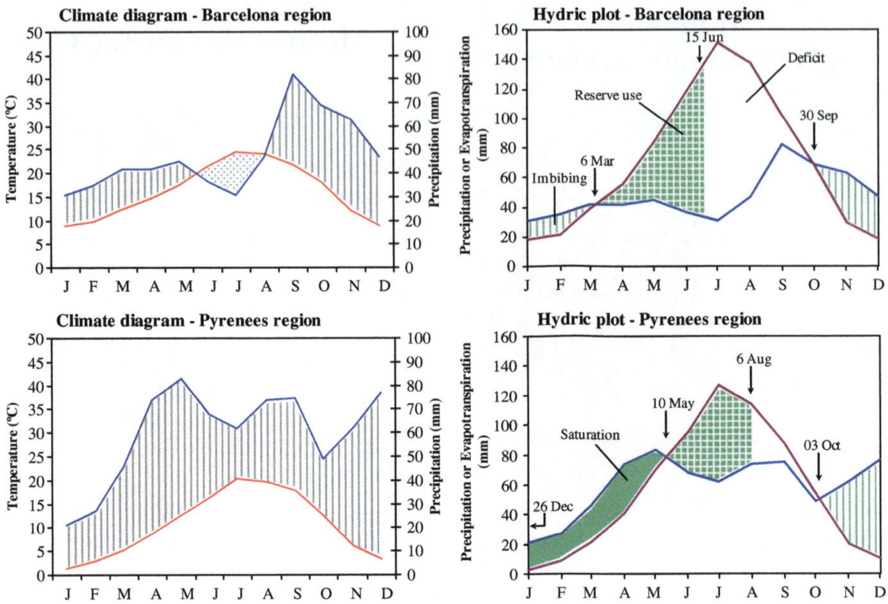

Fig. 6 Climate diagrams and hydric plots for two distinct areas in the Llobregat basin. Data for Barcelona derive from the meteorological station in Sant Boi de Llobregat. Data for the Pyrenees region are from the Sorribes meteorological station. In the climate diagrams, precipitation is the blue trace, and temperature the red one. Arid and wet periods are depicted as a pointed and vertical strip pattern, respectively. In the hydric plots, potential evapotranspiration is the brown line, and precipitation the blue one. Imbibing periods are depicted as green vertical strip pattern, periods of reserve use as green squared pattern, deficit periods as white areas, and saturation as a plain green area. Key dates were added for reference. All data from the Worldwide Bioclimatic Classification System, 1996–2009, http://www.globalbioclimatics.org

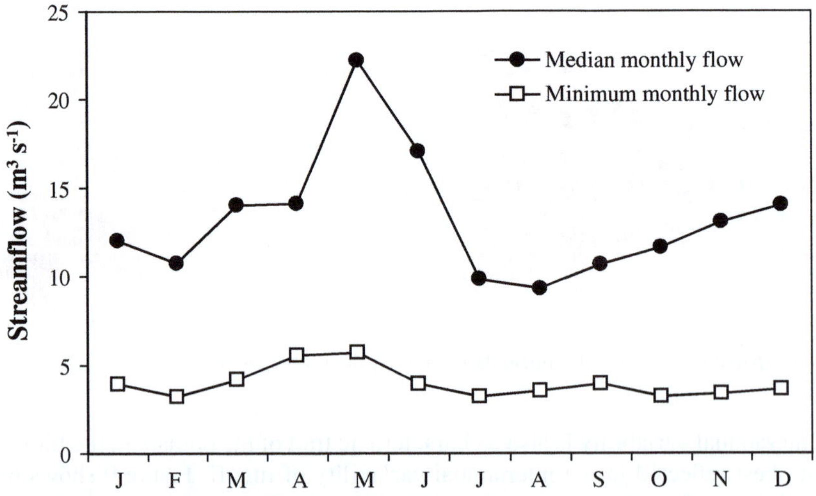

Fig. 7 Median and minimum monthly flow in the Llobregat River near the mouth (Martorell). Data are from the period 1945–1998, provided by the Catalan Water Agency (ACA)

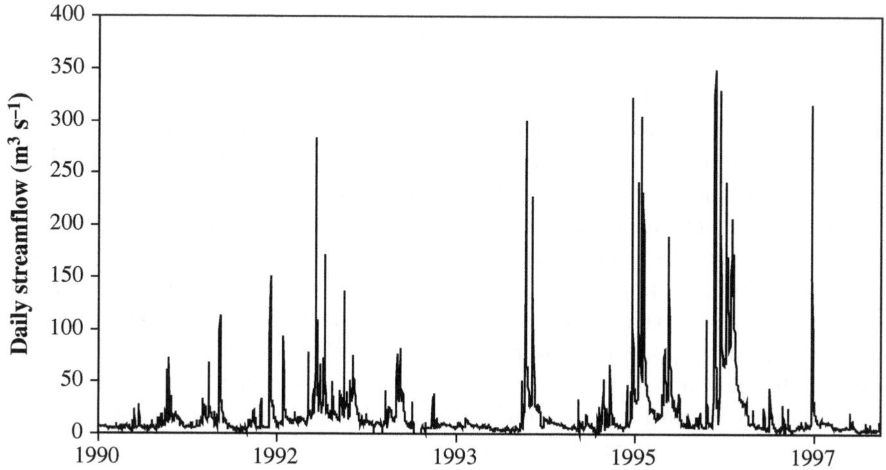

Fig. 8 Daily mean flow in the Llobregat River near the mouth (Martorell). Data are from the period 1990–1998, provided by the Catalan Water Agency (ACA)

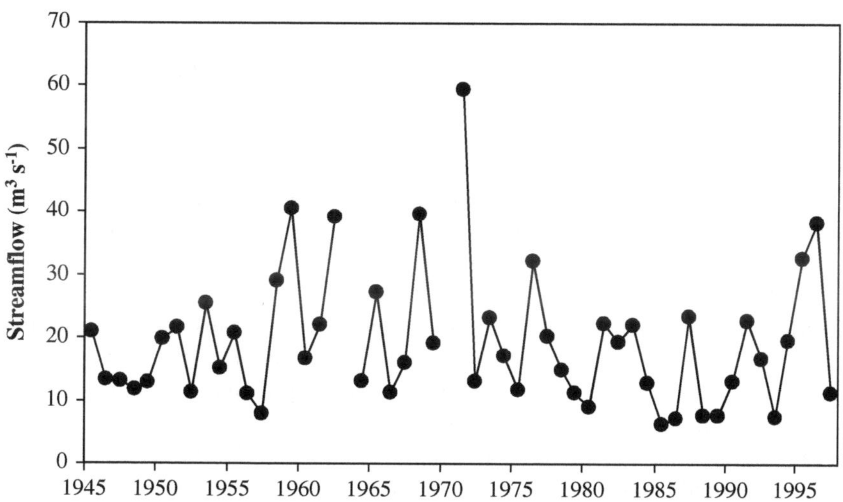

Fig. 9 Mean annual flow in the Llobregat River near the mouth (Martorell). Data are from the period 1945–1998, provided by the Catalan Water Agency (ACA)

variability, with mean flows between 6 and 40 $m^3 s^{-1}$. This is noteworthy considering that the magnitude of wastewater effluents in the whole basin is ~4 $m^3 s^{-1}$. Therefore, the impact of human activities during dry years is severe.

1.4 Hydrological Alterations due to Global Change

Many studies have anticipated future threats related to shifting availability in water resources in Mediterranean countries [6]. Future climatic trends point toward the aggravation of water scarcity that will add to an already precarious situation in many Mediterranean regions [7]. Results in experimental basins throughout the world show that changes in land cover from grassed to forested areas are associated with a reduction in runoff [8, 9]. Northeastern Spain experienced a demographic growth during the late nineteenth and the first half of the twentieth centuries that resulted in the expansion of agriculture, the decrease in forested areas and overgrazing of steppes and natural vegetation [10]. Conversely, social and economic changes in recent decades have reversed this trend and resulted in the abandonment of most old marginal agricultural fields and the decrease of human pressure on forests and pastures [11].

The historical flow records of nearly all of the streams on the southern flank of the Pyrenees show decreasing annual trends at least during the last 60 years [12]. This alarming trend is attributed to three causes of dissimilar weight depending on the watershed: climate shift, increasing irrigation, and the encroachment of forest cover in the headwaters [12, 13].

In the Llobregat River Basin the amount of water currently used for irrigation (~44 hm^3) is a small fraction of the total water load at the mouth (~600 hm^3), and at the annual time scale this value should be regarded as a gross estimation, since an unknown but probably significant volume of the withdrawn water eventually returns to the river. Moreover, most of the irrigation diversions occur in the deltaic region near the mouth (~34 hm^3), and the single significant diversion for irrigation located upstream (the Sèquia at Manresa) dates from the Middle Ages. Thus, we can discard irrigation as a fundamental driver of any streamflow trend detected in the headwaters of the Llobregat basins during the last decades.

Indeed, water flow in the Llobregat and Anoia Rivers showed significant decreasing trends between 1945 and 2005 [12]. The reduction in runoff was particularly conspicuous during the spring months, probably reflecting alterations in snowcover at the Pyrenees. Conversely, no significant trends have been detected during summer. At the annual time step [12], Lorenzo-Lacruz et al. estimated a decreasing trend between 1% and 3% per year.

To unravel the role of climatic drifts and land use changes in the generation of those observed trends Gallart and coworkers [14] applied two complementary methodologies to independently estimate the amount of streamflow trend explained by climate and land use changes in the headwaters of the Llobregat Basin. First, the authors estimated the changes in forested cover in two locations in the basin between 1957 and 1993. Then they applied the Zhang equation [15] as a simple model to translate the differences of land cover types into differences of annual evapotranspiration (see [14] for further details). Under the assumptions of this methodology, forest regrowth in a given basin increases evapotranspiration and, as a result, decreases streamflow in its river courses. Gallart et al. [14] found that the

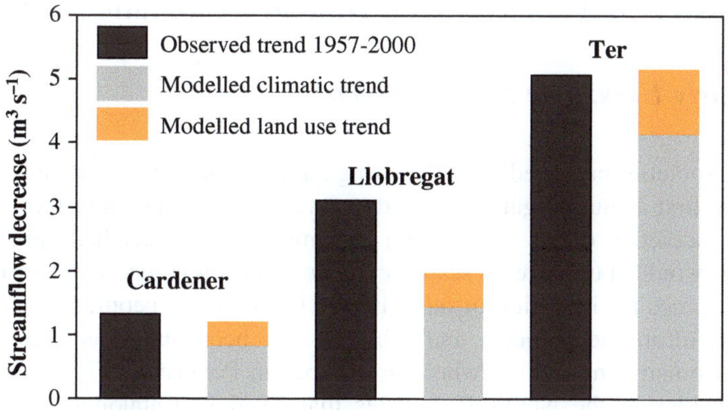

Fig. 10 Magnitude of the observed streamflow decreasing trend in two gauging stations from the Llobregat Basin (Cardener at Olius and Llobregat at La Baells), and in the Ter River at Sau (this last example included for comparison). *Black bars* denote the trend calculated with observed data, and *grey* and *orange bars* correspond to the trends calculated with the SACRAMENTO model and the land use changes (Zhang approach), respectively. If the hydrological model and the Zhang water balance results are fundamentally correct, the result of pooling both trends should resemble the observed trend in each gauging station. Results are adapted from [14] and [16]

most conservative estimate of the streamflow trend promoted by forest expansion was a reduction between 0.36 and 0.55 $m^3\ s^{-1}$ of the mean annual water flow. Independently, Gallart et al. [14] estimated the streamflow trend promoted by climatic drift during the same period applying the hydrologic model SACRAMENTO, which only takes into account climate variability, but not land uses changes. The result was that climate explained a streamflow trend that amounted between 0.84 and 1.43 $m^3\ s^{-1}$, depending on the site.

The question remains on whether or not the sum of the two streamflow trends estimated with the two independent methodologies (climatic trend + land uses trend) reasonably match with the streamflow trend observed in the two gauging stations from the Llobregat Basin considered in those studies (Fig. 10). Considering the uncertainties associated to the estimation of the two hydrologic trends, the good correspondence between observed and estimated trends when the two drivers (climate and land uses changes) are pooled is impressive. In the case of the Llobregat at La Baells (Fig. 10), the lack of fit may be explained by the short record available to properly calibrate the SACRAMENTO model (see [14]). In any case, all results converge to a significant streamflow trend in the watercourses of the Llobregat Basin during the last decades, mainly promoted by a climatic shift, probably related to an increase in the mean air temperature [14]. Land uses changes have a moderate role in this change, account for the 17%–27% of the observed trend. These estimations broadly coincide with results obtained by [16] for the nearby Ter Watershed using an independent approach.

2 A Historical Perspective on Human Exploitation of the River

2.1 Early Uses: The Feared River

Neolithic societies benefited from the rising temperatures and colonized the Iberian Peninsula, first as hunter-gatherers, and eventually as rudimentary farmers. These primitive societies drank water from streams in the watershed, just as other mammals would, but there is scarce evidence that they succeeded in managing water resources for irrigation or other uses. The Neolithic peoples employed very basic agricultural techniques, and while it has been suggested that irrigation predates Roman conquest elsewhere in the Iberian Peninsula [17] this is disputed when considering the Neolithic peoples that lived in Catalonia [18]. There is agreement however that the Phoenicians, Greeks, and Romans had already developed irrigation systems long before the native peoples of Iberia, and they brought this knowledge with them when they traded or settled on the peninsula.

Unlike the societies in Mesopotamia or Egypt, the first settlers in the Llobregat watershed shunned the river, preferring to live at higher elevations instead. Living near a Mediterranean river must have been an inconvenience and danger rather than an asset. Maps of prehistoric settlements show that Neolithic peoples avoided major water ways in the watershed, preferring to live in mountain caves instead. Archeologists have found Neolithic remnants perched far above the Llobregat River, in the caves of Collbató, or in the mountains of present day Sant Feliu de Llobregat and Begues [19, 20]. These original peoples have left us with little material evidence of their society, but their names endure, such as the Palladianus from Pallejà or the Apararia from Abrera [19].

Even if Neolithic tribes knew that irrigation could increase crop production, it is still difficult to imagine that irrigation alone would bring them closer to the water's edge. River banks must have been dangerous and inhospitable for the establishment of permanent settlement due the erratic and flushing nature of Mediterranean rivers. Smaller streams and tributaries were surely preferable sites for habitation. Historians have suggested that the first river dwelling people may have been outcasts, excluded from the secure caves or strategic mountain tops [21]. When forced to migrate to lower elevations, the refugees were more vulnerable to attack from wildlife or rival tribes, and without the technical or social skills to protect themselves from floods, fires, or other threats lurking in the valleys.

We might wish to believe that rivers played a critical role in determining the location of human settlements. As far back as 1476, it has been asserted that rivers played a fundamental role in the development of Catalonia: "Human society has always begun along rivers, because there are three things that lead to growth in this world, water, land and human industry" [22]. And yet in the case of the Llobregat watershed, rivers appear to be largely shunned, more feared than revered.

If Neolithic tribes avoided living near Catalonia's large rivers, the Romans hardly paid any more attention to them. The site for Tarraco, the Roman capital of Hispania Ulterior, was chosen because of the depth of its port, not its meager stream. Roman Barcelona, or Barcino, was founded on the top of a hill, Mount Tàber (16.9 m), flanked by two small streams, the Merdança and Caganell, each named after the human waste they carried [23]. The Roman's were not attracted to Barcino's location between the Llobregat and Besós Rivers, but rather they found the enclave conveniently located midway between Tarraco and Emporiae which provided a useful resting point for marching legions. And when Barcelona grew to become the political and administrative capital of Old Catalonia, neither of the larger rivers played any commercial role in the city's development.

In fact, when Roman soldiers rested in Barcino on their way to Emporiae, the Llobregat Delta was entirely underwater. Two thousand years ago the Llobregat River met the Mediterranean Sea 3 km inland from its current point today, probably near Sant Boi de Llobregat [19]. In the centuries that have passed since then, the rains have eroded the mountains, the sediment has tumbled down the watershed, and pushed back the Mediterranean to create the delta we see today.

The Romans called the Llobregat Rubricatus of Hispania Tarraconensis. Rubricatus translates from Latin as red or purple, although it is unclear if the color refers to the river itself or the dark red soils colored by iron oxide that flank the river between Barcino and the major river crossing in Ad Fines (Martorell). The early Roman geographer Pomponius Mela refers to the Llobregat as Rubricatum majus, suggesting that the river branched out into smaller channels as it passed through the delta and reached the Mediterranean [24]. Rubricatus was not a unique name; since the Romans also spoke of another Rubricatus River in northern Africa, located in modern day Libya [25].

Before humans used the Llobregat River to irrigate their crops, propel their mills, or transport their waste, the river was a source of problems: a threat and a nuisance; an obstacle and a barrier. First and foremost, the river's floods threatened to wash away life and property. The frequent and dangerous floods led the river to be referred to as *flumen terrible et periculosum* [24]. We do not have direct reports on the Llobregat's floods during first millennium after Christ, perhaps in part because there were so few structures to destroy or people to write about them. However, Thorndycraft et al. [35] after the analysis of paleoflood deposits in the Llobregat River described that floods higher than 2,300 $m^3 s^{-1}$ were exceeded in five occasions during the last 2,700 years, reaching discharges of 3,700–4,300 $m^3 s^{-1}$. The first written report of a flood in the Llobregat dates back to 1143, that flood washing away the original Roman bridge at Martorell. In the next 850 years, up until 1900, historians have counted 129 more major floods along the Llobregat, although certainly there must have been more [24]. As human settlements began to develop closer to the river, residents complained that their proximity to its water made it an unpleasant place to live. In 1807 municipal officials from El Prat lamented that the river's frequent floods were the primary cause of its "misery" [24]. And indeed, some of the floods were of biblical proportions. One flood transformed the Llobregat Delta into an inland sea, covering the entire area with water from Montjuic to Castelldefels [24].

2.2 Bridging the River

The Llobregat might have inspired more sympathy had it at least been navigable; but alas, not even that. The Lobregat's shallow and irregular discharge made its waters unforgiving to floating vessels. The river was large enough to obstruct travel and communication, but not large enough to be useful for transport. This was not unusual for this part of the world. Of all of the rivers in northeastern Spain, only the Ebro River (Hiberus Flumen) to the South has any evidence of useful commercial transport. Thus, towns near the river were largely cutting off from communicating with those on the other side, and crossing the river was a major inconvenience for locals and travelers trekking along the Mediterranean rim. The Roman bridge at Ad Fines (Martorell) must have been the first permanent crossings of the Rubricatus. The bridge was a critical piece of the Roman's via Augusta, a protected transportation artery which connected Imperial Rome with the Tarraco, and further south, the silver mines in Carthago Nova, and the aristocratic city of Gades (Cadiz) in Baetica [23]. The bridge crossing at Martorell was a full day's journey from Barcelona.

But the inconvenience of the Llobregat River also provided strategic military defense. For over 300 years, between 801 and 1148, the Llobregat and Cardener Rivers marked the divisionary lines of the Christian and Islamic worlds [23]. The rivers served as a natural moat, albeit sometimes an ineffective one, that protected Catalunya Vella (Old Catalonia) from southern invasions. The threat of southern invasion prompted the Catalan peasantry living near these rivers to move north, inland, or to higher elevations. Once again, just like the Neolithic peoples before them, safety could be secured far from the river, perched in the high lands.

2.3 Approaching the River: Mills and Irrigation

When the North Africans were finally pushed beyond the Ebro River in the twelth century, the Catalan peasantry slowly began to come down from the mountains and populate the lower Llobregat and Cardener valleys. In 1188 the Catalan king gave the first known water concession over the Llobregat to build a flour mill at Molins the Rei. In 1317 a second permanent crossing of the Llobregat was built at Monistrol, giving the church and pilgrims access to the monastery of Monsterrat. A few years later another medieval bridge was built in Castellbisbal [19, 22].

By the middle ages, small scale irrigation along the river's banks was widespread, especially in the upper reaches of the Pyrenees. The medieval period also saw a boom in flour mills as the peasantry became more skilled in harnessing the rivers energy. In the Merlès stream alone, historians have documented 32 medieval flour mills [26]. Many medieval mills have also been found in Capellades along the Anoia River [27]. This tradition of harnessing hydropower along the Anoia would eventually develop into other river related industries such as a paper mills, textiles, and leather tanning.

2.4 River Industries

The Llobregat and Cardener Rivers played a central role in the industrialization of Catalonia. As elsewhere, industrialization began with the production of fabrics and textiles. Spaniards in the eighteenth century produced fabrics by hand; manually separating the seeds from cotton fibers, carding the fibers, spinning the thread, and weaving the fabric. Starting in the early nineteenth century industrialists began to import new technologies from northern Europe in an effort to modernize textile production. One of the first imported technologies was the water frame, which substituted the manual spinning of cotton for a mechanized process with wooden and metallic cylinders. The first water frame appeared in the Llobregat watershed in 1801, and by 1804, these devices spread throughout the Cardener and Llobregat valleys [28].

The mechanization of fabric production was largely driven by hydraulic energy, and this brought entrepreneurs to set up production near rivers. The first company town along the Llobregat was Can Prat, built in the town of at Puig-Reig in 1810. However, hydraulic energy alone was insufficient to massively produce fabrics. It was not until the development of the steam engine in 1833 that industrialists obtained the efficiency necessary to make textile production highly profitable. A few decades later, in the 1860s, the innovation of the modern turbine allowed for the more efficient and widespread use of hydropower [26].

To exploit these technological innovations, entrepreneurs descended upon the Cardener and Llobregat to set up production along the river in the second half of the nineteenth century. The amount of energy produced depended on two critical factors: total river discharge and vertical drop. The more discharge and higher the vertical drop, the more energy produced. The problem with setting up industry next to the river was that they were far from existing towns and infrastructure (in addition of course to the threat of flooding). This led industrialists to build company towns: complete with housing for workers and their families, schools for their children, a church, a theater, and a local store.

The largest and most profitable company towns in the Llobregat were built in the 1870s: Colònia Burés (1873), Vilodomiu Nou (1874), Ametlla de Merola (1874), Colònia Pons (1875), and Colònia Borras (1876). Hydropower was a major source of energy, but it was always complimentary to the steam engine. By 1888 the best spots in the Llobregat were occupied, and industry migrated east to the Ter watershed [27].

With the development of the industrial colonies, one after another, the Llobregat River became to be called the "most exploited river in the world" [27]. Only in retrospect is it easier to appreciate this navel glaring. But the reputation of the Llobregat as a hard working river, producing much with little, stuck in the minds of the Catalans. A "Llobregat mentality" entered the Catalan lexicon as a way of describing someone able to maximize the exploitation of resources, no matter how small, for the maximum advantage.

Industrial activity along the Llobregat peaked in the early twentieth century. At the height of production, between 1900 and 1910, the Sedó company town employed over 2,000 workers [27]. The last industrial colonies were built at the turn of the twentieth century, at which point, electrification gave industrialists the freedom to start production on higher ground, far away from the unpleasant Mediterranean rivers.

2.5 *Dams*

For centuries, floods from the Llobregat had threatened to wash away crops, boats, and bridges. The first efforts to contain the river's waters date back to the fourteenth century when the residents of Sant Boi de Llobregat built levies along the river's edge in the hopes that the structures would protect them from rising waters. But these levies were repeatedly washed away, despite diligent efforts to test building materials and designs that gave the river a generous floodplain [24].

In the 1880s, engineers drafted the first dam project in the Llobregat watershed at a site near Cabrianes [29]. The goal was to mitigate flood damage and provide water for crop irrigation. But the proposal met opposition from the forester Rafael Puig Valls, who advocated for managing forests in the upper segments of the watershed and recognized the natural variability of Mediterranean streams, but emphasized that floods could be mitigated with savvy ecosystem management [29, 30]. The Cabrianes dam project was eventually discarded, but would repeatedly come up again in the various water management plans in the twentieth century.

A plan in 1902 projected five dams in the Llobregat basin: two on the main stem of the Llobregat, one below Montserrat at La Puda, and another at the town of El Pont de Vilomara. The third dam was to be erected along the Cardener River, above Clarianes, the fourth near the confluence of the Anoia with the Montbui, and a fifth on the Anoia between Vallbona and Cabrera. The hydraulic plan of 1902 included similar ambitions and aspirations throughout Catalonia and the rest of Spain, but most of these projects never broke ground [27].

In 1933 engineers drafted another ambitious hydrological plan with the primary objective of increasing hectares under irrigation. Energy and drinking water were secondary objectives. In the plan of 1933, the Cabrianes dam project resurfaced again, as did the proposal to build a dam along the Cardener at Sant Ponç. In 1935, the Catalan Government revised this proposal, and in addition to Sant Ponç, planned for dams at Malagarriga, Sorba, Vilomara, and Boades [26]. However, these plans were interrupted by the Spanish Civil War (1936–1939).

Of all the dam proposals in the Llobregat watershed, the first to be built was at Sant Ponç. Construction began in 1949 and the project was completed 1954. When complete, the reservoir held a capacity of 12 hm^3 [27]. It was the first human made reservoir in the watershed, and later followed by the Baells in 1979 and the Llosa del Cavall in 1997.

2.6 Water Pollution and Salinity Conflict

The first momentous conflict over water pollution concerned salinity concentrations in the Cardener and Llobregat Rivers. Starting in 1926, the owner of a textile mill along the Cardener, Antonia Burés Borràs, noticed that her metallic turbines were corroding. She suspected that this chemical reaction was produced by the unusually high concentrations of salts flowing into the Cardener River from the mining industry upstream. Burés soon initiated legal proceedings against the mining company in order to halt their "impurification" of the Cardener's waters. The Spanish Government began two investigations to study effluent flowing out of Súria, but it was not until the third investigation began in 1931, that the Commission for the Study of the Salinity of the Waters of the Llobregat River (CESALL) was established to study the issue in detail [31]. The CESALL commission used the latest science and technology in order to clarify the fundamental chemistry and hydrogeology, in an attempt to create new rules that would reconcile their water uses among competing interests. The extensive deliberations of the CESALL commission were arguably the first watershed planning process in the Llobregat basin [32]. By the time the CESALL Commission had begun meeting regularly in 1931, Sociedad General Aguas de Barcelona had also raised concerns that the salts flowing into the river from the mines were percolating into the Llobregat aquifer, and as a result, polluting Barcelona's main source of drinking water. The company wells at Cornellà had registered rapid increases in salinity values since mining operations began in 1926. The Municipal Laboratory of Barcelona also became concerned that the mining activity would permanently contaminate Barcelona's primary source of drinking water. The CESALL released their conclusions in 1932 and recommended that salinity effluents be limited to 250 mg/l in Pallejà. The commission also advocated for the building of a brine collector that would transport the mining waste directly to the Mediterranean, and a dam that would regulate flows and dilute salinity concentrations during summer months. The legal limits on salinity in the watershed were debated in the Catalan Parliament and approved by the Catalan Generalitat in 1933 [33], however the structural solutions took several decades before the projects materialized. While the dam at Sant Ponç was completed in 1954, the brine collector was not completed until 1983.

In the first half of the twentieth century, Barcelona drank from the Llobregat but only indirectly, since its waters were pumped up from the aquifer below the Llobregat delta. But starting in 1955, Sociedad General de Aguas de Barcelona began to withdraw surface water from the Llobregat River at a new drinking water facility constructed at the mouth of the watershed in Sant Joan Despí (see Fig. 4). From a water quality perspective, the facility's downstream location could hardly have been worse, but the selection was related to the existing water rights for irrigation or industrial production upstream [32]. With rise of industrial activity along the Llobregat in the 1960s and 1970s, water managers began to detect a cocktail of contaminants flowing down the Llobregat River. According to one former employee, the laboratories at Sant Joan Despí have detected "every possible element in the periodic table." Efforts to protect the source quality of the Llobregat River have been a combination of stop-gap measures. In the 1960s, water managers captured two highly polluted tributaries, the Rubí and Anoia, piped them into a channel, diverting them away from the main stem, and then dumped them back into the river just a few meters below the water treatment facility (see next section). This diversion remains today, in addition to other diversions, such as the water collector of the freeway, which also dumps below the water treatment facility. At the same time, it must be recognized that the water quality of the Llobregat River has largely improved due to the construction of municipal waste water treatment facilities.

3 The Llobregat as an Overexploited River: Interconnectivity in the Llobregat Basin and the Urban Loop

In terms of human exploitation of the river, reservoirs are the most conspicuous in the Llobregat. The Llobregat has three main reservoirs. One is located in the Llobregat itself (La Baells; 109 hm^3 capacity) and other two in the Cardener (Llosa del Cavall, 79 hm^3, and Sant Ponç, 24 hm^3). Their waters are mostly used for irrigation and hydroelectric power, but not so much for local drinking purposes. However, their regulation capacity is fundamental to maintain a sufficient amount of baseflow in the downstream reaches to sustain water supply for Barcelona during the summer months.

Apart from big dams, a total of 101 weirs are located in the basin (Fig. 11), mostly to profit the water for small hydroelectric power stations. As stated above, most of these weirs were built to support local industry during the boom of textile and mining activities. Nowadays many of these small dams still are in use, although their role on the power budget of the region is irrelevant. In some sections of the river network the occurrence of weirs challenges the very definition of the Llobregat as a river: in the middle section the main channel is dammed by weirs once every ~2 km, for a total amount of 45 weirs (Fig. 11).

Further to these medium and large infrastructures, the hydrological complexity of the Llobregat cannot be understood without considering the multiple connections, derivations, withdrawals, and returns that are scattered all over the basin. These connections are implemented since long ago and have received new incorporations to make compatible the multiple uses with the preservation of water quality.

In the Llobregat headwaters, the Berga Industrial Channel derives water mainly for hydropower that will be later returned to the river downstream of La Baells dam. It has a maximum capacity of 2.5 $m^3 s^{-1}$ (Fig. 12), and only a minor amount is used for water supply in Berga. Also in the headwaters, but in the Cardener River, some resources are derived from a small tributary of the nearby Segre River (located to the West). Most of this supply is used in the city of Solsona and returns to the Negre and Cardener Rivers in the form of WWTP outflows. A water channel built in medieval times (Sèquia de Manresa) supplies water to Manresa directly from the

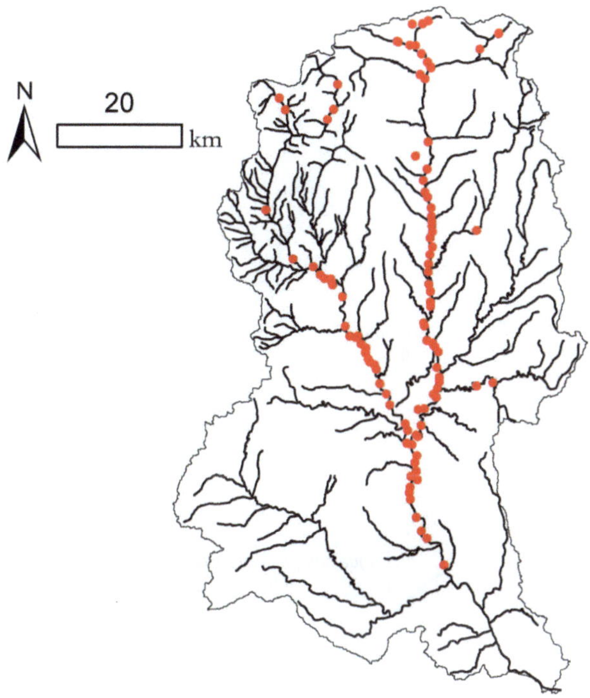

Fig. 11 Location of weirs in the Llobregat River network. Source: Catalan Water Agency (ACA)

Llobregat. It was aimed to provide drinking water as well as water for irrigation in the plains surrounding the city, and later was also used to feed several small hydroelectric plants. Nowadays is the first link of a complex urban schema including natural wetlands, treatment plants, wastewater facilities, and a complex network of pipes (Fig. 12).

As noted above, mining activities enhanced the impact of salt leachates that originate in several areas of the middle section of the river (at the urban nuclei of Balsareny, Sallent, and Súria), causing waters to show conductivity far beyond the quality standards. Therefore, in the 1980s, a salt brine collector was built to directly bypass these salty waters to the sea, avoiding them to enter into the Llobregat waters. The water volume is not so high, but the leachate concentrations are sometimes fivefold the marine concentration. Other industrial salt brine residues are also incorporated in the collector as it develops downstream.

Downstream the city of Manresa and up to the city of Abrera, water is obtained primarily from groundwater sources some of them closely connected to the alluvial aquifers. The aquifer Carmen–Capellades is used to feed the needs of Igualada and some neighboring towns; a large fraction of these waters return later to the river through WWTP effluents. In the Abrera area is placed the second largest Drinking Water Treatment Plant (DWTP Abrera) in the Llobregat. It treats waters from the middle-lower part of the river, and serves a large area including Barcelona. Another

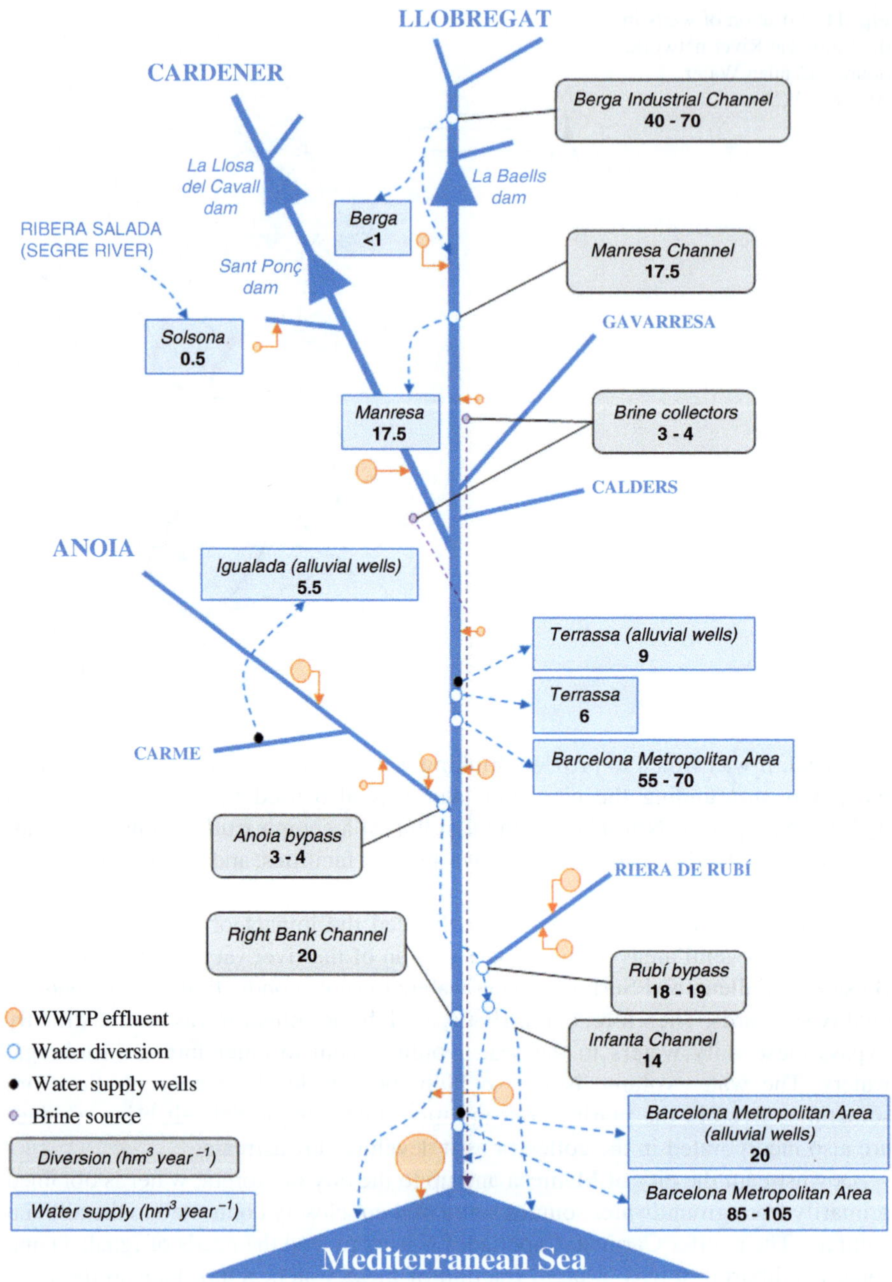

Fig. 12 Scheme of the main withdrawals, diversions, and wastewater effluents in the Llobregat River Basin. Water diversions are denoted by *dashed blue arrows* and named from the origin with a *grey box*. When withdrawals are devoted to urban water supply the *blue arrow* points to a *blue box* indicating the supplied area. Numbers in *grey* and *blue boxes* are hm^3 $year^{-1}$. When the water supply depends on alluvial wells the origin of the intake is denoted by a *black dot*. For WWTP effluents (*orange circles*), the size of the symbol is approximately proportional to the volume of the effluent

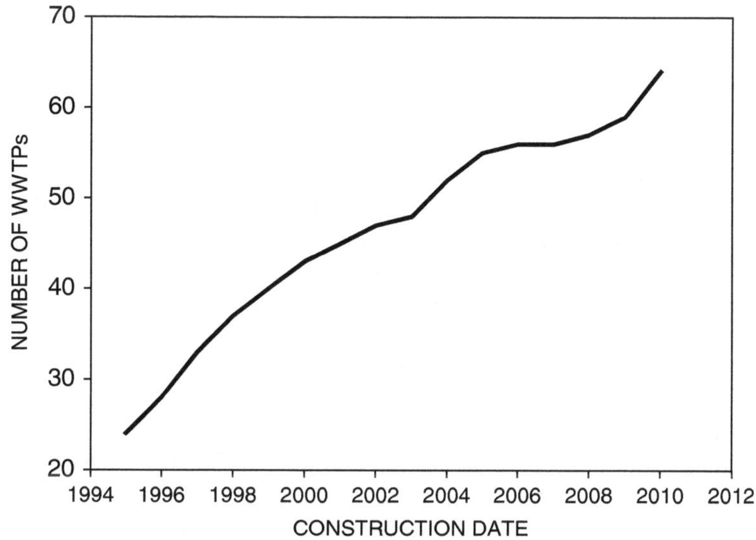

Fig. 13 Evolution in the number of EDARs in operation in the Llobregat River Basin for the last 15 years

and smaller one is the DWTP at Terrassa that treats a lower but significant water volume, from the Llobregat as well as from groundwater alluvial sources.

The last big withdrawal of superficial waters from the Llobregat River is at the DWTP of Sant Joan Despi, located at a short distance from the city of Barcelona near the river mouth. The uptake of this water leaves the river nearly exhausted downstream. When the superficial resources are scarce (summer) or difficult to capture, resources from underground wells (at Cornella and Sant Feliu de Llobregat) are also used.

The treatment of sewage waters in the Llobregat was nearly none before 1981. In 1984 and 1985 the WWTPs of Martorell and Manresa were placed in operation, but it was only in 1995 that the Catalan Parliament approved the Sanitation Plan for the inner watersheds. This Plan defined the fast construction and operation of up to 64 WWTP in the Llobregat watershed (Fig. 13), which presently include a total collector network of 516 km and 113 bombing stations. The present overall treatment capacity of these facilities amounts to 850,929 m^3 per day, and serve more than 2 million people.

The first WWTPs installed at the Llobregat watershed mostly used physico-chemical treatment. However, presently most of the WWTPs in the Llobregat include biological treatment, and in ca. 20% of them there is a biological treatment with capacity to eliminate phosphorus and nitrogen. Three of the WWTPs have a facility for water regeneration (tertiary treatment: Prat de Llobregat, Sant Feliu, Piera). In other two the treatment is based on the use of macrophytes or in water infiltration-percolation (Sant Marti de Sesgueioles and Hostalets de Pierola).

This network of WWTPs is complemented by other infrastructures to help on the sanitation in lower areas of the river. Mostly to avoid further pollution inputs to the superficial waters in the lower part of the river prior to their uptake for drinking purposes, most of the streamflow of the main tributary Anoia is diverted at its mouth to the Infanta Channel, which in part returns to the Llobregat downstream from the intake of the DWTP at Sant Joan Despi. The Anoia waters add to those of the Riera de Rubi and its WWTPs (water volume ca. 0.5 m^3/s), and the waters remain available for irrigation in the Llobregat lowlands. The share in recent years between the two inputs is of 3–4 hm^3 $year^{-1}$ from the Anoia and 18–19 hm^3 $year^{-1}$ from Riera de Rubi. About 8–14 hm^3 $year^{-1}$ of the total are used for irrigation, and the remaining volume is derived through another channel to the Llobregat downstream the DWTP of Sant Joan Despi, and finally reaches the Mediterranean Sea. The Right Bank Channel in the Lower Llobregat derives about 16–22 hm^3 $year^{-1}$ which is also used for irrigation.

Superficial waters in this final section are also allocated in recharge ponds to increase the aquifer reserves of the Delta, which are used as a strategic reserve for drinking water to Barcelona. Some other projects to improve the water quality as well as the volume of the water resources are under planning or execution. The WWTP of El Prat is planned to pump water upstream to maintain environmental flows in the final stretch of the river. Finally, a large desalination plant using reverse osmosis was recently put into operation and is able to produce a maximum yield of 60 hm^3 per year. The plant was defined to allow avoiding critical episodes of water shortages as well as to alleviate the existing pressure on the Llobregat and Ter Rivers, the two main sources for drinking water for the Barcelona metropolitan area.

A significant volume of the water withdrawn from the Llobregat River has already been treated by WWTPs and returned to the river itself, defining a major urban loop in the fate of the water as it travels along the basin. The average volume of water treated and delivered to the river accounts for 205 hm^3 $year^{-1}$ (remember that total streamflow near the mouth is ~600 hm^3 $year^{-1}$). It is estimated that from this huge volume 23 hm^3 $year^{-1}$ are reused at the DWTP in Abrera and 40 hm^3 $year^{-1}$ at the DWTP in Sant Joan Despí. This contribution therefore reaches ca. 8% of the total water volume from the Llobregat supplied to Barcelona Metropolitan Area, but may account ca. 30% under drought situations.

Acknowledgements This research was funded by the Spanish Ministry of Science and Innovation, projects SCARCE (Consolider-Ingenio 2010 program, CSD2009-00065), and CARBONET (CGL2011-30474-C02-01).

References

1. FAO (2010) The wealth of waste: The economics of wastewater use in agriculture. FAO Water Report 35. Food and Agriculture Organization of the United Nations, Rome, Italy
2. Beráslegui Batalla X (2001) Història geològica de Catalunya. In Nuchedel Riveo R (ed.) Patrimoni Geològic de Catalunya. Enresa, Madrid

3. Soler A, Canals A, Goldstein L, Otero N, Antich N, Spangenberg J (2002) Sulfur and strontium isotope composition of the Llobregat River (NE Spain): tracers of natural and anthropogenic chemicals in streamwaters. Water Air Soil Pollut 136:207–224
4. Williams M (2006) Deforesting the earth: from prehistory to global crisis: an abridgement. University of Chicago Press, Chicago
5. Vallès F (2002) El medi natural del terme de Manresa. Centre d'Estudis del Bages, Manresa
6. Milly PCD, Dunne KA, Vecchia AV (2005) Global pattern of trends in streamflow and water availability in a changing climate. Nature 438:347–350
7. Bates BC, Kundzewicz ZW, Wu S, Palutikof JP (2008) Climate Change and Water, Technical Paper of the Intergovernmental Panel on Climate Change. IPCC Secretariat, Geneva
8. Bosch JM, Hewlett JD (1982) A review of catchment experiments to determine the effect of vegetation changes on water yield and evapo-transpiration. J Hydrol 55(1–4):3–23
9. Brown AE, Zhang L, McMahon TA, Western AW, Vertessy RA (2005) A review of paired catchment studies for determining changes in water yield resulting from alterations in vegetation. J Hydrol 310(1–4):28–61
10. Margaris NS, Koutsidou E, Giourga C (1996) Changes in traditional Mediterranean Land-use systems. In: Thornes JB (ed) Mediterranean desertification and land use. Wiley, Chichester, pp 29–42
11. Lasanta T, García-Ruiz JM (1996) Erosión y recuperación de tierras en áreas marginales. Instituto de Estudios Riojanos, SEG. Logroño
12. Lorenzo-Lacruz J, Vicente-Serrano SM, López-Moreno JI, Morán-Tejeda E, Zabalza J (2012) Recent trends in Iberian streamflows (1945–2005). J Hydrol 414–415:463–475
13. Gallart F, Llorens P (2003) Catchment management under environmental change: Impact of land cover change on water resources. Water Int 28(3):334–340
14. Gallart F, Delgado J, Beatson SJV, Posner H, Llorens P, Marcé R (2011) Analysing the effect of global change on the historical trends of water resources in the headwaters of the Llobregat and Ter river basins (Catalonia, Spain). Phys Chem Earth 36:655–661
15. Zhang L, Dawes WR, Walker GR (2001) Response of mean annual evapotranspiration to vegetation changes at catchment scale. Water Resour Res 37(3):701–708
16. Marcé R, Rodríguez-Arias MÀ, García JC, Armengol J (2010) El Niño Southern Oscillation and climate trends impact reservoir water quality. Global Change Biol 16:2857–2865
17. Segura C (1991) Historia de los Regadios en España (...A.C. – 1931). Ministerio de Agricultura Pesca y Alimentación. Madrid
18. Bravo P (1991) Las piezas sueltas. Desde los primeros pobladores hasta las invasiones germanicas. Cataluña. pg 48-50. In: Segura C (ed) Historia de los Regadios en España (...A.C. – 1931). Ministerio de Agricultura Pesca y Alimentación. Madrid
19. Pardo A (1993) El Llobregat (II). Els Rius de Catalunya, Diari Avui
20. Mestres J (2008) Vilafranca abans de la història. In: Arnabat R, Vidal J (eds) Història de Vilafranca del Penedès. Ajuntament de Vilafranca del Penedès
21. Fernández-Armesto F (2001) Civilizations: culture, ambition and the transformation of nature. The Free Press, New York
22. Vila M (1998) Catalunya: Rius i poblament. Publicacions de l'Abadia de Monsterrat, Barcelona
23. Hughes R (1992) Barcelona. Knopf, New York
24. Codina J (1971) El delta del Llobregat i Barcelona. gèneres i formes de vida dels segles XVI al XX. Hores de Catalunya. Esplugues de Llobregat: Edicions Ariel
25. Shelvocke G, Campbell J, Swinton J, Sale G, Psalmanaza G, Brower A (1748) An Universal history: from the earliest accounts of time, vol 18. History of the Numidians, p 124
26. Hurtado V, Mestre J, Miserachs T (1995) Atles d'Història de Catalunya. Edicions 62. Barcelona
27. Latorre X (1995) Història de l'aigua a Catalunya. L'Abecedari, Premià
28. Solà À (2004) Aigua, indústria i fabricants a Manresa (1759–1860). Centre d'Estudis del Bages, Manresa

29. Puig i Valls R (1890) El Llobregat: aguas y montes. Revista de Montes. Año XIV. Núm 325–327. Madrid. Real Academia de Ciencias y Artes de Barcelona. pgs 357–366, 377–388, 427–439
30. Puig i Valls R (1904) El Llobregat: sus cuencas alta, media y baja y obras indispensables que hay que realizar en ellas, para conseguir que las inundaciones sean cada vez menos temibles, y las aguas normales más constantes, con aumentos de riqueza pública y particular. Memorias de la Real Academia de Ciencias y Artes de Barcelona, tecera época, vol. IV, núm. 40. Barcelona p. 524–536
31. CESALL (1932) Comisión para el Estudio de la Salinidad de las Aguas del Llobregat. National Archives of Catalonia. Fons 547. Agència Catalana de l'Aigua. Unitat 479, Barcelona
32. Honey-Rosés J (2012) Ecosystem Services in Planning Practice. PhD Dissertation, Department of Urban and Regional Planning. University of Illinois, Urbana-Champaign
33. Gorostiza S, Honey-Rosés J, Lloret R (in press). La Llei de Salinitat de 1933 i el primer projecte de col.lector de salmorres

Low Llobregat Aquifers: Intensive Development, Salinization, Contamination, and Management

Emilio Custodio

Abstract The most important aquifers of the Llobregat river basin are located in its lower part, namely the Low Llobregat Aquifer System (LLAS). Intensive groundwater development since the late nineteenth century and mostly after the 1950s has dramatically changed its hydrological functioning. Groundwater quality has suffered from salinity increase due to the potash mining area in the centre of the basin and seawater intrusion, as well as agriculture, waste disposal, sand and gravel pit backfilling and leakages. A stricter management is reversing and improving the situation and thus preserving the role of the LLAS as a crucial factor for the water supply to the Metropolitan Area of Barcelona. Besides model-based groundwater quantity management and stricter rules to protect water quality, additional actions include artificial recharge and installing a barrier to halt seawater intrusion by injecting highly treated reclaimed waste water. A groundwater user community was created and plays an important role in moving towards a sustainable use.

Keywords Barcelona, Groundwater, Low Llobregat aquifer system, Quantity and quality management

Contents

E. Custodio (✉)
Department of Geo-Engineering and International Groundwater Hydrology Centre, Technical University of Catalonia (UPC), c/Gran Capità s/n, Ed. D-2, 08034 Barcelona, Spain
e-mail: emilio.custodio@upc.edu

S. Sabater et al. (eds.), *The Llobregat: The Story of a Polluted Mediterranean River*,
Hdb Env Chem (2012) 21: 27–50, DOI 10.1007/698_2011_138,

1 Introduction

In the Llobregat river basin (5,000 km^2 in size), there are three main aquifer systems: the limestone formations, at the headwaters; the limestone and dolostone formations of the Anoia, at the central western side, and the low Llobregat alluvial deposits, which are divided into two upstream small alluvial formations, and the lower valley and the delta. This chapter deals mostly with the intensively exploited, partly urbanized, lower valley and delta aquifers, which together are crucial for the water supply to the Metropolitan Area of Barcelona.

In the central part of the Llobregat basin, there are extensive, cropping out, and deep evaporite formations of Oligocene age, mostly gypsum and halite. They contain important resources of potassium salts (potash) that have been exploited in four deep mines since the 1920s, although currently operations have greatly decreased. The area is surrounded by large salt dumps, whose leaching by rainfall and surface water, jointly with the disposal of brines from the mines and the mineral processing plants, is the origin of an important salinity contamination of the river water. Since the mid 1980s, a large part of dissolved salts is collected into a brine pipeline and brought to a sea outfall, although part of the problem remains. See [30] for a detailed explanation.

Population concentrates in the valleys and the inter-range plains, especially in the lower part, where the large conurbation of Barcelona and its Metropolitan area are located, about 4.5 million inhabitants. A large fraction of the Low Llobregat alluvial areas that were highly productive irrigated agricultural areas are now transformed into urban and suburban areas, and roads, motorways, and railroads. The average rainfall is about 650 mm/yr. The average river flow is about 18 m^3/s, but with large seasonal and year to year variations. A large part of the river flow originates in the high, more humid and colder headwaters. There are occasional large floods peaking at more than 3,000 m^3/s. In droughts, the river flow may be reduced to as low as 1 m^3/s, with river tracts that were occasionally dry under natural conditions. The river and the related aquifers are important water supply sources and reserves.

2 Hydrogeological Characteristics of the Low Llobregat Area

The low Llobregat valley (Fig. 1) contains permeable alluvial sediments resting on and bounded by low permeability Palaeozoic shales and Miocene and Pleistocene clays. Alluvium resulting from the large climate changes along the Quaternary fills the bottom of the low part of the basin [1]. There are various encased river terraces

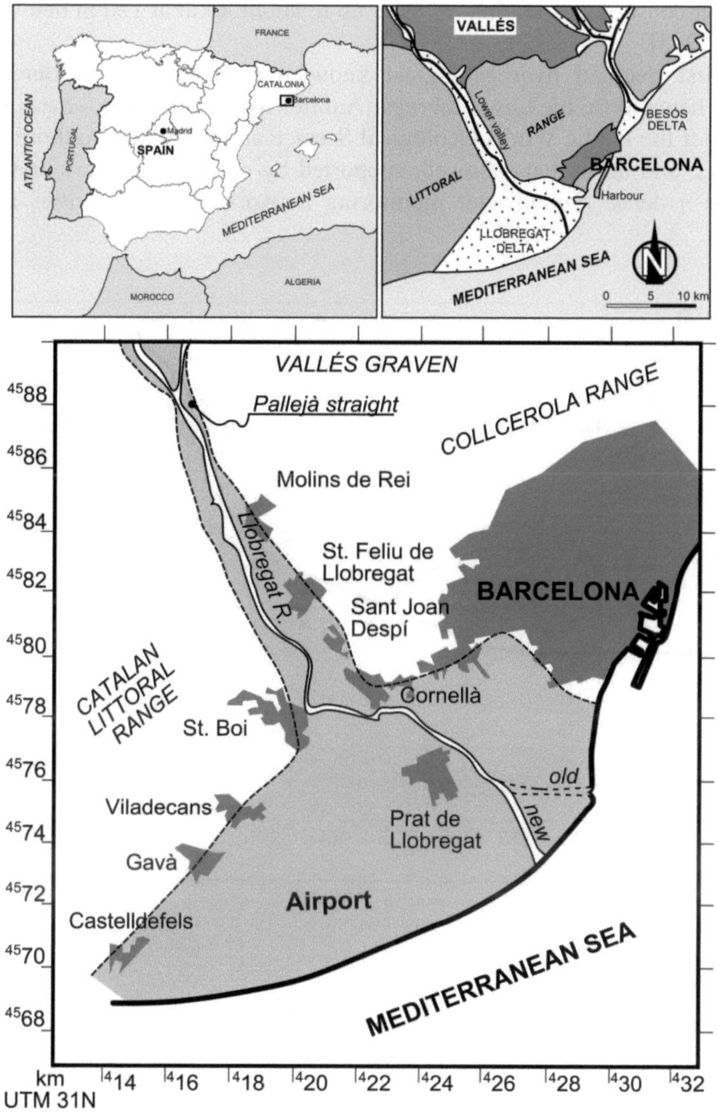

Fig. 1 The low Llobregat area and its situation

connected to the river, formed by silt, sand, and gravel. The thickness increases downstream, up to more than 50 m at the valley's lower end. These flat formations are 800–2,000 m wide, with narrow passages when crossing the ranges. In the lower part, the terraces show the influence of sea level changes along the Quaternary. In the lowest part, these sediments become deltaic sediments deposited at different times, which are partly eroded and encased. The present delta, rounded by along-the-shore marine sediment transport, contains up to 60 m of Holocene sediments at the current coastline, extending offshore up to about 4 km at 120 m below present sea level [2–5].

The first detailed hydrogeological knowledge of the Low Llobregat area aquifer – henceforth, the Low Llobregat Aquifer System (LLAS) – dates from the 1960s, after the Public Works Geological Service and the Water Authority studies and works, and those that followed, supported by extensive surveys and drilling, and quite a dense groundwater monitoring network. These activities and their

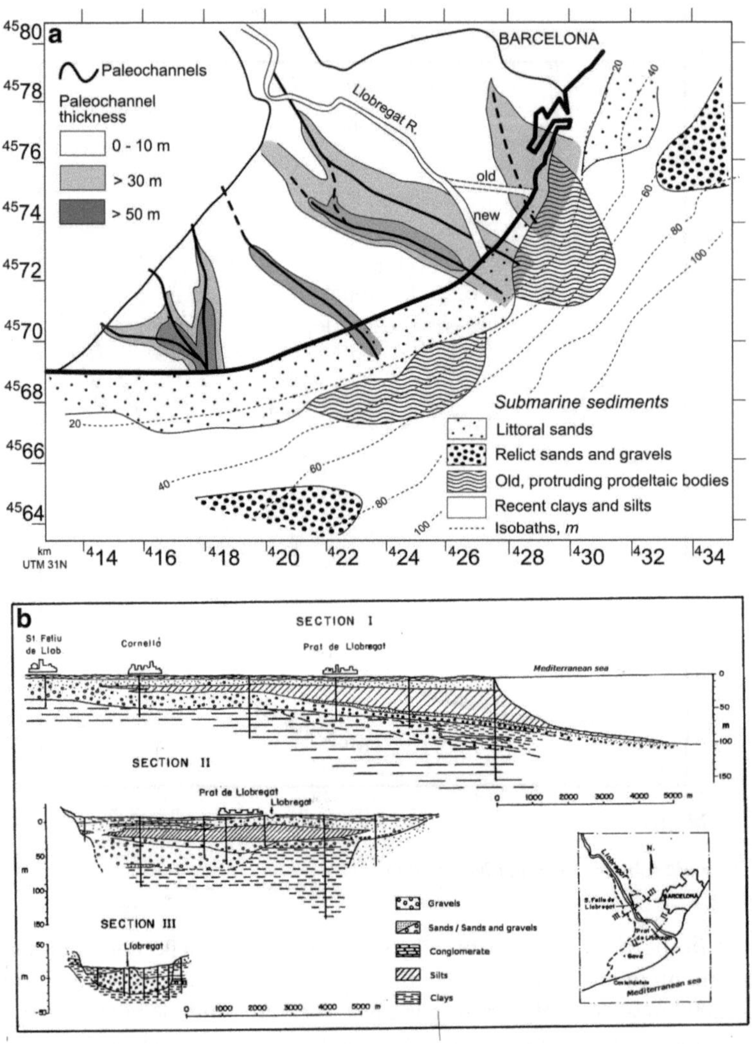

Fig. 2 The Llobregat delta. (**a**) The area with indication of submarine formations and the depth of the sea bottom (in [4]), and the position of the known late Pleistocene river channels that are part of the deep aquifers (modified from [5]). (**b**)Typical hydrogeological cross sections (in [8])

results fostered further research activities and studies that still continue, especially with regards to the delta aquifer hydrogeology [6, 7]. Relevant references will be given when dealing with the different aspects.

The Lower valley alluvium holds a water-table aquifer of encased river terraces that contain coarse gravels. The delta holds a two-aquifer system. The delta upper aquifer formed after the sea level recovery about 10,000 years ago. It consists of a complex arrangement of river and side tributary channels, beach and eolian sands, organic matter-rich silts and clays filling old marshes, and a silt and clay cover deposited by floods. Sandy formations dominate, forming a water-table aquifer, slightly confined locally, and up to 20 m thick.

The deep delta aquifer contains Late Pleistocene sand and gravel sediments – even very coarse gravels – from the river and side tributaries, when sea level was at about 120 m and the coastline about 4 km offshore, where they crop out partially (Fig. 2). This deep formation, 5–15 m thick and spatially variable, is confined below an up to 40 m thick wedge of clay, silt, and fine sand sediments deposited under Early Holocene estuarine conditions, grading upward to the upper aquifer. It forms an aquitard that thins out offshore and also toward the delta boundaries, where the upper and deep aquifers meet. Below these delta sediments, remnants of close-to-the-coast and offshore older deltaic formations exist. Nevertheless, they have much poorer hydrological characteristics than the deep aquifer. The basement consists mostly of very low permeability, Pliocene clay formations.

The deep delta aquifer is in hydraulic connection with the lower valley aquifer, forming together a unit, the main aquifer (MA). Henceforth, what is presented refers to the LLAS, comprising the lower valley and the delta, with special reference to the MA. Surface areas are 40 km^2 for the lower valley and 80 km^2 for the delta.

3 Groundwater Development

Groundwater development in the Low Llobregat area started in the late nineteenth century for the supply of Barcelona by means of large diameter, deep wells in the lower end of the valley and through the drilling of small diameter, deep boreholes in the delta that at that time were flowing wells. This fact, together with the high yield of the wells, attracted water demanding factories, especially textiles, paper mills, and chemical industries. Development accelerated in the 1930s and especially during the 1950s and 1960s, peaking in the 1970s (Fig. 3) at about 120 hm^3/year (1 hm^3 $= 10^6$ m^3). This intensive development has progressively decreased, due to the increased cost of water abstraction, the dramatic loss of quality by seawater intrusion, the progressive urbanization of the area and the pressure from the public administration and civil society to reduce pumping.

After the opening of a river water treatment plant for urban supply in the late 1950s, groundwater abstraction in the lower valley for supplying Barcelona decreased and progressively the aquifer became an important reserve for dry periods

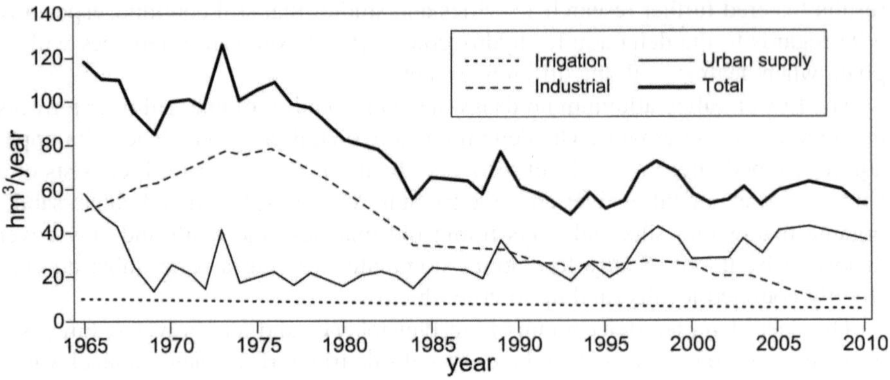

Fig. 3 Evolution of groundwater development in the LLAS since 1965

and for occasional moments of poor river water quality. Current exploitation is about 50 hm^3/year, with variations according to the year and the water management decisions taken by the water authority. Under the current circumstances, recharge activities, and management, this exploitation is considered a sustainable average value.

4 Hydrogeological Conceptual Model Under Natural Conditions

The lower valley alluvium elevation varies from +25 to +11 m relative to mean sea level along 10 km. This is a rather large slope due to the coarse materials. Recharge was due to direct rainfall infiltration on the alluvial area and lateral contributions. Under natural conditions, the Llobregat River recharged water during dry periods in the upper reaches and was permanently effluent downstream, with important outflows in the lower end, at the delta inlet. This natural condition was altered by the introduction of irrigated agriculture in the mid-nineteenth century, in both sides, by means of two canals fed by the river that also irrigated the eastern and central areas of the delta.

In the delta upper aquifer, recharge derived mostly from rainfall and lateral contributions from the mountaneous sides, plus the canal losses and excess water after irrigation was operating.

In the delta deep aquifer, the river-controlled head at the delta inlet was about +11 m. At about 4 km offshore (12 km from the delta inlet), the deep aquifer crops out at about −120 m; there, the equivalent freshwater head is 3 m, according to the Badon Ghijben–Herzberg ratio [9]. Thus, seaward groundwater flow and discharge at the sea bottom outcrop was possible. This deep flow was slow and variable, according to the aquifer characteristics, but was enough to flush out the original marine water in a large part of the delta deep aquifer in a few centuries. Thus, in the central and eastern sector, the delta deep aquifer contained mostly recent fresh water transferred from the lower valley.

The delta inlet is shaped as an alluvial fan with an elevation of +11 m at the apex. Elevation is +5 to +7 m at the boundaries. The remaining area, most of the delta surface, has an elevation between +0.5 and 3 m. Thus, most of the deep aquifer head was above land surface, thus allowing the drilling of flowing boreholes in the period between the end of the nineteenth century and the early twentieth century.

The aquitard wedge has a very low vertical permeability in a large part of the coastal area. Inlandward the wedge thins out, becomes more permeable, and finally it grades into sand close to the inner boundaries.

Under natural conditions, the lower valley aquifer contained water derived from local rainfall and some river water, which was later on increased after the irrigation activities in the lower valley took place. The delta upper aquifer was fed by rainfall and lateral inflows and affected by river water from the irrigation canals in the central and eastern parts.

Some data point to a natural river chloride content of about 80 mg/L. After the late 1920s, the saline pollution from the salt mines started, propagated to groundwater through the river bed and the irrigated areas, and moved into the delta aquifers.

5 Hydrogeological Conceptual Model Under Disturbed Conditions

The main consequence of intensive groundwater development has been a generalized water head drawdown and the loss of flowing conditions in the main aquifer. Henceforth, the conditions preferably accounted for are those of the 1980s when abstraction was at its maximum. Figure 4 is a simplified longitudinal cross section, comparing the undisturbed (natural) and disturbed conditions.

In the lower valley aquifer, the water table was permanently below river level down to 20 m in some critical periods. This resulted in the river being perched since the infiltration rate was less than that needed to sustain a saturated zone below. Under this situation the river recharged along the whole valley, although more readily upstream due to the higher vertical head gradients and coarser materials there. In this case, floods play a key role in recharge due to the lateral spread of water – now greatly reduced by the embankments – and the scouring effect of the river bed by the increased water velocity. To compensate for the decreasing flood effects and the siltation of the river bed, infiltration has been artificially enhanced by leveling and scrapping periodically the bed in a river tract. Indirect river water recharge was produced in the canal-irrigated areas. It was important due to the large water depths applied in the fruit tree fields to kill weeds, which also prevented the use of herbicides. The progressive reduction of irrigated areas and the pressure to

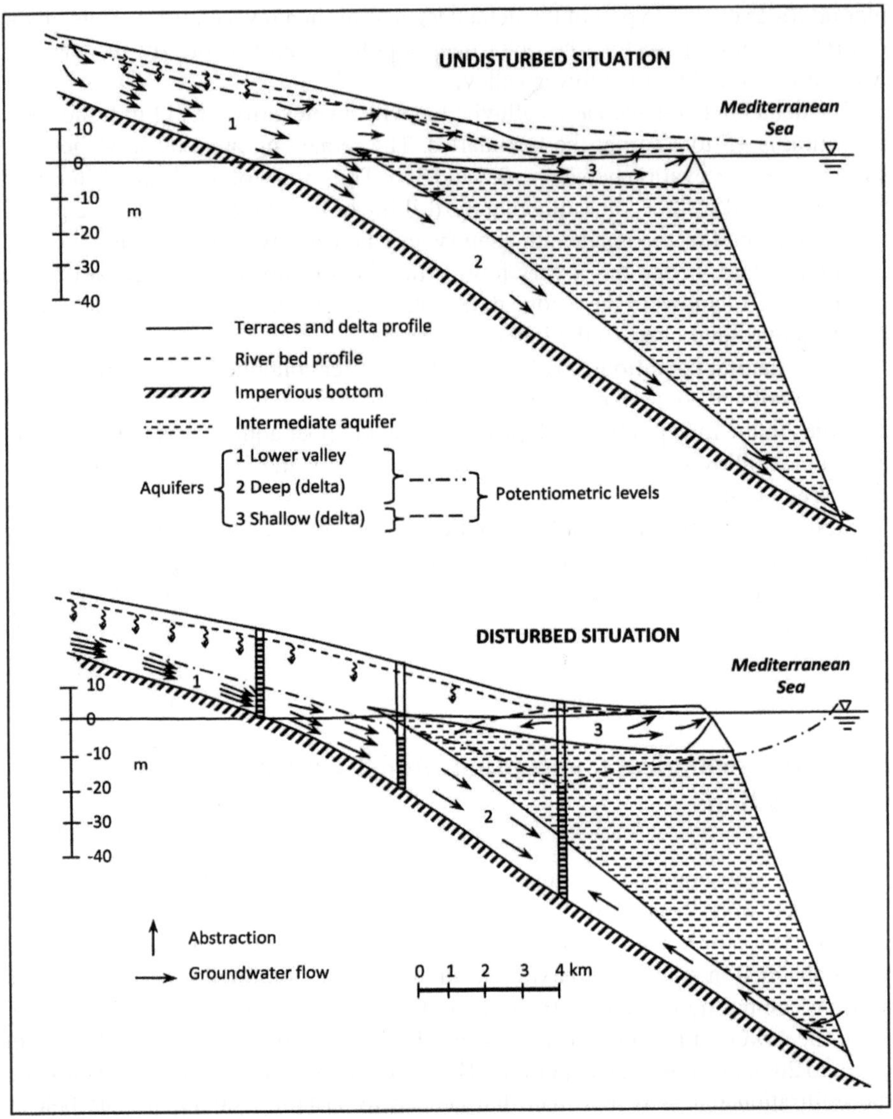

Fig. 4 Simplified hydrogeological longitudinal cross sections of the LLAS, showing the undisturbed (natural) situation before 1900, and the disturbed situation in the 1980s. Current piezometric levels of the deep aquifer are much higher, although not above surface level; the lower valley water table is also higher but not attaining the river channel bed (after [6])

reduce irrigation water depth is against this positive effect on aquifer recharge. Under these circumstances most of the groundwater in the lower valley aquifer is derived directly or indirectly from the river, as the water chemical and isotopic composition shows.

In some years, in the central area of the deep delta aquifer, the piezometric level was as low as −15 m, and down to −8 m along the central coastal area. This increased groundwater transfer from the lower valley, and from the upper aquifer through the inner boundaries. A large part of the deep aquifer water up to the head minima at the main pumping centers, was replaced by very recent, river-derived water. The low piezometric levels induced seawater penetration from the offshore outcrops. The show up of this saline intrusion was delayed about 20–40 years. This was first noticed around the mid 1960s [10]. Afterward, the intrusion developed fast, advancing toward the pumping centers (Fig. 5), mostly along two main paths that correspond to more permeable features of the deep aquifer [11]. Since groundwater gets radially into the wells, those in the saline front abstract a mixture of freshwater with a lower proportion of seawater due to the decreasing seaward permeability and aquifer thickness. Wells that continued operating after being salinized – mostly for industrial cooling purposes – protected inlandward wells, but when they ceased operation, which happened progressively, the intrusion advanced toward other inlandward wells. The saline water penetration changed with piezometric pumping pattern conditions in the lower valley aquifer. Salinity due to seawater intrusion did not attained the supply wells for Barcelona

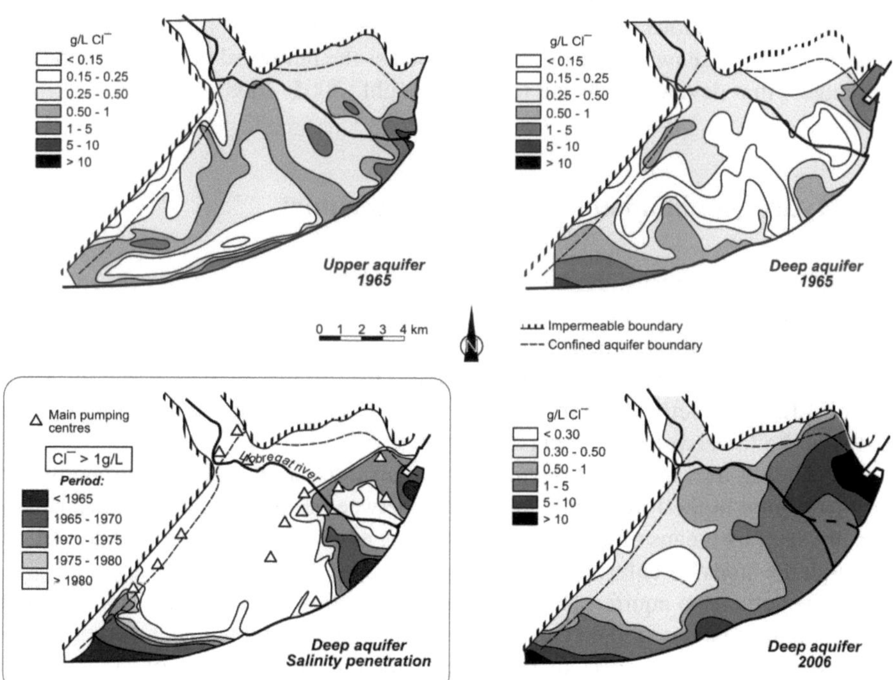

Fig. 5 Chloride content in the delta aquifers, after the first survey in 1965, in the upper aquifer and the deep aquifer, and in the deep aquifer in 2006, with indication of seawater intrusion advancement from prior to 1965 to 1980, which is the period of faster penetration

placed at the end of the lower valley, although in 2006, during a serious drought, one of them was slightly affected for a short period of time. The Prat de Llobregat supply wells have been badly affected. The situation worsened after the Harbour Authority of Barcelona started the excavation of inland docks in the early 1970s, in a place where the aquitard thins out and becomes sandier.

The situation in the delta upper aquifer underwent less hydrogeological modifications since there is only a small exploitation due to aquifer conditions and the poor quality of water. An important change was the transformation of the whole eastern area and a large part of the central area, formerly canal irrigated, into urban and industrial areas. Sea water intrusion was not significant, except for very local situations or the existence of still unflushed marine water in the recent, westernmost area. The drainage of the airport runaways and their extension have caused a general lowering of the water table, but the main negative effect occurred on the existing coastal wetlands. However, in the delta boundaries, where the deep and upper delta aquifers meet, there is an important drawdown, enhanced recharge of contaminated local water and some areas with saline increase from deep, still unflushed aquifer layers.

The current situation (2011) in the LLAS is quite different due some wet years (2009–2011) and the management efforts. Groundwater abstraction decreased, especially in the delta area, and a seawater intrusion barrier is in operation [28]. The most conspicuous result is a groundwater head recovery that greatly reduces and pushes back seawater intrusion and disperses the landward brackish groundwater fingers that developed during the last drought (2005–2008).

6 Hydrogeochemistry and Mass Transport

The formulation and validation of the conceptual models have largely relied on hydrogeochemical and environmental isotope studies, and on modelling. These works started in 1964, when the LLAS was already quite largely disturbed [10] and the river water had a large saline content. This salinity increase was transferred to the LLAS, directly through the river bed and at the rather extensive irrigated areas in the lower valley existing at that time.

Depending on water-table depth and total abstraction, the average groundwater turnover time in the lower valley aquifer is between 1 and 3 years. Similar values are obtained by means of tritium data, assuming exponential mixing [12], as well as through the study of chloride content fluctuations.

In the delta deep aquifer, the transfer time of a conservative solute from its inlet to the pumping areas in the centre of the delta (Prat de Llobregat) is variable. The chloride content fluctuation in recharge water due to the cease of potash mine activity during the 1936–1939 Spanish Civil War, and resumption of full activity in 1940, shows a 10-year period of time (Fig. 6), which corresponds to the 1950s conditions [13]. Tritium data yield similar results for the 1960s. This time period is calculated to be about 30 years under natural and also under current conditions and

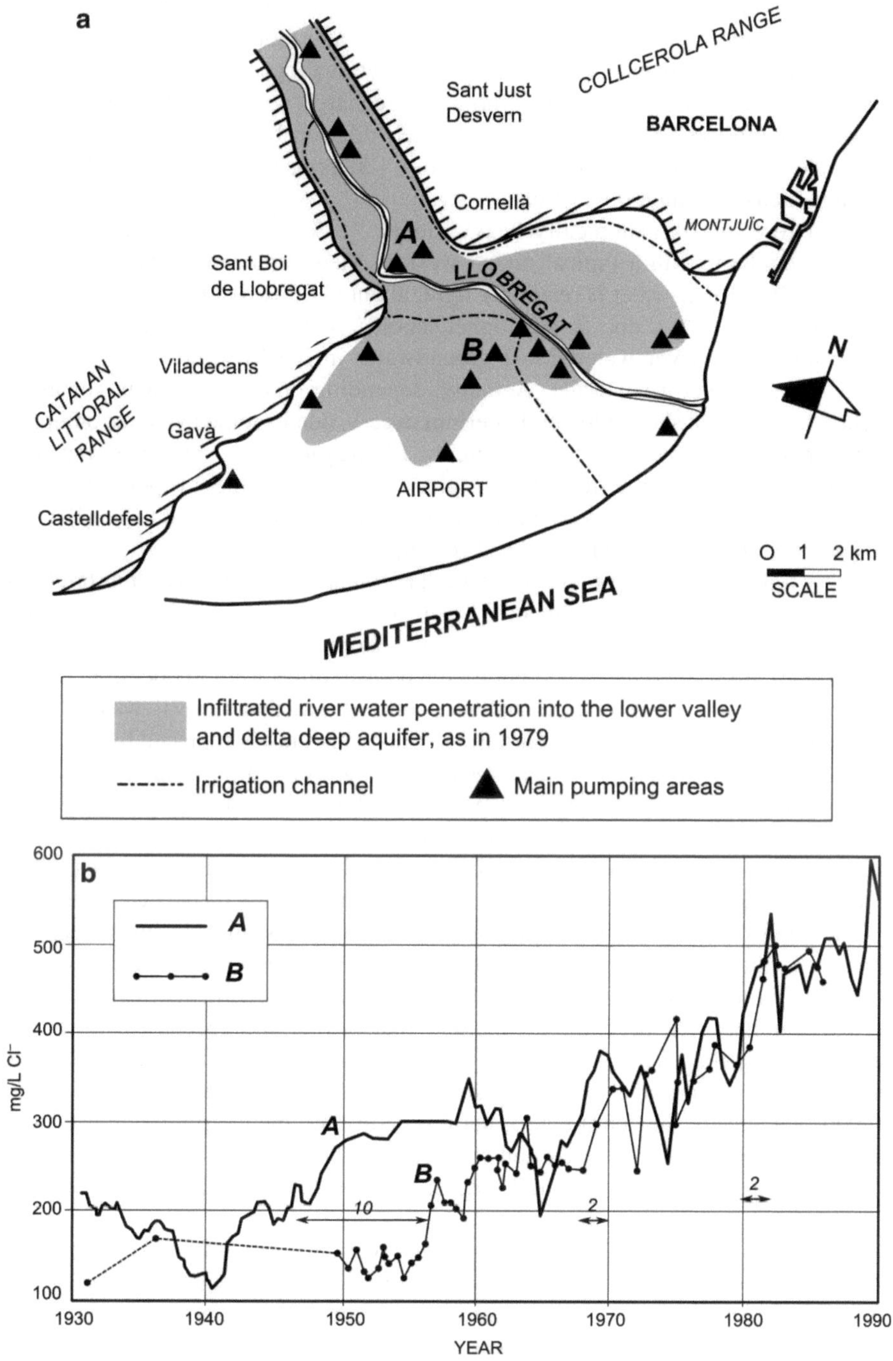

Fig. 6 Replacement of old ("pre-salt-mines") water in the deep aquifer by transfer from the lower valley aquifer, progressing towards the main pumping areas in the central delta. (**a**) Situation in 1967. (**b**) Chloride content evolution in wells A (delta inlet) and B (deepest part of the drawdown cone in central El Prat de Llobregat, where new water was mixed with old water). The *double arrows* and the figures above them indicate the flow time (after [10, 13]; and other sources)

as low as 2 years in the 1980s. This depends greatly on total groundwater abstraction and its pattern, and the average depth of the water table in the lower valley.

Quite high ^{14}C values in some wells in the lower valley in the early 1970s, up to 150 pmc (percent modern carbon), show that recharge was partially produced by direct river infiltration in its bed, with limited dilution by dead carbon from the carbonate-dominated alluvium. Dilution to <100 pmc was expected below the irrigated areas, as found downstream.

Local recharge is isotopically heavy, about −5‰ $\delta^{18}O$, with regards to the SMOW standard, with a typical mixed western Mediterranean deuterium excess of about 14‰. River water is relatively light, about −7.5‰ $\delta^{18}O$, with a deuterium excess of about 10‰, due to the higher, cooler, and Atlantic origin of a large proportion of the river water. Most groundwater in the MA shows the isotopic content of infiltrated river water. Actually, depending on circumstances, it can be slightly heavier and show a lower deuterium excess, due to some evaporation in the many small hydropower reservoirs along the Llobregat valley – and more recently at the three upstream large storage reservoirs – plus some further evaporation in the irrigated areas, where water was mostly applied by inundation of the fields (Fig. 7). Despite this, it is still lighter than local recharge.

These light water isotopic characteristics are found in the delta deep aquifer, both in recent and old – predevelopment – freshwater, due to the same origin; but old water has low Cl, is tritium free, and has prenuclear radiocarbon content. This points to river-derived waters, recharged before the potash mines were in operation and in their way toward the submarine outflow, although at that time it was moving backward due to groundwater exploitation. They were slowly replaced by a mixing of fresh and marine water due to the large transport dispersion. Tritium, radiocarbon, and inorganic ion contents show a sharp transition between groundwater inlandward from the central delta pumping areas and groundwater seaward from them. Currently most of the predevelopment groundwater has been depleted and substituted by dispersed seawater, except in some low permeability parts of the deep aquifer.

The delta deep aquifer is heterogeneous, with more permeable linear features corresponding to the Late Pleistocene river channels shown in Fig. 2 – or the coarser sediments at the eastern boundary – and less permeable sediments in between, mostly old beaches, dunes, and marshes. Thus, groundwater flow is variable along the coast, with areas slowly renovating. ^{14}C data in these areas show this, with water ages of a few thousand years, compatible with the less than 10,000 years age of the confining formations. However, the interpretation is unsure due to the complex inorganic carbon chemistry in those highly reducing ambients, in which 10,000–6,000 years old organic matter-derived carbon is added to groundwater.

Recharged river water can be easily characterized by its high Cl and Na contents, with relatively low SO_4 and moderate HCO_3, and quite high K and Mg. Br concentration is in excess of what can be expected from halite dissolution, since part of the brines derive from processing the Br-enriched, K and K – Mg salts. Locally recharged water is less saline, but high in SO_4, lower in K, and somewhat high in Br. This can be followed through the aquifer system.

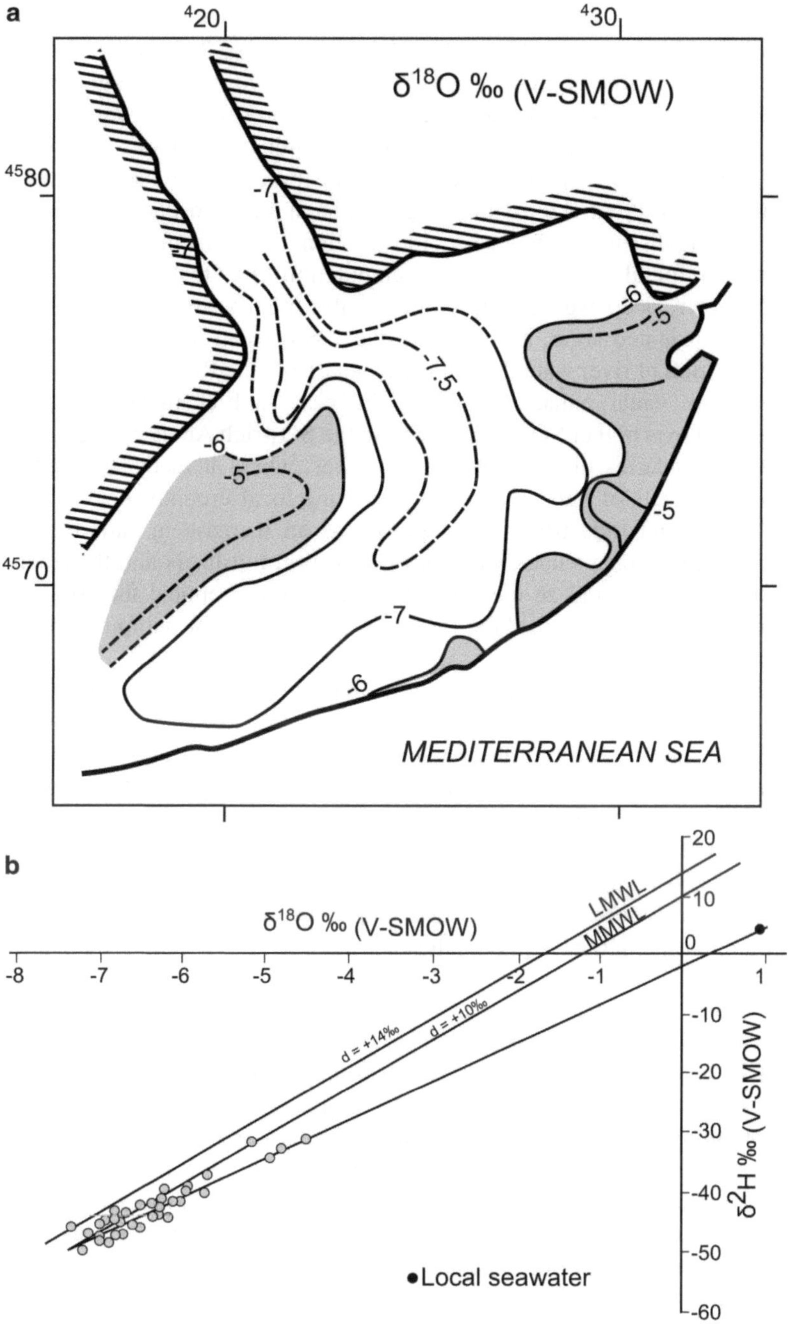

Fig. 7 Water isotopes in the LLAS as in the mid 1970s. (**a**) Distribution of $\delta^{18}O$‰, showing that new and old water correspond to the light river water (a), but with some starting mixing with sea saline water (b), and lateral inflow of heavier local water (c). (**b**) $\delta^{18}O$–$\delta^{2}H$ plot of data that show the deuterium excess (d) of river water, with a slight evaporation effect, and the mixing with seawater (after [6, 11, 12])

River water has a TDIC (total dissolved inorganic carbon) content of about 230 mg/L HCO_3 and about 500 mg/L $CaCO_3$ in hardness. This is transferred to the aquifer when recharge is produced at the river bed, with only a slight increase as dissolved organic matter oxidizes. However, the lower valley aquifer shows a higher TDIC of 350–425 mg/L HCO_3 and a hardness of about 650–700 mg/L $CaCO_3$. This is due to part of recharge being produced in the irrigated areas. There the soil CO_2 partial pressure is higher than in the atmosphere, and infiltrating water is able to dissolve carbonates in the alluvium. This manifests in the carbonate encrustations found in pipes and the accumulation of CO_2 in the bottom of idle, large diameter, deep wells. This points to the relative importance of recharge through the irrigated areas, at least until the 1980s when they began to decrease. The substitution of river water by polluted water from the Rubí and Anoia Rivers in the left bank canal enhanced salinity increase and chemical changes, now accompanied by a higher SO_4/Cl ratio due to the SO_4-rich Anoia river contribution.

At the boundaries of the delta deep aquifer, where it meets the delta upper aquifer, the effects of intensive agriculture using local groundwater for irrigation and of urban and industrial areas, appears as an increase in salinity by water recycling, with a conspicuous increase of SO_4 from fertilizers and the leaching of local Triassic rocks, and an increase of TDIC. Some chemical fingering indicate that this area is able to contribute to the deep aquifer main pumping areas in the central delta (Fig. 8).

Near the coast, the delta deep aquifer is under strong reducing conditions due to the organic matter-rich sediments just covering it. This affects freshwater, old saline water, and recently intruded marine water. The result is SO_4 reduction, Fe dissolution after dissolved SO_4 is depleted, CH_4 formation, and TDIC increase. SO_4 reduction agrees with increased $\delta^{34}S$ and $\delta^{18}O$ in remaining SO_4 [15]. Intruded saline water shows the expected Na/Cl ratio decrease, although the $^{87}Sr/^{86}Sr$ ratio corresponds mostly to aquifer carbonate dissolution by organic matter-derived CO_2, instead of what is expected from closed simple mixing seawater and freshwater. In the saline areas, where still poorly flushed, old seawater is found, the Na/Cl ratio increases.

The hydrochemical characteristics of the delta aquitard were not known before the 1960s. At that time, a geoelectrical survey showed the existence of saline water in the half delta close to the coast. This was also indirectly shown by some small diameter, old deep wells, with corroded casings; short after starting the pump a transient salinity peak appeared. Several, carefully cored boreholes were drilled, and interstitial water was studied for chemistry and tritium content. The result was prenuclear water. Salinity profiles (Fig. 9) correspond to seawater in a process of upward diffusive/dispersive displacement by deep freshwater. Far from the coast, the original seawater is almost flushed out, but is scarcely displaced close to it. Ionic ratios show the progressive upward replacement of seawater by freshwater. In a borehole, from which pore water was obtained by pressure extraction, chemical and water isotope analyses confirm previous findings and show a chromatographic upward displacement [14, 16].

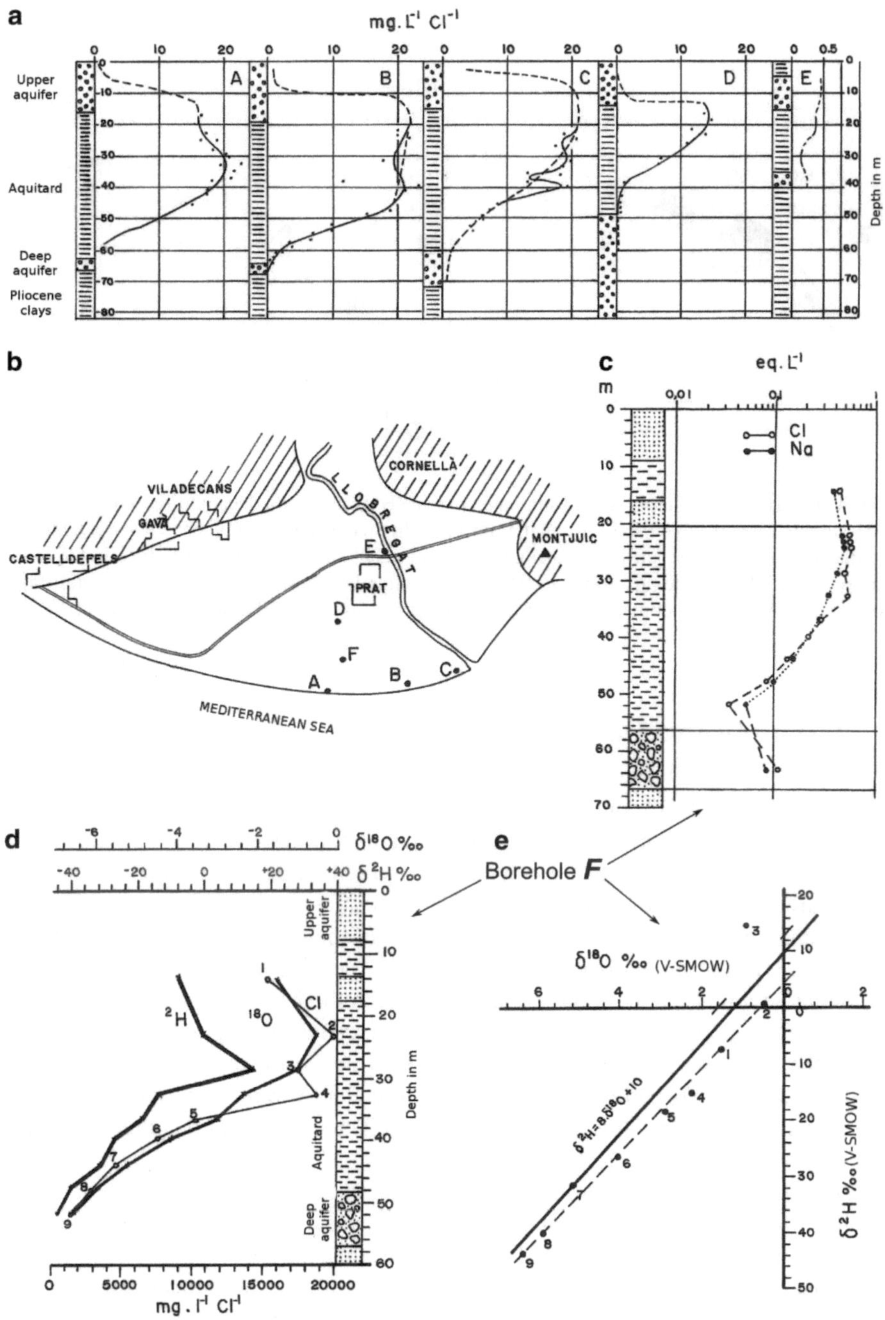

Fig. 8 Saline characteristics of the delta aquitard. (**a**) Vertical profiles showing chloride content of pore water in the locations shown in (**b**). (**c**) Shows that below 40 m there is clear Na^+ deficit relative to Cl^- produced by cation exchange when fresh water is displacing marine water. (**d**) compares Cl, $\delta^{18}O$ and δ^2H, which also show the marine origin of pore water; in the $\delta^{18}O$–δ^2H plot they follow the expected mixing line; the slight displacement is due to slight evaporation of the dripping some water samples during core squeezing (after [11, 13, 14])

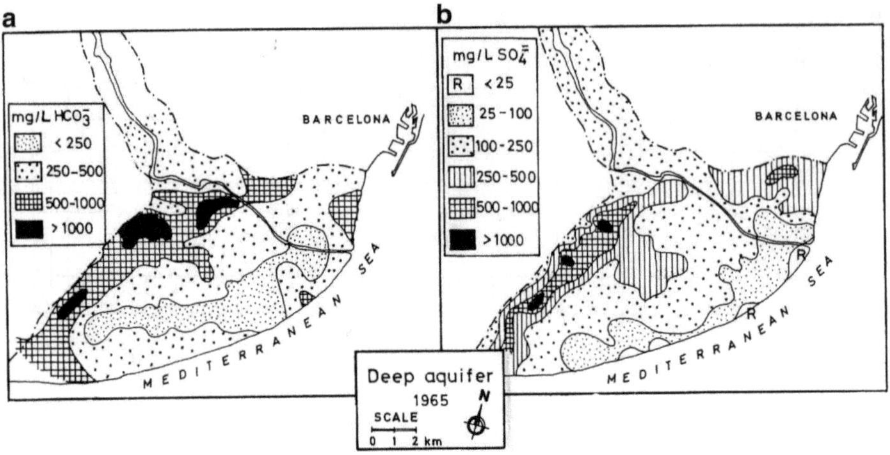

Fig. 9 Bicarbonate content (A) and sulphate content (B) of the delta deep aquifer with data of the first survey carried out (1965). At that time clear human water quality changes were already produced. (**a**) Old water (pre-salt-mines), (**b**) chloride-contaminated water recharged in the lower valley, (**c**) highly reducing ambient (R) and first signals of arriving seawater, (**d**) unflushed formation seawater, (**e**) contamination from local agricultural and urban sources (after [10])

7 Groundwater Contamination

The most conspicuous contamination sources of the LLAS come from river water salinization from the upstream potash mines and from the groundwater extraction-induced seawater intrusion in the delta, as commented before. Further contamination sources are the two highly polluted tributaries, just in the upper part of the area, supporting large urban and industrial areas, one of them also contributing relatively high SO_4 concentrations from natural sources. These tributaries produced serious problems of ammonia and high organic content in the Llobregat river water. So, in the late 1970s they were diverted to the irrigation canal running along the left bank. Since the remaining irrigated agricultural areas in the lower valley use this poor quality water, a SO_4 and hardness increase in recharge water occurred. Downflow, this recharge mixes with the other groundwater. Thus, the mineral content of part of the supply wells to Barcelona increases.

The nitrate content in groundwater is high in some areas of the delta upper aquifer due to intensive agriculture and urban wastewater infiltration, especially near the western boundaries, with some areas that may exceed 100 mg/L NO_3. Likewise occurs in the side terraces of the lower valley, where local recharge may be dominant. However, most of groundwater is rather low in nitrate, as it is in river water, including the possible increase by oxidation of NH_4 in the river water. In the deep delta aquifer, nitrate shows at low or zero concentration. This is due to the redox potential decrease after becoming confined. Close to the coast, their intense reducing conditions favor natural NH_4 in relatively large concentrations, as well as dissolved Fe(II).

Except for the coastal part of the delta deep aquifer, Fe^{2+} is only found in low concentrations since the redox potential is not low enough, but there are exceptions. In the alluvial basins upstream of the LLAS, Fe(II) may be present due to more reducing conditions. Currently, this seems to be propagating to the upstream part of the LLAS, which had not shown this kind of behavior before. Mn(II) is not widespread problem, contrary to what happens in other close-by deltaic areas. No general arsenic problems are known. However, occasional As may be currently found in areas with a previous highly reducing ambient, which now receive recharged water from the seawater intrusion injection barrier.

The highly anthropized area, with large urban and industrial areas and intensive irrigated agriculture – although decreasing and removed in some areas, jointly with many years of poor care for the environment and groundwater quality, explain the many sources and forms of localized groundwater contamination. This goes in parallel with the intense activity in the 1970s and 1980s to extract sand and gravel from the valley and delta, with poorly controlled operation and backfilling of the pits. In some cases, permits were issued for disposal of urban refuse and industrial wastes in these pits. Disposal places are poorly known and documented. Studies were rare, with very scarce means. Current powerful study tools were not available at that time. So, what follows is mostly descriptive and covers only some cases known to the author.

Out of river water salinization due to potash mining and seawater intrusion, and a few occasional breakdowns of brine pipelines, some conspicuous salinity pollution events were produced. They are linked to water softening in local factories by means of ion exchange and resin regeneration with NaCl brines. Since the disposal of saline water into the river was not allowed, and it was not possible to incorporate them into brine pipelines until recently, the common, illegal, disposal way was just casting away the brines into local dry ravines, where they infiltrated and incorporated into groundwater flow, where salinity was diluted, but with a nonnegligible effect downflow. Since the water table is deep due to groundwater development, the unsaturated zone was able to hold a large brine storage, slowly moving down toward the aquifer. During a large river flood, water inundated the creeks and produced an important recharge event that pushed down the brines stored in the unsaturated zone and also flushed them from the lower part due to the water-table rise. The result was a local salinity peak in the aquifer that moved and faded out downflow [17]. The scarce-existing data show that the acute effect disappeared in a few weeks, but with damage to close-by groundwater uses and the factories themselves. This problem is currently under control.

A factory producing hexavalent chromium from chromites operated in the area for many years. Wastes contained rather high contents of soluble Cr(VI). Total production wastes were estimated at 300,000 tons until a controlled disposal site was made available. These wastes were used to backfill sand and gravel pits in the lower valley and delta, and also as a filling to reinforce the local Llobregat River embankments. Three decades after ceasing these activities, the Cr(VI) was partially flushed out by groundwater and reduced by organic matter, but the buried wastes still contain some Cr(VI). It is not known how much remains. Other Cr(VI) problems come from small platting factories where their wastes were infiltrated into the soil to avoid treatment. One such case affects sporadically some water supply wells, depending on the exploitation pattern, the water-table elevation, and the operation of other nearby wells; the factory closed down in the early 1970s, the contamination was noticed in the 1980s, and the effects are still noticeable.

High boron concentrations were detected in the 1970s in the western boundary of the delta after farmers complained about crop poisoning. The origin was traced to an area of backfilled sand and gravel pits where dumps included industrial wastes incorporating boron-rich cleaning agents.

Liquid-chlorinated solvents were stealthy disposed into sand and gravel pits mostly in the 1960s and 1970s. Several problems of groundwater contamination were reported. A large one affected the water supply of El Prat de Llobregat. Water treatment facilities were to be introduced to reduce the contaminant content in the supply water. The problem was noticed in the 1980s and slowly faded out, but it can still be observed at small concentrations. The source can be traced to some sand pits about 3 km away, in a site where groundwater transfer to the deep aquifer was possible through the fine-sand layers separating them. The disposal site was not known, thus preventing remedial action.

Many oil stations have leakage problems, although not large contamination cases are known. A major oil spill accident happened in 1991, just a year before the Olympic Games event in Barcelona in 1992, which prompted special action by the authorities. About 70 m^3 of gasoline leaked into the lower valley aquifer, not far from several supply wells, that soon showed the contamination. The spill was the result of a terrorist action on an oil pipeline and the failed attempts to repair the damages. Action was taken to recover the gasoline floating on the aquifer and impregnating the unsaturated zone, at that time about 15 m thick [18]. The problem was controlled although not fully solved in spite of a long and costly cleaning up process.

Sand and gravel pit backfilling was assumed to be done with inert material. In the best situations, this was done by dumping excavation earths and building demolition debris, the latter partly incorporating gypsum from the wall coatings of the houses. This increased the SO_4 content in groundwater. The worst case was the backfilling with urban refuse practiced in the 1970s by municipal services, including those of Barcelona, at a time where the municipal refuse tips were filled up and a new one was still not available. In the areas where most backfilling was done, no groundwater studies are known. During the excavation in 1991 of a Olympic rowing canal, one of these backfilled pits was intersected, yielding groundwater heavily loaded with organics and sulphides. In addition, many of these pits received occasional disposal of toxic wastes, including chromite wastes, oils, and chlorinated solvents. After several decades, these contaminants still remain in the unsaturated zone. In the lower valley, when the water table goes up after recharge events, the slowly downward moving contaminated water is rapidly incorporated into groundwater flow, thus producing sudden contamination events. After 3–4 decades this still happens, although in a more diluted manner, and it is a major concern for the operation of supply wells. When these pits are found by excavations, they have to be dredged out and the material hauled to a controlled disposal site, at a high cost. This is currently the case at the surrounding of the new artificial recharge infiltration basins under construction.

Problems with organics, chlorinated compounds, pesticides, and emergent contaminants, widely present in groundwater due to the many anthropic influences, are under study and are the subject of [31–35].

8 Groundwater Management in the Low Llobregat Aquifer System

In the LLAS, the clearly unsustainable situation of the 1970s and 1980s, with low piezometric levels, important and increasing salinization, and some pollution events, prompted action to manage the aquifer system, which is crucial for the water supply to the whole Metropolitan Area of Barcelona. Fortunately, the scientific and applied knowledge was quite advanced and numerical simulations models were available.

At that time the Water Act of 1876 was in force according to which groundwater was the property of the abstractor. This partly explained the scarce official interest of the public water administration on them and that the economic resources and legal support were scarce, besides groundwater development being poorly developed at that time. This was theoretically changed by the Water Act of 1985, which included groundwater in the public domain. However, a transitional article to avoid the cost of expropriating existing rights resulted in practice in maintaining groundwater within the private domain. However, the need for protection and management of groundwater was introduced as a principle. This has been further reinforced by the European Water Framework Directive [19] and the complementary "Groundwater Daughter Directive" [20], both incorporated into the current Water Act.

Social aspects are highly relevant in the LLAS. In the 1970s, the results of the Water Authority and the Technical University of Catalonia studies and modelling results convinced local industrialists and suppliers of the need to create a users' association to deal with existing problems, to push the Water Authority for action, and to preserve the groundwater resources they were using. This association was created in 1975 [21] and approved by the Water Authority under the old Water Act of 1876, bringing forward the effects of groundwater intensive exploitation on public water domain. The association was later on extended to the whole LLAS as the Groundwater Users Community of the Low Llobregat (CUADLL). Combined efforts with the Water Authority have halted uncontrolled sand and gravel mining in the area, controlled existing and new wells, self-limited groundwater abstraction, improved monitoring, and promoted correction measures. The CUADLL is an important Civil Society entity able to back, redress, or oppose the Water Authority decisions. It is involved in groundwater monitoring and uses the same mathematical models of the Water Authority to draw its own results and check those of the others.

Conjunctive use of surface water and groundwater as a mean to increase the availability and security of water supply was early considered by the Water Supply Company of Barcelona and by the Water Authority in the 1960s [22, 23]. They were based on the hydrological, hydrogeological, and hydrogeochemical studies carried out at that time, supported by mathematical models [24]. Tools improved later on. In the 1990s, models considered marine water intrusion [8], and later on included mass transport [25, 26]. The most recent model is used by the Water Authority for management purposes.

Another technical management tool is artificial recharge. In order to increase groundwater reserves in the lower valley aquifer, in the late 1940s the Water Supply Company of Barcelona started enhanced river bed infiltration by scrapping, and recharge of potable water in single- and dual-use wells in the late 1960s [27]. In order to compensate for the decreasing recharge area due to new infrastructures, artificial recharge by means of infiltration basins is underway, as presented in [36]. To halt seawater intrusion and try to reverse it, maintaining a given level of groundwater abstraction and with the possibility of using groundwater reserves in droughts, a coastal barrier is operated by the Water Authority, in which highly treated and partly desalted reclaimed water from the large wastewater treatment plant at the Llobregat delta coast is injected into the delta deep aquifer. Total exploitation cost of injected water once in the aquifer is about 0.29 €/m^3, with a total construction cost of 26 M€, and for design flow of 150 L/s [28].

The Water Authority started groundwater monitoring in the 1960s with rather a dense network of multilevel piezometers, although now aging and slowly degrading. The Water Supply Company of Barcelona has data from the 1920s and some factories from the 1950s until they closed down. Now the Water Authority is carefully involved in management and correction measures [29].

Environmental aspects were poorly taken into account in the past, but are now considered in those cases in which there are areas to be preserved and restored. They refer mostly to wetlands in the delta coastal area, with mixed surface water and groundwater dependence.

Groundwater management has some restrictions. One of them is that high water-table levels become a serious problem for existing tunnels and buildings, mostly constructed when they were low. Recharge efforts can be lost by drainage or require expensive waterproofing efforts. High water-table levels also mobilize contaminants existing in the previous unsaturated zone of many poorly known areas.

9 Concluding Remarks

The urban, industrial, and irrigation water supply of the Metropolitan Area of Barcelona is complex and demanding. There are notable achievements in demand reduction, since the per capita domestic consumption is about 125 L/day, which is compatible with quite reasonable living standards. Also no water restrictions have been applied in spite of recent serious drought periods. This is the combination of diverse supply sources, the prolonged and joint use of surface and ground water, and more recently the reclamation of carefully treated low quality groundwater, the use of reclaimed waste water for nondrinking purposes and for aquifer recharge, and a seawater desalination plant. In such a complex system, the role of the LLAS has evolved toward a strategic water reserve of up to 150 hm^3 for the critical moments of droughts, major failures, and contamination events, now combined with the operation of the seawater desalination plant as a complementary backing in moments of water stress. The aquifer storage is crucial since it is not possible to get such a reserve volume in the nearby area of influence at a reasonable cost. The preservation of the LLAS as a continuous water source and as a reserve includes the need for artificial recharge and conjunctive use, and preservation and improvement of their quality.

Groundwater quantity aspects are nowadays reasonably addressed, but quality aspects have still to be solved in a densely populated area that suffers from some old deficiencies, water salinity from the mining area, seawater intrusion, and the results of past moments of degradation and poor care. Steps forward are artificial recharge, efficient operation of brine and saline water pipelines, partial treatment of the surface and groundwater by reverse osmosis and electrodialysis, and the seawater intrusion barrier.

The management of the LLAS depends on assuring recharge, agreeing on the spatial and temporal abstraction quantity and pattern, and reducing seawater intrusion by maintaining high enough piezometric levels in the coastal area. But this means difficulties to the injection barrier since it cannot continue to operate by gravity, and the wells have to be prepared for injection under pressure. High water tables are desirable to maintain a high groundwater storage, but the inundation problems and the drainage of underground infrastructures favors sustaining a given level of continuous groundwater abstraction, with the appropriate pattern and timing.

Other LLAS quality problems are more difficult and pervasive. Some of them refer to the slowly degrading wastes introduced and buried in the past. New threads appear as the area holds expanding human activities. This has obliged to carefully treat part of the pumped groundwater before distribution. Current stricter regulations, although positive, may further stress the situation. However, there are improvements; as wastes are more controlled, some substances are restricted or banned, and the more potentially dangerous activities have ceased or are under close control. This is the result of a more conscious and well-staffed water administration, more social understanding and involvement, and improved monitoring. It should be remarked that the ethical behavior of many people at key levels has been essential for the system being salvaged from a serious and irreversible degradation.

References

1. Llamas MR, Molist J (1967) Hidrología de los deltas de los ríos Besós y Llobregat [Hydrology of the Besós and Llobregat river deltas]. Rev Agu, Barcelona 2:139–154
2. Checa A, Díaz JI, Farrán M, Maldonado A (1988) Sistemas deltaicos holocenos de los ríos Llobregat, Besós y Foix: modelos evolutivos transgresivos [Holocene delta systems of the Llobregat, Besós and Foix rivers: transgressive evolutive models]. Acta Geol Hispánica 23:241–255
3. IGME (1989) Mapa geológico de la plataforma continental y zonas adyacentes [Geologic map of the continental platform and neighbouring zones]. Hoja 35/42. Barcelona 1:200 000. Instituto Geológico y Minero de España. Madrid
4. Manzano M, Peláez MD, Serra J (1986) Sedimentos prodeltaicos en el delta emergido del Llobregat [Prodeltaic sediments in the emerged delta of the Llobregat]. Acta Geológica Hispánica 21–22:205–211
5. Simó JA, Gámez D, Salvany JM, Vázquez–Suñé E, Carrera J, Barnolas A, Alcalá FJ (2005) Arquitectura de facies de los deltas cuaternarios del río Llobregat, Barcelona, España [Facies architecture of the Llobregat river Quaternary deltas, Barcelona, Spain]. Geogaceta 38:171–174
6. Custodio E (1981) Sea water encroachment in the Llobregat delta and Besós areas, near Barcelona (Catalonia, Spain). Sea Water intrusion meeting: intruded and fossil groundwater of marine origin. Rapp. och Meddelanden. Sveriges Geologiska Undersökning, Uppsala 27:120–152
7. Custodio E (2008) Acuíferos detríticos costeros del litoral mediterráneo peninsular: valle bajo y delta del Llobregat [Coastal detritic aquifers of the peninsular Mediterranean litoral: Llobregat low valley and delta]. Monográfico: Las Aguas Subterráneas. Rev Asociación Española Enseñanza de las Ciencias de la Tierra Madrid 15(3):295–304
8. Iribar V, Carrera J, Custodio E, Medina A (1997) Inverse modelling of sea water intrusion in the Llobregat delta deep aquifer. J Hydrol 198(1–4):226–244
9. Custodio E, Bruggeman GA (1987) Groundwater problems in coastal areas. In: Studies and reports in hydrology, vol 45. UNESCO, Paris, pp 1–576
10. Custodio E (1967) Études hydrogéochimiques dans le delta du Llobregat, Barcelona (Espagne) [Hydrogeochemical studies in the Llobregat delta, Barcelona (Spain)]. General Assembly Bern. Intern Assoc Sci Hydrol Publ 62:135–155
11. Iribar V, Custodio E (1992) Advancement of seawater in the Llobregat delta aquifer. In: Custodio E, Galofré A (eds) Study and modelling of saltwater intrusion into aquifers, 12th salt water intrusion meeting. CIMNE, Barcelona, pp 35–50
12. Custodio E, Iribar V, Manzano M, Skupien E (1992) Utilización de isótopos ambientales en el Valle Bajo y Delta del río Llobregat (Barcelona, España) para resolver problemas de flujo y de transporte de masa en los acuíferos [Use of environmental isotopes in the Llobregat river Low Valley and Delta (Barcelona, Spain) to solve flow and mass transport aquifer problems]. In: Water resources development. IAEA, Vienna, pp 385–414
13. Custodio E, Bayó A, Peláez MD (1971) Geoquímica y datación de aguas para el estudio del movimiento de las aguas subterráneas en el delta del Llobregat (Barcelona) [Geochemistry and water dating to study groundwater movement in the Llobregat delta (Barcelona)]. 1er Congreso Hispano–Luso–Americano de Geología Económica Madrid–Lisboa 6:51–80
14. Manzano M, Custodio E, Carrera J (1992) Fresh and salt water in the Llobregat delta aquifer: application of ion chromatography to the field data. In: Study and modelling of saltwater intrusion into aquifers, International 12th Salt water intrusion meeting. CIMNE, Barcelona, pp 207–228
15. Custodio E, Solà V, Soler A (2010) New data on seawater intrusion into the Llobregat delta aquifer, Barcelona, Spain. In: Proceedings of the 21st Salt Water Intrusion Meeting (SWIM–21). Universidade das Açores, Ponta Delgada, São Miguel:15

16. Xu T, Samper J, Ayora C, Manzano M, Custodio E (1999) Modeling of non-isothermal multi-component reactive transport in field scale porous media flow systems. J Hydrol 214:144–164
17. Custodio E, Galofré A (1987) Fast release of salinity after a flood in the Llobregat valley (Catalonia). In: van Dujnverbooden W, Waegeningh HG (eds) Vulnerability of soil and groundwater to pollutants. Elsevier, Amsterdam, pp 563–573
18. Carrera J, Custodio E, Sánchez–Vila X, Medina A, Manzano M, Galarza G (1993) Modelling a large oil spill in the aquifer of the Llobregat lower valley, Barcelona. Environmental pollution: science, policy, engineering. Eur Centre Pollut Res 1:369–376
19. WFD (2000) Directive 2000/60/EC of the European Parliament and of the Council establishing a framework for common action in the field of water policy. Water Framework Directive, Brussels
20. GWD (2006) Directive 2000/118/EC of the European Parliament and of the Council: Groundwater Directive. GWD, Brussels
21. Codina J (2004) Las aguas subterráneas: una visión social. El caso de la Comunidad de Usuarios del Llobregat [Groundwater, a social visión. The case of the Llobregat Users Community]. Real Acad Cien Exact Fis Nat (Esp) Madrid 98(2):323–329
22. Llamas MR (1969) Combined use of surface and ground water for the water supply to Barcelona (Spain). Bull Int Assoc Sci Hydrol 14(3):119–136
23. Llamas MR, Vilaró F (1967) Die Rolle der Grundwasserspeicher by der Wasserversorgung von Barcelona [The role of groundwater in the water supply of Barcelona]. Das Gas–und Wasserfach, Wasser–Abwasser 34(15):945–953
24. Cuena J, Custodio E (1971) Construction and adjustement of a two layer mathematical model of the Llobregat delta. Mathematical models in hydrology. In: Proceedings of the IASH–UNESCO–WMO, Warsaw Symposium, vol 2, pp 135–155
25. Abarca E, Vázquez–Suñé E, Carrera J, Capino B, Gámez D, Battle F (2006) Optimal design of measures to correct seawater intrusion. Water Resour Res 42. doi: 10.1029/2005WR004524
26. Vázquez–Suñé E, Abarca E, Carrera J, Capino B, Gámez D, Pool M, Simó T, Batlle F, Niñerola JM, Ibáñez X (2006) Groundwater modelling as a tool for the European water framework directive (WFD) application: the Llobregat case. Phys Chem Earth 31(17):1015–1029
27. Custodio E, Isamat FJ, Miralles JM (1976) Twenty-five years of groundwater recharge in Barcelona (Spain). Artificial groundwater recharge. DVWK Bull 11(I):171–192
28. Ortuño F, Molinero J, Garrido T, Custodio E, Juárez I (2010) Seawater intrusion control by means of an injection barrier in the Llobregat delta, near Barcelona, Catalonia, Spain. XXXVIII IAH Congress, Krakow. In: Zuber A, Kania J, Kmiecik E (eds) Groundwater quality sustainability. Extended abstracts, vol 267, pp 2263–2268 (CD printing)
29. Niñerola JM, Queralt E, Custodio E (2009) Llobregat delta aquifer. In: Quevauviller PI, Fouillac A-M, Grath J, Ward R (eds) Groundwater monitoring. Wiley, New Jersey, pp 289–301
30. Aguilera R, Sabater S, Marcé R (2012) In-stream nutrient flux and retention in relation to land use in the Llobregat River Basin. Hdb Env Chem doi: 10.1007/698_2012_143
31. Eljarrat E, Barceló D (2012) Occurrence and behavior of brominated flame retardants in the Llobregat River Basin. Hdb Env Chem doi: 10.1007/698_2011_139
32. Petrovic M, Barceló D (2012) Inputs of pharmaceuticals and endocrine disrupting compounds in the Llobregat River Basin. Hdb Env Chem doi: 10.1007/698_2011_141
33. García Galán MJ, Díaz Cruz MS, Barceló D (2012) Occurrence and fate of sulfonamide antibiotics in surface waters: Climatic effects on their presence in the Mediterranean region and aquatic ecosystem vulnerability. Hdb Env Chem doi: 10.1007/698_2011_140
34. Llorca M, Pérez F, Farré M, Picó Y, Barceló D (2012) Perfluorinated compounds' analysis, environmental fate and occurrence: The Llobregat River as case study. Hdb Env Chem doi: 10.1007/698_2012_147
35. Postigo C, Mastroianni N, de Alda ML, Barceló D (2012) Illicit drugs and metabolites in the Llobregat River Basin. Hdb Env Chem doi: 10.1007/698_2012_146
36. Sanchez-Vila X, Armenter JL, Ortuño F, Queralt E, Fernández-Garcia D (2012) Managed artificial recharge in the Llobregat aquifers: Quantitative versus qualitative aspects. Hdb Env Chem doi: 10.1007/698_2012_154

Managed Artificial Recharge in the Llobregat Aquifers: Quantitative Versus Qualitative Aspects

Xavier Sanchez-Vila, Josep-Lluïs Armenter, Felip Ortuño, Enric Queralt, and Daniel Fernàndez-Garcia

Abstract The area involving the Lower Valley and Delta of the Llobregat River has a long-standing tradition in Managed Artificial Recharge (MAR) activities. For more than 50 years different MAR activities have taken place, and at present the area provides a catalog of methodologies put into practice. This chapter provides an overview of artificial recharge activities in the river and riverbanks classified according to types, objectives, water sources, and focus. Regarding types, both surface and deep MAR activities can be found at different locations. In terms of objectives, we can distinguish between using opportunity water, resources increment, or improving groundwater quality. Regarding water sources, recharge water may come from the river, a diverted channel, or treatment plants. Finally, regarding focus, activities can be seen as used for scientific purposes, demonstration, or actual industrial activity.

Keywords Artificial recharge, Infiltration, Pumping well, River scarification, Water quality

X. Sanchez-Vila (✉) and D. Fernàndez-Garcia
Universitat Politècnica de Catalunya – BarcelonaTech, Jordi Girona 1-3, UPC Campus Nord, 08034 Barcelona, Spain
e-mail: Xavier.Sanchez-Vila@upc.edu

J.-L. Armenter
Agbar – Aigües de Barcelona, Barcelona, Spain

F. Ortuño
ACA – Agència Catalana de l'Aigua, Barcelona, Spain

E. Queralt
CUADLL – Comunitat d'Usuaris del Delta del Llobregat, Barcelona, Spain

CUACSA – Comunitat d'Usuaris de la Cubeta de Sant Andreu, Barcelona, Spain

S. Sabater et al. (eds.), *The Llobregat: The Story of a Polluted Mediterranean River*, Hdb Env Chem (2012) 21: 51–68, DOI 10.1007/698_2012_154,

Contents

1 Introduction

1.1 The Need for MAR into the Llobregat River Aquifers

Most of the Mediterranean coastal areas are under water stress conditions. Increase in population combines with the decrease in the amount of water resources caused among other things by climate change. An additional factor is the decrease in quality associated to water resources. This decrease is itself caused again by overpopulation, as well as by agricultural and industrial practices.

The Llobregat River is not an exception. The river and its aquifers have been and are still heavily exploited to provide water to a number of cities and villages, amongst them the ones that constitute the Barcelona Metropolitan Area. At present, the water supply in the Barcelona city and its surroundings has four sources:

(a) Surface water coming from the River Ter (water transfer to the Llobregat-Besòs system)
(b) Surface water from the Llobregat River
(c) Groundwater from the aquifers located in the Llobregat Lower Valley and its Delta
(d) Small amounts from different sources, among them the Barcelona aquifer or the Besòs aquifer

Pressure on resources is function of supply and demand. Water consumption per capita has been significantly and progressively reduced from about 130 L/day in 1996 to 107 L/day in 2009. This means that both the water company has been efficient in reducing water losses in the system, and also the population has positively reacted to the different calls made by the administration to reduce water use, mainly related to the big drought in 2006–2008. The total amount of supply water in 2008–2010 ranged between 6.5 and 7 m^3/s.

The average flow in the river is about 19 m^3/s, but as a consequence of the Mediterranean climate, during long periods of time the actual flow is comparable to the amount of water that needs to be diverted into the area. For this reason, water management in the Barcelona region calls for a combination of surface and subsurface water. While surface water is the most often used water source (under so-called normal conditions), aquifer abstractions take place when the river cannot supply the system or when the quality of the river water is not adequate.

The hydrogeological setting of the Llobregat Lower Valley and Delta has been well known since the studies carried out by the Water Administration in the 1960s, followed by many others (see [1] and references therein).The Delta is formed by two sand and gravel aquifers separated by a silty-clayey wedge deposited under an estuarine environment. The upper unconfined aquifer is about 15 m thick, and lies below current land surface, of Holocene age. The lower one, named "main aquifer," is 10–20 m thick, of late Pleistocene age. This latter aquifer is confined and very conductive, with transmissivity values ranging between 1,000 and 5,000 m^2/day. The silty wedge is of variable thickness, decreasing upward until it disappears close to the opening of the Delta. This situation extends into the Lower Valley, where we find a single aquifer, unconfined and highly transmissive. A comprehensive summary on the details of the geological and hydrogeological settings can be found in previous chapters of this book.

Historically, groundwater extractions started around 1890 in the Delta confined aquifer. The industrial development and the increase in population led to a sustained increase in the groundwater abstraction that peaked in 1970s with values close to 120 MCM/year. These extractions produced undesired effects in groundwater heads, but mainly a deterioration of the water quality due to seawater intrusion. Exploitation is expected to be sustainable at a rate of 40 MCM/year. Yet, still in 2009, abstractions were slightly higher, on the order of 55 MCM/year.

Altogether we picture a situation where groundwater is a key component in the water supply system in the Barcelona region. Yet, natural recharge seems not to be sufficient to guarantee a sustainable exploitation of the system. This situation was already recognized more than 50 years ago, and about that time, artificial recharge activities started taking place. This situation has evolved in such a way that nowadays a number of different recharge activities are being simultaneously conducted in different parts of the system and with different methodologies and purposes. Furthermore, while first initiatives were only based on a good knowledge of the system and best practices – adopted from other parts in the world – nowadays a MAR system involves a careful planning, design, modeling, construction, and finally monitoring. Bear in mind that any such facility has a very strong potential impact on water management, not only locally but also in an integrated basin-scale framework.

This chapter provides an overview of managed artificial recharge activities taking place in the Lower Valley and Delta of the Llobregat River, with emphasis on types, objectives, water sources, and focus, with some historical background on the reasons behind each facility and the expectations for the future. The activities presented in this chapter are:

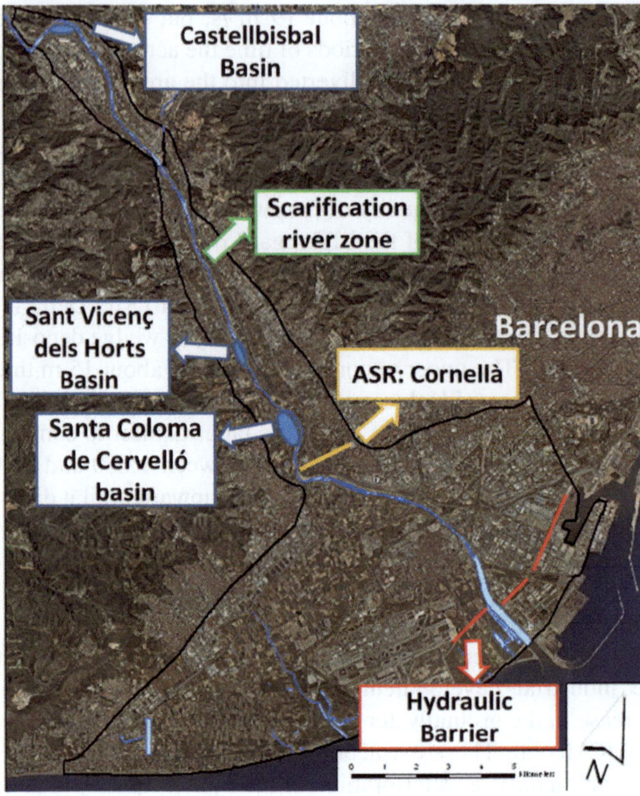

Fig. 1 Location of the different MAR activities taking place in the Llobregat River area (Lower Valley and Delta)

- Scarification activities at the river bed (Lower Valley)
- Deep injection from drinking water quality surpluses (Delta)
- Surface ponds in Castellbisbal (St. Andreu Basin)
- Surface ponds in St. Vicenç dels Horts (Lower Valley)
- Surface ponds in Sta. Coloma (Lower Valley)
- Seawater positive barrier (Delta)

and their location can be found in Fig. 1. In all six cases we will deal with a bit of history, original motivation, type and source of water, volumes affected, monitoring issues, and measured or expected impacts on the subsurface.

1.2 A Summary of Managed Artificial Recharge Practices: Objectives and Methodologies

Artificial recharge is one of the many components involved in integrated water resources management. Integration means that conventional and non-conventional sources should be studied together. Conventional sources involve rivers, reservoirs, and aquifers. Non-conventional sources include desalination, wastewater reuse, rainfall harvesting, fog harvesting, subsurface dams, Aquifer Storage and Recovery (ASR), and Aquifer Storage Transfer and Recovery. These last two activities can be merged into the concept of Managed Artificial Recharge.

MAR may be defined as man-made controlled operations aimed at transferring water from ground surface into underlying aquifers. The issue of "managed operations" distinguishes this concept to that of natural replenishment or natural recharge. In a MAR facility location, flow rates and quality of the infiltration water are all decision variables, and thus need to be included into the design of the facility and subsequent operation. Moreover, being decision variables they can be included in an optimization process.

From a general perspective, MAR practices may have one or a number of objectives. In the Llobregat area the following objectives have been considered:

(a) Storage of water: In quantitative terms this is the most common objective. The idea is to supplement the difference between the groundwater demand and the natural aquifer replenishment. Water is stored in the void space of aquifers, and due to the relatively slow movement of water, most of it can be eventually recovered by appropriate operations. Both long-term and short-term storage may be practiced. In years with excess surface runoff, water may be stored in aquifers for use in dryer years. Short-term storage may be practiced in order to make a more efficient use of the water supply lines.
(b) Control of regional hydrological regime: By artificially recharging an aquifer, water levels, or piezometric heads, are raised. By manipulating these levels, a number of elements can be controlled: rate and direction of flow, water of inferior quality (e.g., seawater) intrusion, spring discharge, land subsidence, and seepage to or out of adjacent water bodies (rivers and lakes). The amount of water to be infiltrated to produce a significant rise in head elevations depends on aquifer type. In confined aquifers, because of the very low value of storativity, small volumes of recharged water are required. In unconfined aquifer volumes should be much larger, since heads are controlled by specific yield.
(c) Control of water quality: The indigenous water mixes with the recharged one as a result of hydrodynamic dispersion. The quality (in terms of dissolved matter) of the resulting water can be controlled by a combination of MAR practices and by carefully designing a pumping scheme within the aquifer. Due to the very slow movement of water in the aquifer, months to years may elapse between the time water is introduced into and extracted from an aquifer. During that time, chemical reactions (natural attenuation, sorption, ion exchange, and filtering) may take place. Thus, the aquifer might improve the quality of the injected water. This is why reuse of reclaimed sewage water is often implemented in conjunction with MAR.

The last point to be considered in the design of a facility is the existence of a number of methodologies for artificial recharge. In each particular facility, the selected method depends on the type of aquifer and existing soil and the topographical and geological conditions, as well as on social, economic, and legal restrictions.

We distinguish three families of methods, all taking place in the Llobregat River area:

1. Methods for enhancing infiltration: The objective is to increase infiltration by various agro-techniques, which affect ground surface roughness, slope, vegetation cover, etc. The purpose is to extend the time and area through which infiltration from precipitation and surface runoff take place. An example is the removal of soil or obstacles in areas of high infiltration capacity such as sand dunes. Another example is riverbed scarification.
2. Surface spreading methods: Here water is diverted into specially constructed ponds, basins, or trenches, dug along ground surface contours to favor the infilitration of water through a high permeable surface bottom. These techniques should be implemented only in phreatic aquifers where no impervious layer of significant areal extent between the bottom of the infiltration basins and the water table is present. As infiltration proceeds, the bottom of the pond becomes gradually clogged, so that maintenance tasks should be performed periodically. The economy of artificial recharge by surface techniques depends to a large extent upon the availability of land and the capacity to maintain high infiltration rates with time. On the other hand, infiltration is a natural process driven by gravity, so that very little energy is consumed.
3. Artificial recharge through wells (deep recharge): either recharging or dual-purpose wells. Operation is costly since it involves a large amount of energy to inject the water from the surface into the deep aquifer. The main concern of such facilities is the reduction of infiltration capacity with time due to well screen clogging. For this reason, deep recharge involves the use of high-quality (often drinking quality) water. Artificial recharge through wells is practiced either in confined aquifers, when extended impervious layers are present between the ground surface and an underlying phreatic aquifer, or else when land is expensive or unavailable.

In the following sections we deal individually with the different MAR practices that have been implemented in the Llobregat River and related aquifers along the years.

2 Scarification Activities at the Riverbed (Lower Valley)

In recharging rivers, the amount of water infiltrating from the river toward the aquifer reduces with time due to a combination of sediment transport and deposition of fine material as well as a number of geochemical and biological reactions. Scarification is a method that consists in removing, by means of a scrapper, the silty sediments of the riverbed that are progressively developed in the river bottom, removing surface biofilms (see Fig. 2). The immediate consequence is a recovery of the initial infiltration rates in the riverbed after scarification, since sands and gravels outcrop again. All removed material is put back into suspension in the river and carried downgradient. For this reason scarification is always performed with the scrapper moving in the direction of river flow.

Fig. 2 (**a**) Close look to the superficial clogging layer before scarification; (**b**) Machine used first to even the river bottom, allowing for (**c**) the scrapper (a smaller machine) to perform the actual scarification

In the upper part of the Llobregat Lower Valley, scarification practices have been taking place since the 1950s, operated by the Barcelona Water Supply Company (Agbar). They are usually done in spring and fall when the river flow ranges between 10 and 35 m^3/s and the turbidity lower than 100 NTU. The minimum flow has been established in order to avoid the impact on river flow; the maximum value implies that water flows too fast (with too much energy) so that scarification practices cannot be performed in a safe environment for both machinery and personnel. Additional restrictions involve ammonia and chloride concentrations below 1 mg/L and 350 mg/L, respectively.

Scarification in the Llobregat River can be seen as an opportunity method, taking place only when the river carries water of sufficient quality. The estimated recharge capacity is 40,000 m^3/day considered as the difference between natural recharge and total recharge after scarification. Activities take place along 1 km of the river section. The cost of the method is really low and has been estimated in 0.03 €/m^3 of infiltrated water.

3 Deep Injection from Drinking Water Quality Surpluses (Delta)

Historically, in the middle of the twentieth century the water supply systcm was proved insufficient to fully supply the demand of the Barcelona Metropolitan Area. This was the reason why an additional transfer of water from the Ter River was established. Under such new conditions, the Water Purification Plant located in

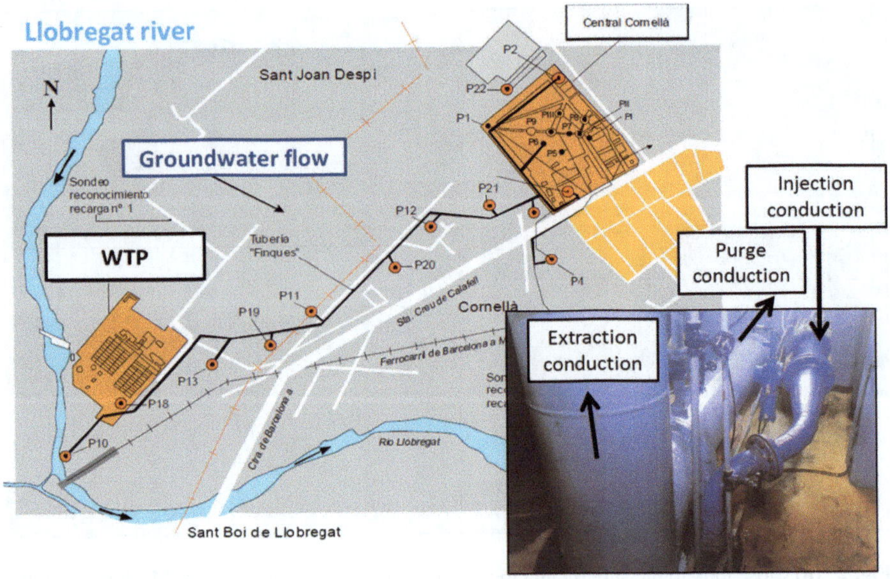

Fig. 3 Location of the deep injection practices in Cornellà (*left*). Picture of the pipes and valves in the reversible wells (*right*)

Sant Joan Despí did not have to work at full capacity except on peak demand periods.

Yet, it was best in economic terms to keep the system at full capacity, producing an excess of water. Thus, in 1969 a decision was taken to use this surplus water in a combined short-term management system to inject it into the aquifer by means of recharging wells. The stored water could later be extracted whenever the river conditions (quantity or quality) forced groundwater resources to take over.

The actual system is located in the Agbar Water Company premises. The system was designed to make use of seven existing wells, completed with five additional wells. These new wells were capable of handling larger infiltration rates. They were drilled exclusively for recharge purposes and were located in a transverse section of the aquifer, perpendicular to the natural flow direction (Fig. 3). The recharge area is located where the two existing aquifers in the Delta are unified into a single unconfined aquifer. The well arrangement causes an extended decrease in the head gradient in the Pallejà/Cornellà area.

The recharge capacity of the system has been evaluated in 75,000 m^3/day. The unit cost of the infiltrated water depends on the actual infiltrated volumes, but the average cost is 0,25 €/m^3. Actually, the amount recharged has varied largely with time. The maximum recharge occurred in the period 1989–1996. Since then, the infiltrated volumes have been very small in quantitative terms.

As stated before, the main risk of any recharge operation is clogging. While clogging also occurs in surface facilities, those involving deep wells may be potentially fatal since it may cause problems in very expensive installations.

Thus, the design of the facility must ensure the effective cleaning of the borehole and its surroundings. This involves setting rigorous standards in the recharged water quality. The threshold values established to consider water of enough quality for infiltration purposes in the Cornellà wells are the following: turbidity, 0.2 NTU; pH 6.5–8; and concentration restrictions in many elements such as sulfate, 200 mg/L; magnesium, 30 mg/L; sodium, 200 mg/L; aluminum, 0.15 mg/L; nitrate, 20 mg/L; ammonium, 0.1 mg/L; TOC, 3 mg/L; detergents, 0.1 mg/L; and trihalomethanes, 0.035 mg/L.

A control of the degree of clogging around the wells is carried out periodically to evaluate the evolution of the infiltration capacity. Such control consists of measuring the infiltration cones around the well after a cleanup operation. Only once, in 1981, an excessive degree of clogging around one of the wells was observed. A cleanup operation took place involving compressed air until the initial infiltration capacity was restored.

4 Surface Ponds in Castellbisbal (St. Andreu Basin)

The Cubeta de Sant Andreu has been heavily exploited since the 1960s, mainly due to industrial development. In fact, in this area, industry has largely displaced agriculture to become the main water consuming activity. Until the 1990s, the water supply to the villages of Castellbisbal, Corbera, and Sant Andreu de la Barca was achieved through wells located within the basin. All these populations have grown substantially in recent years. From the summer of 1984, there was an important drop in groundwater levels due to a severe drought. The maximum drawdown was recorded in February 1985. To solve the problem in mid-May 1985 a number of surface infiltration ponds were designed and built close to the riverbed, in the area known as Can Albareda. The actual location of the recharge ponds took advantage of geologically favorable conditions.

The initial facilities consisted of a settling reservoir, a water channel leading to four infiltration basins. Each pond was separated from the next one by a bar of clay materials 1.5–2 m wide. The initial recharge area was some 7,000 m^2 in extension. Later a new channel was built along the right margin. This channel ensured that the system does not deteriorate with river flows up to 150 m^3/s.

With the construction of a highway in the vicinity of the recharge facility, a number of hydraulic works took place in the area, including the enlargement of the river section. The surface ponds were put again in function until the occurrence of an extreme flood event above the design value destroyed some of the walls, and the ponds operation was discontinued.

The Catalan Water Authority (ACA) and the water users' community of the St Andreu Basin (CUACSA) signed an agreement in 2002 to rebuild the old ponds with a new design (Fig. 4) and to transfer their management and maintenance from the former to the latter. The ponds were inaugurated in April 2010, and consisted of a wetland extending some 14,000 m^2 and an infiltration pond over 6,000 m^2.

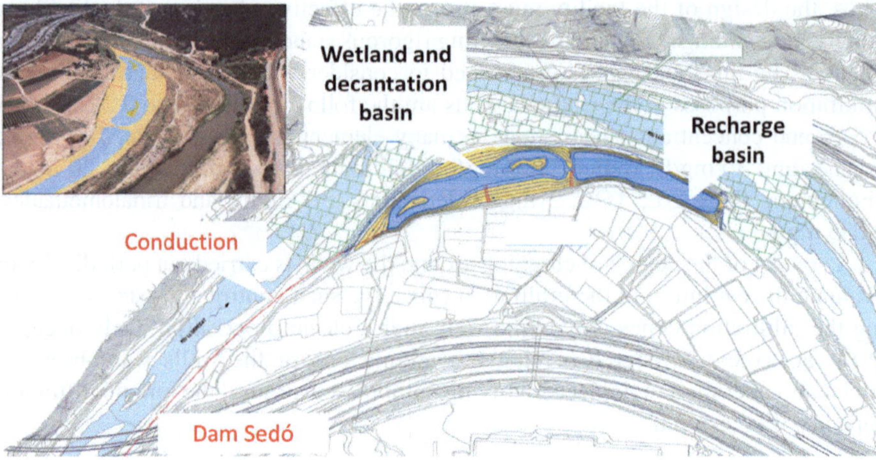

Fig. 4 Plan view of the surface infiltration ponds in Castellbisbal

The total amount of recharge predicted was 1.8 MCM/year, a value computed assuming an infiltration rate of 1 m^3/m^2/day. The flow diverted into the ponds is a function of the river flow (qr); when qr exceeds 5 m^3/s, 500 L/s are diverted. This value reduces to 50 L/s when qr is larger than 2 m^3/s. Below this last value no water is diverted into the infiltration ponds.

Quality is controlled automatically. The recharged water is collected from the river and in the interconnection of the two basins there is a control station programmed to automatically take water samples at a preset frequency to check the water quality regarding three indicators: concentration of ammonium, conductivity, and turbidity. When the water quality exceeds a predefined threshold for any of these three indicators, the station is programmed to automatically close the interconnection gate, shutting down the system. The actual thresholds are set to 2 mS/cm for electrical conductivity, 1.5 mg/L for ammonium, and 100 NTU for turbidity. Clogging of the recharge ponds due to the suspended matter of river water occurs approximately every 30 days.

5 Surface Ponds in Sant Vicenç dels Horts (Lower Valley)

A recharge pond system of 1 Ha was constructed in Sant Vicenç dels Horts as a compensatory measure for the decrease in the recharge area in the vicinity of the Llobregat River caused by building a High Speed Train railroad.

The system is composed of two interconnected ponds. The sedimentation pond is upstream with a surface of 4.000 m^2 and the infiltration pond, of 5.000 m^2, is downstream. There are three potential sources for the recharged water. It can come directly from the river or from two sources of reclaimed water: the Reversible

Electrodialysis Plant located in Sant Boi, or the Waste Water Regeneration Plant located in El Prat. Whenever reclaimed water can be poured directly into the infiltration pond, the sedimentation process becomes unnecessary and both ponds can be used for infiltration purposes.

The main objective of this facility is to use it as an experimental field laboratory. The area has been densely monitored both in terms of vadose zone and aquifer. This instrumentation includes suction cups at different depths and piezometers drilled to different depths. A well and a number of additional piezometers were drilled around the pond.

A number of experiments have been carried out in the system and its surroundings. Small infiltration tests using a double-ring infiltrometer were first performed. Six tests were run, showing a high average infiltration capacity of 5.7 m/day, with a standard deviation of 5.5 m/day. Starting March 2009, the system was put into operation for three months. This can also be seen as a three-month infiltration test at the full pond scale. During the test the recharging water was diverted from the river through a pipe. The volume provided was 0.42 MCM in 78 days, resulting in an average estimated infiltration rate of 1.08 m/day.

One conclusion of this test is that most of the turbidity of the water is removed before reaching the sedimentation pond due to deposition occurring in the inlet pipeline. When supply is discontinued and the system is back to operation we might expect remobilization of this fine material and a peak in turbidity at the pond. Another interesting finding was that in order to obtain a constant rate of infiltration, a higher water column in the infiltration pond was needed over time. That is to say that the infiltration pond is affected by a clogging process. Figure 5 shows how the infiltration (normalized by the water column) decreases over time, by more than half in 3 months.

To evaluate the impact of clogging nine double-ring infiltrometer tests (including the same six locations already tested, plus three new selected locations) were performed after the pond operation was discontinued. The resulting average infiltration capacity was now 2.1 m/day, with a standard deviation of 2.0 m/day. The net reduction is consistent with that observed in Fig. 5.

The surface ponds in Sant Vicenç dels Horts are still very active in terms of research activities being carried on. At present the focus is water quality and, in particular, evolution of organic and emergent compounds along the infiltration process, as well as studies in bioclogging formation.

6 Surface Ponds in Santa Coloma de Cervelló (Lower Valley)

The recharge ponds in Santa Coloma de Cervelló are expected to become the largest MAR facility in the Llobregat area. The aim of this facility is water storage for the Barcelona Area to be used as complementary water supply. The ponds are an initiative of a number of public and private entities, including CUADLL, ACA, Agbar, the Environmental department of the Catalan government (DMAiH), and

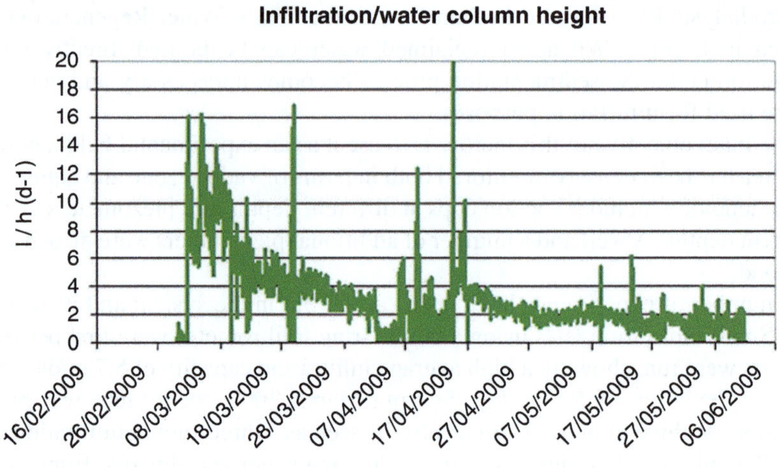

Fig. 5 Infiltration rate normalized by water column (in day^{-1}) observed during a 3-month infiltration test. Rate decreases by more than half due to a clogging process

the Environmental Entity of the Metropolitan Area of Barcelona (EMSHTR). All the preliminary studies needed to create the project have been completed and the ponds will go into construction in 2012.

The geological and geophysical characterization is one of these aforementioned preliminary studies. It was conducted by means of tomography and test drillings interpretation. More than 30 test drillings were done, and 22 of these test drillings were kept as piezometers, configuring the monitoring network for the test and for the future recharge system. As a result of these studies, a geological interpretation of the medium was obtained concluding that the clay and silt layer present in the Santa Coloma area are thicker in the downstream direction (Fig. 6). For that reason the basins are projected upside down: infiltration ponds will be located upstream of the area and the sedimentation ponds downstream where the vertical hydraulic conductivity is smaller.

Two different kinds of infiltration tests were done. The first one was by means of infiltration rings (a total of seven tests) at a depth of 2–3 m aimed at understanding the behavior of the top surface layers. The second kind of tests involved building pilot ponds at the same depth of the designed ones. Two small ponds of 150 m^2 were dug in different places. The tests results were input into the hydrogeologic model to calibrate hydraulic parameters of the vadose zone, especially vertical hydraulic conductivity. The infiltration rate deduced from these tests ranged from 1 m/d (in the first pilot pond) to 10 m/d (in the second). The big difference is due to the location of each pilot pond, indicating that the exact location of the future infiltration ponds is very important.

Two numerical models were constructed, involving two different sections, one transverse and another longitudinal with respect to the river flow direction. From these models, the minimum infiltration rate was estimated in 8 MCM/year [2].

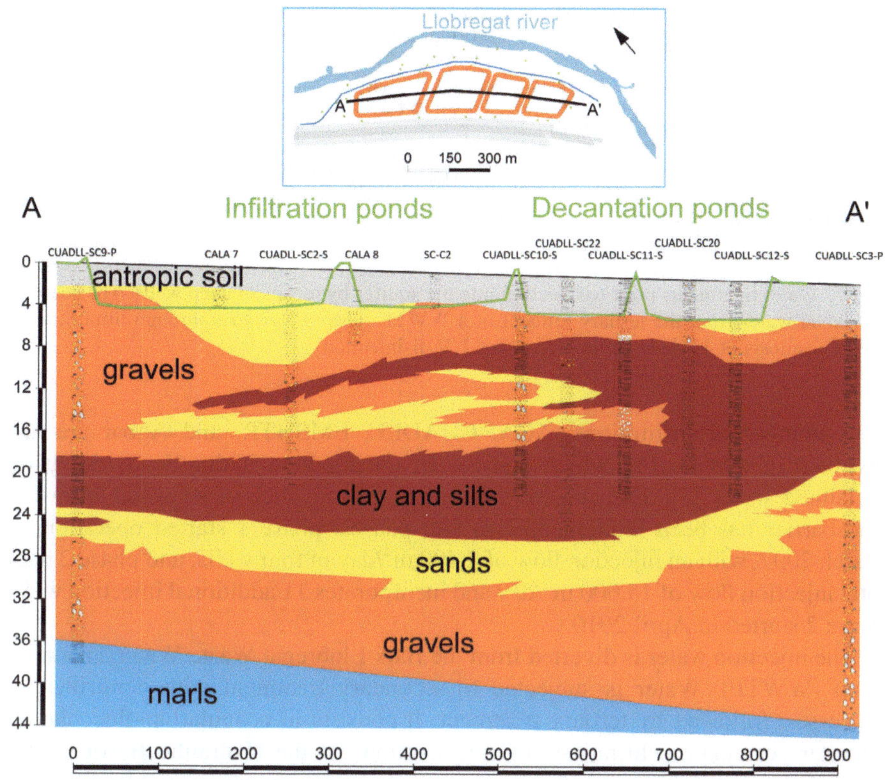

Fig. 6 Tomography and geological cross section with the basins profile overlapped. The two sedimentation ponds (S.P) are located downstream since the clay and silt layer is thicker in that direction

This rate is between the two values obtained from the pilot ponds. The models also foresee that only 5% of the infiltrated water would flow toward the river.

7 Seawater Positive Barrier (Delta)

Seawater intrusion processes have affected the main Delta aquifer since the 1960s. The intensive exploitation of groundwater resources, along with the excavation of part of the confining layer in the eastern coastal corner, has led to the progressive deterioration of groundwater quality (see [3] and references therein). Two main seawater intrusion fronts exist, one in the central coastal area and the other in the easternmost part. Current groundwater abstraction is about 54 MCM/y, but it exceeded 100 MCM/y in the 1970s. The sustainable exploitation rate to avoid further groundwater deterioration is estimated in 40 MCM/y [4].

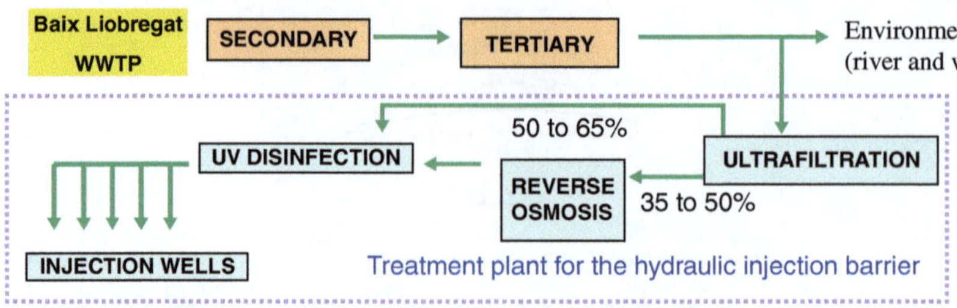

Fig. 7 Water treatments prior to injection into the aquifer by means of deep wells. The source is reclaimed water from the tertiary effluent of a WWTP. Then water goes through ultrafiltration, reverse osmosis of 35–50% of the water, and UV disinfection

A number of institutions, ACA, CUADLL, EMSHTR, and Agbar are also involved in this facility which consists of an injection hydraulic barrier using reclaimed water [5]. The objective is to halt the advance of seawater intrusion. The barrier has been implemented in two phases: phase 1 started operations in March 2007, with an injection flow of 2,400 m^3/day in four wells, and phase 2 has a total injection flow of 15,000 m^3/day and incorporates 11 additional injection wells. Phase 2 started in April 2010.

The injection water is diverted from the Baix Llobregat Waste Water Treatment Plant (WWTP). Water is subjected to secondary treatment with denitrification processes, followed by tertiary treatment. It consists in coagulation–flocculation, lamellar decantation, filtration, and disinfection. At the Hydraulic Barrier Plant, prior to the distribution into the injection wells, water experiences ultrafiltration, reverse osmosis, and UV disinfection (see Fig. 7 for the full scheme). Ultrafiltration is performed for bacteriological purposes to avoid the introduction of bacteria and nematode eggs, to prevent physical clogging, and to extend the functional life of reverse osmosis membranes. Reverse osmosis is necessary to reduce salinity, since the tertiary treatment water usually contains more than 600 mg/L of chloride due to upstream saline water discharges to the Llobregat River basin. Chloride content in the areas of the aquifer unaffected by seawater intrusion is about 350 mg/L, which is the target value. The reverse osmosis concentrate is disposed into the sea through submarine outfall of the WWTP. Water is not chlorinated before injection due to the potential appearance of trihalomethanes in the aquifer.

The water quality control is carried out in compliance with the Sanitary Authority requirements. Some injection water parameters (electrical conductivity, pH, temperature, ammonia, and turbidity) are obtained continuously and automatically, and some bacteriological and chemical parameters (nematodes, *Escherichia coli*, P, N, Cl, NO_3, NH_4, and TOC) are monitored weekly or biweekly. Major elements, metals, and volatile organohalogenated compounds are monitored on a bimonthly basis. A complete analysis of all the Drinking Water Quality Regulation parameters is performed once a year.

Injection at Phase 1 began on March 26, 2007. By May 2011 around 4 MCM of reclaimed water had been injected. Average values of injected water are 347 mg/L

Fig. 8 *Left*: Injection wells of the full hydraulic barrier (phases I and II) and location of the WWTP. *Right*: Reduction in chloride content in the aquifer between 2007 and 2009, indicating the movement of water injected by the hydraulic barrier in Phase I

for Cl^-, 1.85 ms/cm for electrical conductivity, 7.43 for pH, and <0.09 NTU for turbidity. Injected water temperature measured in the injection wells ranges between 14°C in winter and 29°C in summer. To date coliforms, *E. coli*, *Clostridium*, nematodes, or organochloride compounds have not been detected, and injection water complies with the Drinking Water Quality Regulation requirements.

Fifteen injection wells were drilled for the hydraulic barrier (4 in Phase 1 and 11 in Phase 2). They are located 1-km inland from the shore, along a 6-km line. The wells are 70 m deep and penetrate the 6–10-m thickness of the main confined aquifer along the barrier. The location of the injection wells can be found in Fig. 8 (left). Wells are equipped with flow-meters, automated electrovalves to maintain constant flow, and downhole sensors to control injection rates. Injected volume, water temperature, water electrical conductivity, and well hydraulic head are recorded every hour, and wells are controlled remotely.

Cleaning is necessary to avoid clogging of the well screen and surroundings. This is done by means of backflush with submersible electropumps or compressed-air pumping devices. It is carried out periodically during a short period of time in which the water discharge is three times that of injection. No decrease of well efficiencies (flow/head increase) has been detected so far. This is attributed to the combination of high quality of the injected water, as well as to the strict cleaning program.

Water quality and quantity monitoring involves existing boreholes and wells, plus 17 new boreholes that were constructed in the framework of the barrier project. During Phase I, 8 monitoring points were available, covering 3 km^2. Currently, there are 37 monitoring points (17 new piezometers and 20 existing points) covering 30 km^2. All monitoring points are sampled, and chemical analyses are carried out to determine major elements, nitrogen compounds, metals, and organochlorinated compounds. The 17 new piezometers have also a permanent sensor at the midpoint of the aquifer thickness that records hourly water level, temperature, and electrical conductivity of the aquifer.

The analytical results and field monitoring of the aquifer during Phase 1 of the project show highly positive results [6]. Since 2007, there is a progressive decrease in chloride (Fig. 8, right), sodium, calcium, magnesium, iron, and ammonium contents. Only a slight nitrate increase has been observed, since it is present in the injected water. Bicarbonate content is constant. The area influenced by recharged water ranges from 1 to 2 km around the injection wells and is irregular with respect to the two sides of the barrier. This is because groundwater flow from the seaside is halted or greatly reduced, while recharged water readily moves inland toward the pumped areas. The barrier has an oxidizing effect on the otherwise highly reducing aquifer ambient. So far, a clear redox perturbation is only observed in the close vicinity of injection wells.

The hydraulic barrier is operated to maintain the aquifer dynamic head in the range of 1–3 m above sea level to control the progression of seawater intrusion. Each well has its own injection flow rate that depends on local aquifer characteristics and the static aquifer head. Current total operation costs (fixed and variable) are estimated in 0.15–0.18 €/m^3, depending on total injection flow.

The injection barrier was discontinued in 2011, not for technical reasons, but rather for economic reasons.

8 Summary and Final Remarks

Three artificial recharge pond facilities are either planned or have already been put into operation in the lower valley and the Delta area of the Llobregat River with the objective to increase available water resources in the Barcelona region. Each of these ponds is currently in a different operational stage: Castellbisbal recharge ponds were inaugurated in April 2010; St Vicenç dels Horts ponds are ready to recharge but are still being used as a large scale recharge laboratory; finally, in Sta Coloma de Cervelló all the preliminary studies have been completed and the ponds will be constructed soon.

On top of that, a number of MAR practices exist in the Llobregat River area. This includes a hydraulic barrier against seawater intrusion with two phases already put into operation, scarification practices in the river bed, and recharge of drinking quality water surpluses into the Delta deep aquifer by means of injection wells.

Altogether the water volumes recharged into the aquifer are highly variable in time. Once all activities are in operation, a total of 22.5 MCM/year could be eventually recharged. Table 1 summarizes the target recharge rates associated

Table 1 Target rates for each of the MAR practices taking place in the Llobregat River area. The maximum recorded water volumes recharged in a given year is also displayed

Recharge methodology	Target rate (MCM/year)	Historical maxima (MCM/year)
Scarification	2	5
Castellbisbal Basin	2	0.3
St. Vicenç dels Horts ponds	2	0.9
Sta. Coloma de Cervelló ponds	8	–
Deep injection wells	3	14
Hydraulic barrier	5.5	2.5

with the different sites and methodologies employed in the Llobregat River area. It is important to highlight from this table that the different facilities could accept larger amounts than the target ones. This is also shown in the table, which contrasts the target rates against the maximum recorded water volumes recharged in a given year. Importantly, this indicates that both scarification and deep injection could increment the target value in a given year if the hydraulic conditions were favorable.

9 Conclusions

Water management in the influential area of the Llobregat River is quite complex due to an increasing urban and industrial development, the resulting large number of existing infrastructures (roads, railroads, airport, harbor enlargement, etc.), and the typical changing climate characteristics associated with the Mediterranean coastal areas. These developments have had a clear negative effect on the total amount of natural aquifer recharge that can be normally attributed to the area, increasing surface runoff and modifying groundwater flow and solute transport. A suite of Managed Artificial Recharge practices have been adopted in the past 60 years to palliate this drop in water quantity. As a result, the Llobregat River area is probably nowadays one of the places in the world where a larger variety of recharging methodologies have been implemented in a smaller area, ranging from scarification to deep injection wells and artificial recharge ponds.

Remarkably, albeit these artificial recharge practices have been insufficient to completely restore the resources of water in the area, the simultaneous application of multiple artificial recharge methodologies in the same area has been proven to be an efficient approach to decrease water scarcity beyond what a single technology can achieve.

References

1. Gámez D (2007) Sequence stratigraphy as a tool for water resource management in alluvial coastal aquifers: application to the Llobregat delta (Barcelona, Spain). PhD Thesis, UPC-Barcelona Tech. 1–177 + An
2. Luna M, Salas J, Molinero J, Queralt E, Rull M, Colomer V, Trevisan L, Ruiz E, Guimerà J, Niñerola JM (2009) Evaluación de la capacidad de infiltración e impacto hidrogeológico de las balsas de recarga artificial del Baix Llobregat (Santa Coloma de Cervelló). Jornadas sobre la Investigación de la zona no saturada del suelo –ZNS'09. Vol IX
3. Abarca E, Vázquez-Suñé E, Carrera J, Capino B, Gámez D, Batlle F (2006) Optimal design of measures to correct seawater intrusion. Water Resour Res 42:W09415. doi:10.1029/2005WR004524
4. Vázquez-Suñé E, Abarca E, Carrera J, Capino B, Gámez D, Pool M, Simó T, Batlle F, Niñerola JM, Ibáñez X (2006) Groundwater modelling as a tool for the European Water Framework Directive (WFD) application: the Llobregat case. Phys Chem Earth 31(17):1015–1029

5. Ortuño F, Niñerola JM, Teijon G, Candela L (2008) Desarrollo de la primera fase de la barrera hidráulica contra la intrusión marina en el acuífero principal del Delta del Llobregat. IX Simposio de Hidrogeología, Elche (Spain)
6. Ortuño F, Niñerola JM, Armenter JL, Molinero Huguet J (2009). La barrera hidráulica contra la intrusión marina y la recarga artificial en el acuífero del Llobregat (Barcelona, España). Boletín Geológico y Minero, IGME, 120(2):235–250, ISSN: 0366–0176

In-Stream Nutrient Flux and Retention in Relation to Land Use in the Llobregat River Basin

Rosana Aguilera, Sergi Sabater, and Rafael Marcé

Abstract The Llobregat as well as other streams and rivers elsewhere transports and transforms nutrients, regulating nutrient export downstream. The most common approach is to study nutrient in-stream processes at the reach level, but a correct understanding of their role on nutrient cycling in the watershed requires taking the perspective of the entire river network. Considering the complexity and the spatial scale of the Llobregat River network, modeling tools are useful to determine the sources and the processes by which nutrients are transported at the basin scale. In this chapter, we apply a hybrid process-based and statistical model (SPARROW, SPAtially Referenced Regression On Watershed attributes) to the Llobregat River basin in order to estimate the annual nitrate and phosphate loads that reach the drainage network. The model allows emphasizing the description of the in-stream processes that affect nutrient transport and retention in the Llobregat, and it might be useful in other impaired rivers as well. Although forests and grasslands are the predominant land cover types in the Llobregat basin, activities related to the presence of agricultural land and human agglomerations cause impairment in water quality and alter the natural flow. Increased exported nutrient load and decreased in-stream nutrient removal coincide in the downstream direction, probably due to the significant chemical and geomorphological changes in the lower Llobregat River basin.

Keywords Basin scale, In-stream removal, Modeling, Nutrients

R. Aguilera (✉), S. Sabater, and R. Marcé
Catalan Institute for Water Research (ICRA), Scientific and Technological Park of the University of Girona, Emili Grahit 101, 17003 Girona, Spain
e-mail: raguilera@icra.cat

S. Sabater et al. (eds.), *The Llobregat: The Story of a Polluted Mediterranean River*, Hdb Env Chem (2012) 21: 69–92, DOI 10.1007/698_2012_143,

Contents

1 Introduction

Undesirable levels of in-stream nutrient concentrations generally are the major cause behind proliferation of algal masses in slow-moving and nonturbid rivers. As they transport nutrients and other pollutants along the fluvial network, eutrophication of downstream water bodies (also including lakes, reservoirs, and coastal areas) is an issue of major concern [1]. Overall, the existence of high nutrient concentration does not only affect the structure and function of the aquatic ecosystem but may also be detrimental to human health. The identification and quantification of nutrient inputs and their sources is vital to achieve good water quality since it provides the basis for efficient management of nutrient sources.

Nutrient inputs are usually the most effectively managed among the varied stressors that affect water quality in a basin [1]. Analyzing the influence of nutrient sources on water bodies, such as land use and wastewater effluents, often requires the evaluation of nutrient cycling processes at a larger spatial scale than the reach level (e.g., at the watershed level) [2]. Considering that a reach is part of a stream or river which, in turn, belongs to a network of streams within a catchment, it becomes apparent that conditions found at the reach level are affected by those found at larger spatial scales [3]. Appropriate tools and techniques are needed in order to identify the stressors that alter water quality and, most importantly, to link the stream's response to the responsible stressor [4]. The relative importance of nutrient sources associated with human activities is better expressed in terms of the processes that occur within the rivers compared to merely analyzing measurements [5]. Modeling tools become useful in determining the sources and the processes by which pollutants are transported at the basin scale [5, 6].

One of the most common problems encountered in European river basins is nutrient enrichment. In Mediterranean basins, water pollution acts as an additional stressor on an already scarce water supply and aggravates the damage to stream health in the region [7]. Such is the case of the streams and rivers in Catalonia, where damming, water extraction, and urbanization worsen the effect of current pressures on the already vulnerable water cycle [8]. Particularly, the Llobregat

River basin is characterized by the presence of many weirs along the main channel as well as the occurrence of high concentration of pollutants, which tend to increase as water travel downstream [9].

Studies of the water quality status of the Llobregat River basin have established the presence of high levels of pollution that gradually increase from upstream to downstream direction, overlapping with the density distribution of industrial and urban nuclei [10]. Biological indicators have also confirmed that the streams and rivers in the basin are characteristic of an impaired fluvial system. A study determined that there is a general decrease of species richness of diatoms and invertebrates downstream, as well as an increase in the species tolerant to organic and chemical pollution [11]. Moreover, another study concluded that the invertebrate community structure in the soft sediment of the Llobregat River basin mirrors the pollution gradient observed from upstream to downstream sites [12]. Coincidently, these studies found that salinity (due to salt mine debris), high nutrient concentration, and low water flow are among the main accountable factors for the prevalence of biologically poor communities.

In this chapter, we evaluate the patterns in nutrient sources and in-stream retention characteristics. The data cover a 7-year period (2000–2006) on water quality in the Llobregat River basin reaches. Nutrient apportionment is achieved by means of a modeling tool, as well as by the examination of nutrient removal capabilities within the river sections under major anthropogenic stress exerted along the network. Our results suggest that customary modeling approaches are not appropriate for describing in-stream nutrient processes in basins including impaired reaches. In order to consistently upscale nutrient retention across scales, a deeper knowledge of the governing drivers of lumped biological activity in reaches must be obtained.

2 Modeling In-Stream Nutrient Transport and Retention

2.1 SPARROW Model

When selecting an appropriate modeling tool, a balance between model complexity and data availability must be reached. Most modeling efforts at the basin level have employed mechanistic models, which require enormous amount of water-resources data at a highly detailed temporal and spatial scale, but that in turn are able to supply dynamic responses at a fine temporal and spatial resolution [13]. However, detailed input data may not be readily available and a greater number of assumptions must be made in that case. On the other hand, conventional empirical models frequently come across the challenge of upscaling the biogeochemical processes that affect nutrient fluxes along stream networks and reservoirs within a basin, with no guarantee that measurements performed at one particular scale will be valid to parameterize processes at other scales.

The hybrid statistical and process-based model SPARROW (*SP*Atially *R*eferenced *R*egression *O*n *W*atershed attributes) attempts to address the aforementioned issues by spatially linking watershed characteristics to the stream network and by using mechanistic functions based on empirical observations to describe nutrient dynamics. SPARROW also provides robust measures of uncertainty as a result of the statistical estimation of the regression parameters [14]. Furthermore, the robustness of the parameter estimates can be evaluated by applying a bootstrap algorithm incorporated in SPARROW.

The fundamental component of the SPARROW model is the stream reach and its associated incremental catchment (Fig. 1). Conceptually, the nutrient load leaving a reach is the sum of two components, the load generated within the upstream reaches and transported to the reach via the stream network and the load originated within the reach's incremental watershed and delivered to the reach segment [14]. Nutrient flux generated within the incremental watershed is subject to attenuation in the terrestrial phase (represented by land-to-water delivery factor) and to in-stream and reservoir decay within the reach. The load generated upstream undergoes additional in-stream attenuation in the reach being evaluated. Regarding the temporal dimension, the SPARROW model works at the annual time-step; this may be sufficient to locate high risk areas within the basin that are vulnerable to particular sources of pollution [13]. For instance, the implementation of a statistical modeling approach based on SPARROW formulations in a French river system resulted in the demarcation of critical areas where the contribution of agricultural practices to nutrient loads was predominant [15].

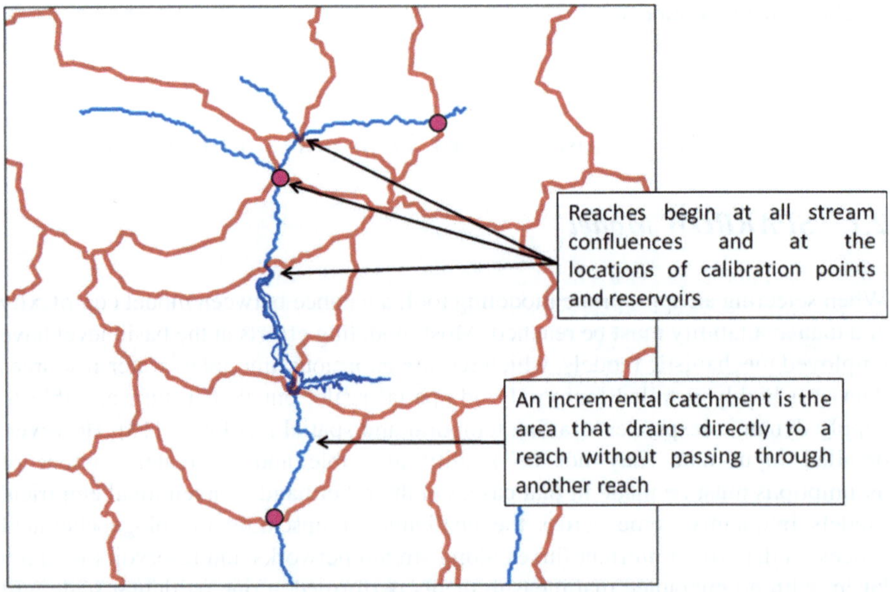

Fig. 1 Fundamental units of the spatial scale in the SPARROW model

The SPARROW model, developed by the United States Geological Survey (USGS), has been previously applied to the regional assessment of water quality in the United States of America [16] and in the evaluation of nutrient loads in New Zealand [17]. As an example, the delivery of nitrogen to the Gulf of Mexico (where eutrophication and hypoxic areas are a recognized problem) and the relationship with nutrient aquatic processing and stream size was characterized by applying the SPARROW model in the Mississippi River basin [18].

2.2 *In-Stream Decay Specification*

Streams and rivers are pathways that transport nutrients arriving from land and upstream sources and thereby serve as spots for nutrient transformation, storage, and removal. In other words, river networks become regulators of exported material to downstream aquatic ecosystems. Understanding the coupling between terrestrial nutrient sources and the aquatic transport within a basin is vital for the management of receiving water bodies such as lakes, estuaries, and marine coastal environments [19, 20].

Nutrient transport and retention processes have been mainly studied at the reach level, and they are frequently characterized by metrics provided by the Nutrient spiraling concept [21]. According to this conceptual framework, the complete cycle of a nutrient atom is examined as it travels downstream, including its transformation by biological uptake and its subsequent release to the water column. Overall, nutrient spiraling metrics combine both the hydrological and biological factors that are involved in nutrient processing in streams. Current modeling tools provide a valuable means for unraveling such processing in systems larger than headwater streams. Models assist in the evaluation of nutrient cycling from the stream network perspective via upscaling, to a basin point of view, the most relevant mechanisms that occur at the reach scale.

The models developed to simulate nutrient transport and retention at the watershed level usually rely on reaction rate expressions in order to describe nutrient loss in terrestrial and aquatic ecosystems. Most of the rate expressions implemented in modeling assume first-order kinetics in the behavior of nutrient uptake, which means that the rate of nutrient loss is proportional to the load (or concentration) of the constituent(s) being modeled [18, 20]. The proportion of nutrient removed in a given time-step (R) has been expressed on a volumetric basis (depth dependent) as an exponential function of the rate k and the time-of-travel τ (or residence time) of the solute in a given stream:

$$R = 1 - \exp(-k\tau). \tag{1}$$

Alternatively, the mass-transfer coefficient, v_f, could be used to describe the nutrient migration to the benthic compartment. This depth-independent measure, also known as uptake velocity, quantifies the vertical velocity of the solute expressed as an exponential function related to the ratio of the water residence time and the mean depth d:

$$R = 1 - \exp\left(v_{\mathrm{f}} \times \frac{\tau}{d}\right). \quad (2)$$

SPARROW models have implemented a first-order decay equation to quantify the proportion of constituent being removed in a particular reach [18]. This implies that nutrient removal rates would increase linearly with nutrient concentration, with no signs of potential saturation. However, it has recently been argued that the concentration of nutrients in streams determines the rate of uptake, which could eventually reach partial saturation. Data of nutrient spiraling metrics have shown that the efficiency of process rates relative to increasing concentration eventually declines. According to a recent study which evaluated the saturation of nitrogen cycling in a broad range of streams in Kansas (USA) [22], N uptake increases with NO_3^- concentration, but at a slower rate than the augment in N concentration thus resulting in a loss of nitrogen processing efficiency. This study concluded that an Efficiency Loss (EL) model is able to describe the empirical relationship between biological nitrogen uptake and increasing NO_3^- concentration. Within the EL concept, log-transformed areal uptake rates (U) increase with log-transformed nutrient concentration, where the slope of the relationship is less than one. Several subsequent studies have also observed that the relationship between areal uptake rate and NO_3^- concentration in streams shows a partial saturation effect [23–25].

The formulation for reach nutrient attenuation in SPARROW has been modified [26] to include a wider range of stream uptake dynamics (including first-order kinetics and EL). The new model specification was based on an empirically derived power law related to available nutrient concentration. The fraction (R) of the nutrient load in a given reach that is removed as a function of stream channel characteristics is therefore expressed as:

$$R = 1 - \exp(-v_{\mathrm{f}} \times H_{\mathrm{L}}^{-1}). \quad (3)$$

The inverse of the hydraulic load (H_{L}^{-1}) is the ratio between reach time-of-travel (water residence time) and water depth in a given reach. The uptake velocity (v_{f}), which represents the average downward velocity at which nutrients are removed from the water column by migrating into the benthic compartment, was calculated as follows:

$$v_{\mathrm{f}} = a \times C^{b}, \quad (4)$$

where a and b are power law parameters (intercept and slope, respectively) and C is the available nutrient concentration in each reach. Biological factors affecting nutrient removal are represented by the uptake velocity variable, and the hydraulic load reflects the hydrological conditions of the reach being evaluated.

2.3 Watershed Characteristics

In the water quality modeling exercise presented in this chapter, nitrate and phosphate flux measurements in different points along the streams and rivers were linked to sources related to natural areas and anthropogenic activities. The principal nutrient sources in the Llobregat River basin were both diffuse from agricultural activities and end-of-pipe inputs derived from wastewater discharges in urban and industrial areas. Diffuse nutrient sources were represented by main three different land uses (Table 1) based on multi-temporal data captured by the Thematic Mapper sensor of the Landsat satellite during the year 2002, provided by the Cartography Institute of Catalonia.

Point source loads for both nitrate and phosphate were obtained from wastewater and industrial effluents data (Fig. 2) provided by the Catalan Water Agency (ACA). These nutrients are considered to be discharged directly into the river and therefore are not subjected to any transport or retention processes within the landscape.

The average nutrient concentrations of the selected 23 ACA monitoring points during the period of 2000–2006 were 11.2 mg NO_3^- L^{-1} (or 2.6 mg N L^{-1}) and 0.9 mg PO_4 L^{-1} (or 0.3 mg P L^{-1}) (Fig. 3). The spatial distribution of highest and lowest concentration values slightly differs among nutrients. High average concentration of nitrate ($>$15 mg L^{-1}) is found in the middle section of the Llobregat River (vicinity of Manresa) and in reaches of the Anoia River and the Cardener River (particularly at Clariana de Cardener). The concentration of phosphate is also high in these reaches, the highest average value being at the lower part of the Anoia River.

The river Riera de Rubí is a particular case of impairment in the basin. This tributary is situated on the left margin of the Llobregat River, and carries most inorganic nitrogen in the form of ammonium (NH_4^+, the least toxic of the ammonia-nitrogen forms at low pH). Ammonia-nitrogen (NH_3-N) has a more toxic form at high pH (un-ionized ammonia), which is also present in this stream in smaller concentrations. The discharge of ammonia-nitrogen and its subsequent oxidation can seriously reduce the dissolved-oxygen levels. This is the case at the Papiol monitoring station (near the confluence of the Riera de Rubí and the Llobregat River), where the average ammonium concentration was 25.5 mg NH_4^+ L^{-1} (average concentration of ammonia was 2.1 mg NH_3 L^{-1} between 2003 and 2006), and the value for nitrate was 5.4 mg NO_3^- L^{-1}. Phosphate concentration was also recorded to be high in this point (average 2.7 mg PO_4 L^{-1}). The high pollution level of this tributary is caused by the high density of industries and

Table 1 Classification of main land use types found in the Llobregat River basin

Land use types	% of total basin area	Predominant land cover type (% of total basin area)
Urban	6	Housing (2.9)
Cultivated	24	Dryland crops (19.6) and vineyards (2.39)
Natural	70	Woods and meadows (28.3) and needle-leaved forest (31.6)

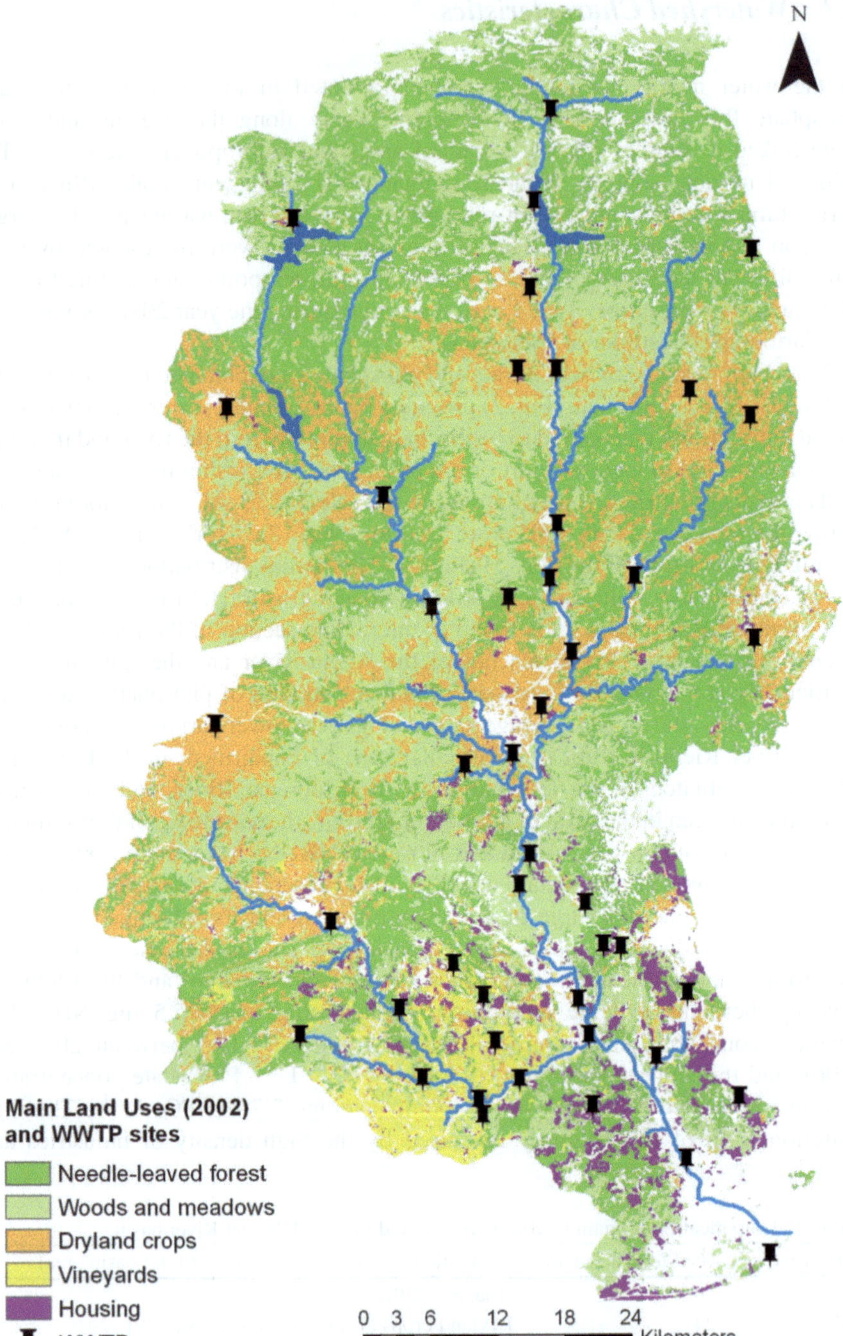

Fig. 2 Land use distribution in the Llobregat basin with location of WWTP effluents

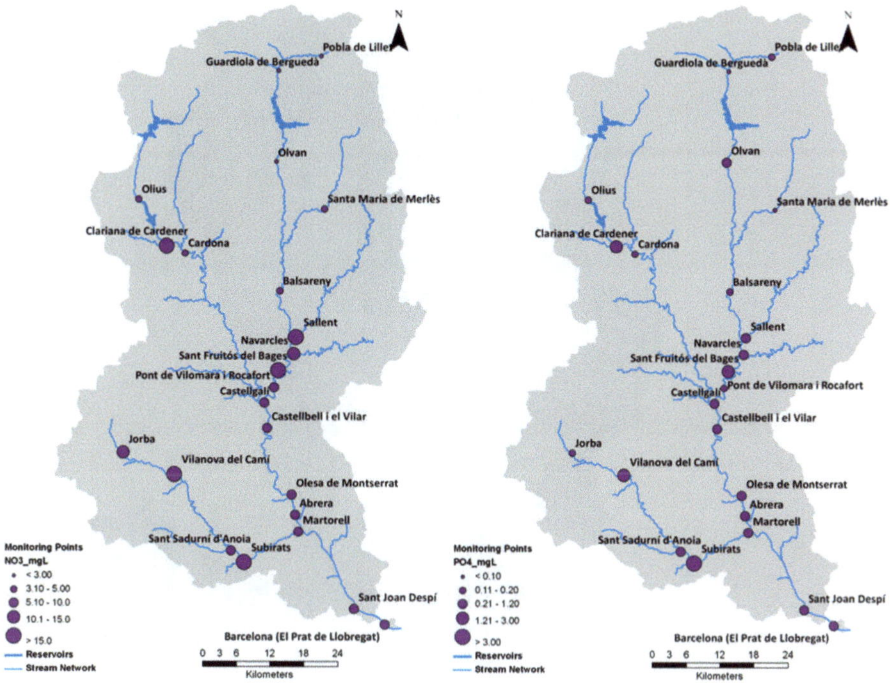

Fig. 3 Mean nitrate and phosphate concentration values at selected monitoring sites

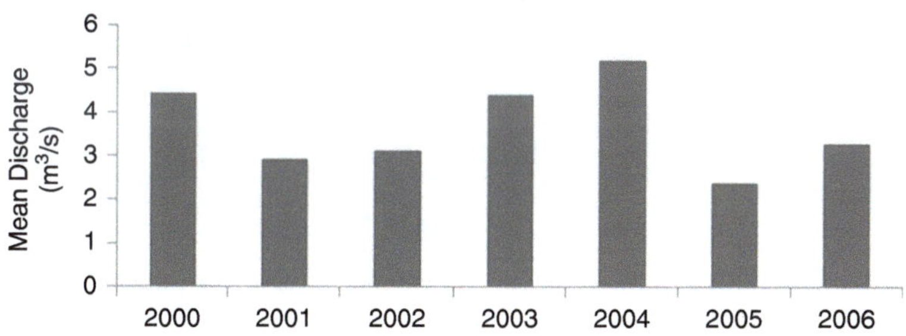

Fig. 4 Mean basin discharge variability between the years 2000 and 2006

wastewater spills occurring in the area [27]. Since nitrate was not the predominant source of inorganic nitrogen and most of the streamflow does not reach the Llobregat River (waters are derived through a derivation channel to the lower Llobregat River; see [36], this volume), the monitoring data gathered from the El Papiol station (near the city of Castellbisbal) were not included in this modeling exercise.

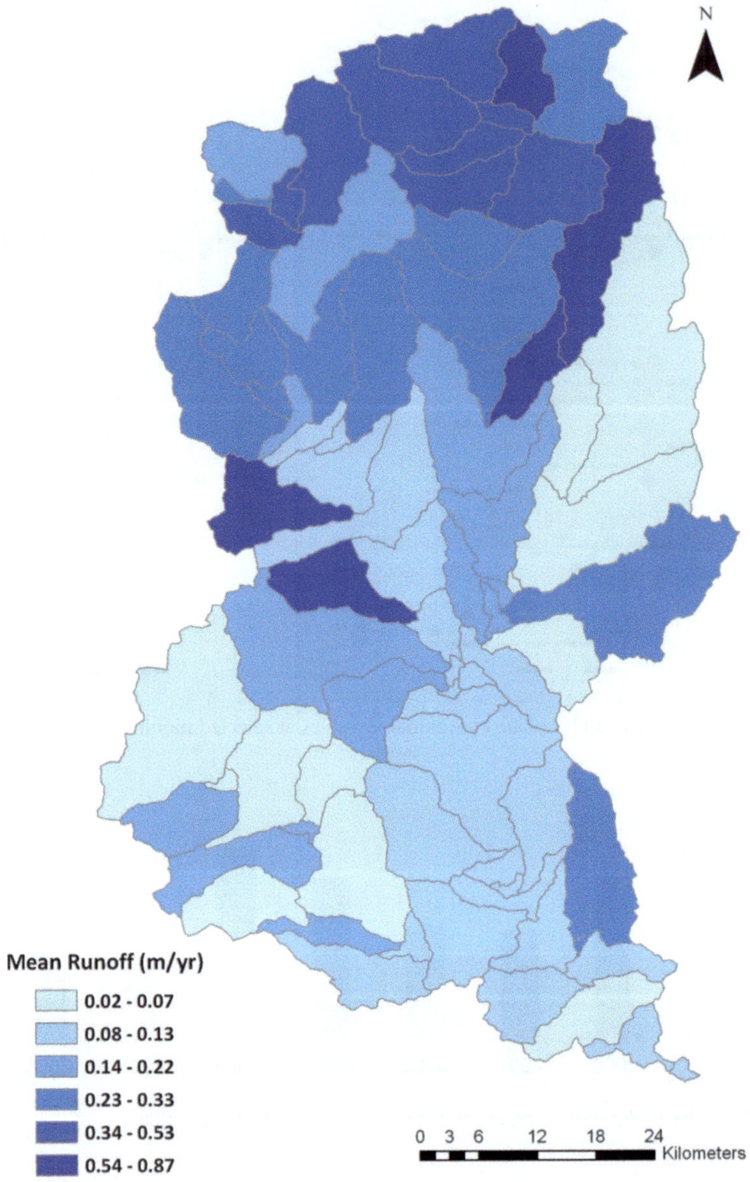

Fig. 5 Spatial distribution of mean runoff generated in each sub-watershed for the period comprised between 2000 and 2006

Inter-annual climate variability was depicted in the mean loads observed at the monitoring points and in the values of average discharge and runoff assigned to each of the sub-watersheds. Mean annual discharge in the Llobregat River basin during 2000–2006) is both variable between years (Fig. 4) as well as highly variable between sub-watersheds (Fig. 5).

3 Nutrient Loads and In-Stream Retention

The nitrate model calibration results provided a reasonable fit between measured and predicted loads. The nitrate model was considerably efficient as shown by its coefficient of determination ($R^2 = 0.86$; RMSE of 0.55). The Yield-R-Square value indicated that 66% of the variance of observed nitrates yield is explained by the estimated model. In the case of the phosphate model, the values of R^2 and Yield-R-Square were 0.80 (RMSE of 0.73) and 0.60, respectively. Figure 6 illustrates the relationship between observed load and that predicted by the SPARROW model calibration for each constituent.

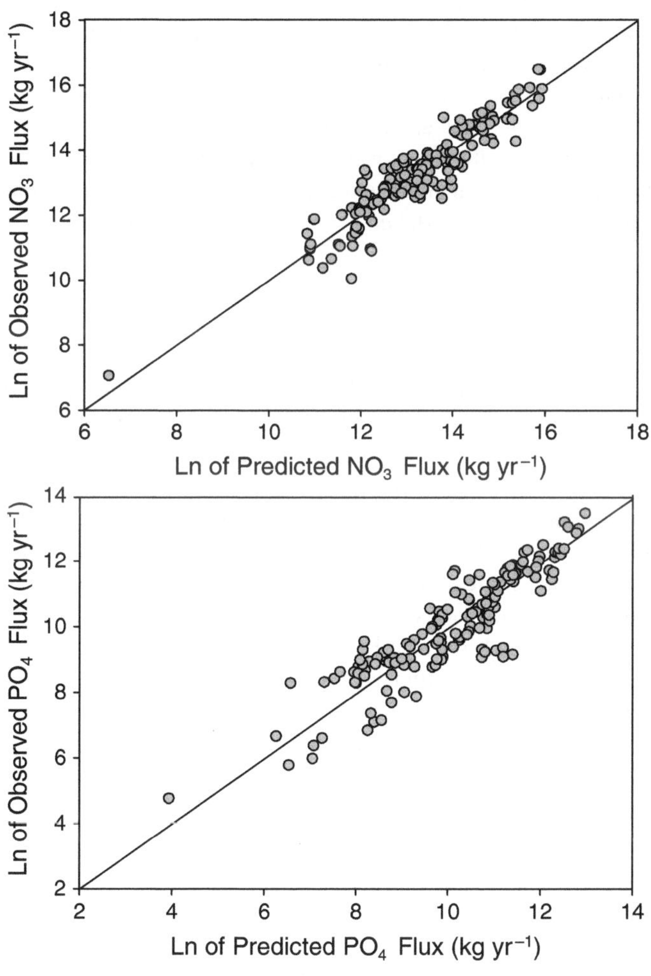

Fig. 6 In-stream observed annual loads compared to model predicted loads at calibration sites (*top*: nitrate, *bottom*: phosphate)

3.1 Nutrient Export

3.1.1 Urban Land

The explanatory variable related to diffuse sources from Urban land included in the SPARROW models reflected the amount of nutrients transported by urban runoff. However, after the parametric and bootstrap estimation exercises, diffuse pollution from urban runoff was discarded after running both nutrient models, resulting in coefficients values for Urban land equal to 0. The behavior could probably be explained by the distribution of point sources (WWTP and industrial spills) in the basin, which coincide with the spread of the land use areas identified as Urban land. Thus, point sources appear to take preference in explaining nutrient flux deriving from developed lands. A recent comparison of the regional models implemented in the United States [16] stated that even though diffuse urban sources can affect water quality in streams, wastewater spills could take a greater importance in explaining urban loads due to the large masses of nutrients that are discharged directly into streams. In the case of the Llobregat River basin, the urban runoff sources may be considered to be included in the point source inputs that streams receive, also bearing in mind that sewage and rainwater separation is not undertaken in some areas of the territory.

3.1.2 Cultivated Land

In terms of nutrient sources, those associated with agricultural uses played a major role in explaining nitrate yield; the export coefficient value yielded an annual average of 7.43 kg N ha^{-1} $year^{-1}$, although cultivated land occupies only 24% of the basin area (versus 70% covered by Natural land). A recent modeling study in two agricultural European catchments [6] reported average nitrogen export coefficients between 21 and 35 kg N ha^{-1} $year^{-1}$. The extrapolation from a European model of nitrogen export [28] indicates that the Llobregat River basin is characterized by 0–2 and 3–12 kg ha^{-1} of N annual diffuse emissions, which agree with the results obtained in the SPARROW model of the Llobregat River basin.

The phosphate export coefficient attributed to agricultural land (0.83 kg P ha^{-1} $year^{-1}$) is significantly lower than that of nitrates. A publication on nutrient export [29] reported a median export coefficient for nonrow crops of 0.8 kg P ha^{-1} $year^{-1}$ (range 0.7–1.6), which agrees with the estimated average for the Llobregat River basin.

3.1.3 Natural Land

Nitrate export from natural land (which include forests, grass, and shrub lands) in the Llobregat River basin, estimated by the SPARROW model, held a slightly higher value (3.34 kg N ha^{-1} $year^{-1}$) than the median observed in forested regions of North America (2.5 kg N ha^{-1}). However, the natural land export value fell within the range attributed to forest land export (2.0–3.5 kg N ha^{-1}) [29]. Nutrients derived from this type of land use correspond to diffuse anthropogenic sources and/or natural background concentrations of the nutrients under evaluation. Natural land export of nitrates potentially includes atmospheric deposition, particularly in the sub-watersheds where higher precipitation and runoff values are observed, which in turn coincides with the area where the largest percentages of forest and grass lands are found. Overall, atmospheric deposition can be a negligible contributor of nitrogen in impaired systems like the Llobregat River basin, but where other nutrient sources such as those included in this modeling exercise are not present deposition is usually the largest source of nitrogen that reaches streams [16].

Also, the export coefficient attributed to natural land in the phosphate model (0.21 kg P ha^{-1} $year^{-1}$) fell within the range previously reported in the literature [29]. In previous SPARROW model applications, forest land was identified as a statistically significant source of the phosphorus exported to water bodies (coefficients ranged from 2.04 to 14.7 kg TP km^2 $year^{-1}$) [16]. As a potential explanation, the authors stated that phosphorus accumulated in stream bed sediments can become an important net source, given the annual time-step that SPARROW is based on.

3.1.4 Point Sources

For sources that are expressed in units of loads, parameter estimates should approximate the value of 1. Any deviation from this value expresses a systematic bias in the databases used to calculate these loads, which is in fact the rationale to include such a parameter in this model. Point source coefficients related to nitrate pollution were greater in the upper part of the basin (0.76), whereas wastewater spills in the lower section (downstream Abrera station) could have been overestimated prior to model calibration, which could explain the low coefficient value of 0.15 [-]. In the case of phosphates, the point sources took a greater portion than in the nitrates model, *upstream* (1.54) and *downstream* (0.79), approximately twofold and fivefold higher than the values obtained in the nitrate model, respectively.

3.2 *Nutrient Apportionment*

One of the advantages of the SPARROW modeling technique is that the effects of statistically significant nutrient sources can be separated and evaluated in a spatial context, thus identifying areas where each source contributes the largest portions of nutrient loads [30]. The maps in Fig. 7 show the spatial distribution of the share of the three major nutrient sources in sub-watersheds of the Llobregat River basin. Natural sources for both nutrients tend to predominate in the upper part of the basin, whereas agricultural activities are more relevant in the middle section and the upper Anoia River. Point sources are especially important in the lower part of the basin,

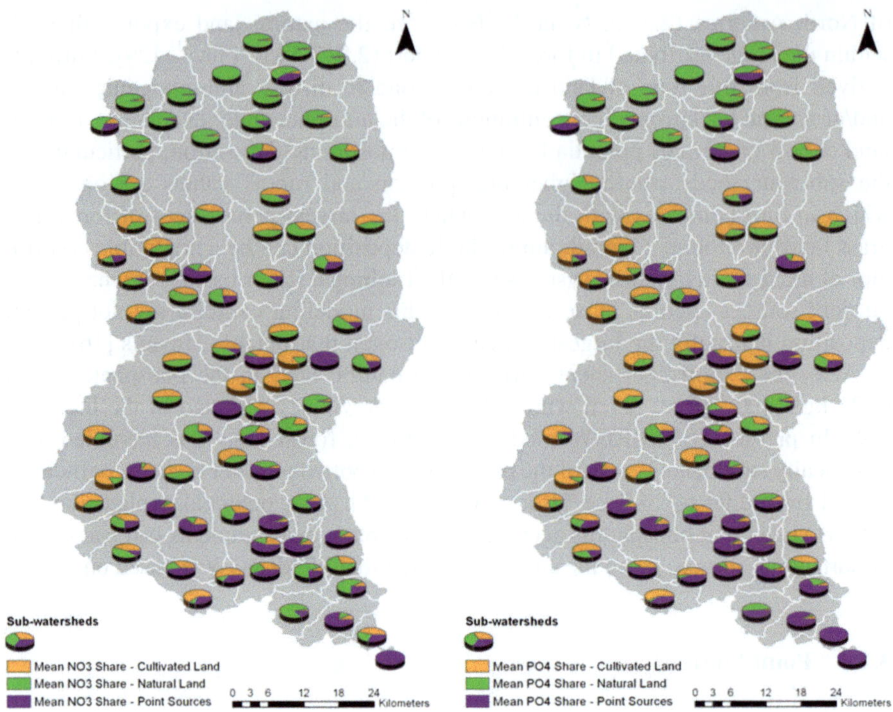

Fig. 7 Mean source shares of nitrates (*left*) and phosphates (*right*) between 2000 and 2006 at each sub-basin considered in the model

where they dominate the apportionment of nutrient loads, particularly that of phosphates.

Overall, the mean predicted yield for total flux leaving the reach generated in SPARROW was 17.5 kg NO_3 ha^{-1} $year^{-1}$ (or 4.03 kg N ha^{-1} $year^{-1}$) for the nitrate model and 0.8 kg PO_4 ha^{-1} $year^{-1}$ (or 0.26 kg P ha^{-1} $year^{-1}$) for phosphates (predicted incremental yield for total flux generated within the sub-watershed of 89.8 and 12.8 kg ha^{-1} $year^{-1}$, respectively). The predicted flow weighted concentration values obtained by the model were 19.4 mg NO_3 L^{-1} (4.5 mg N L^{-1}) and 0.8 mg PO_4 L^{-1} (0.3 mg P L^{-1}).

Total nutrient loads (Fig. 8) include the effect of land-to-water delivery transformations but not those of in-stream attenuation processes. In other words, these are the loads that reach the streams from the surrounding land and upstream sources, where no in-stream decay has taken place. The greatest values of nutrient flux are found mainly in the vicinities of the Llobregat River middle and lower sections (starting approximately downstream of the La Baells reservoir and increasing in the downstream direction), as well as in the lower reaches of the Cardener and the Anoia rivers.

Diffuse sources in the basin, particularly natural land loads, accounted for the major part of nitrates arriving to streams (Fig. 9). In the phosphate model, point sources took a greater portion of the total load generated in the basin when compared

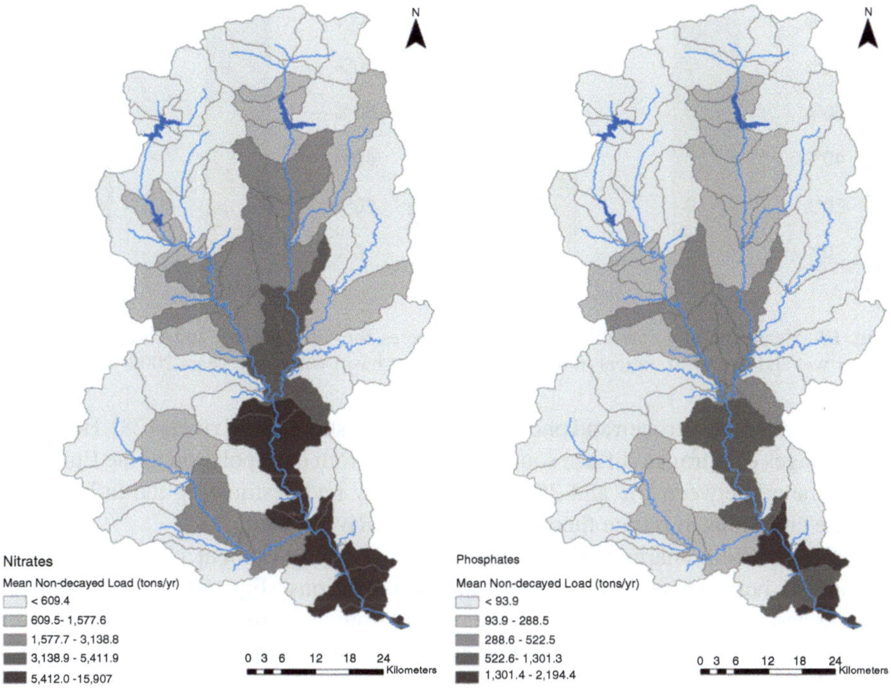

Fig. 8 Average nondecayed nutrient loads generated in the basin (predicted total flux leaving the reach if there were no in-stream or reservoir attenuation processes)

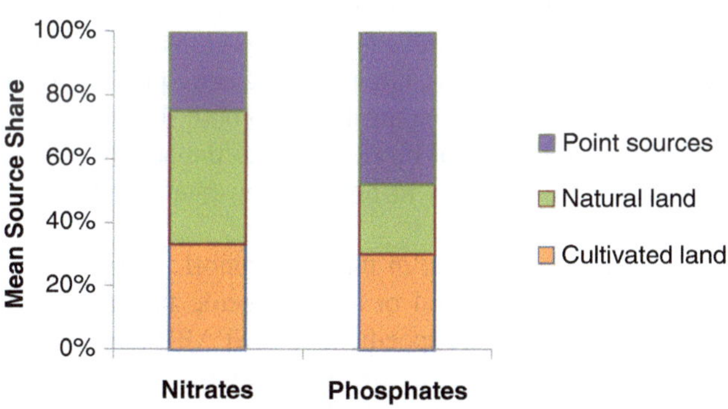

Fig. 9 Mean share (%) attributed to diffuse (Cultivated and Natural land) and point sources in the Llobregat River basin for the period of 2000–2006

to percentages found in the nitrate model. Nevertheless, diffuse sources were also important in explaining phosphate load, particularly those originating in areas of Cultivated land (as opposed to Natural land in the nitrate model). Studies performed in large European basins [13, 28] have confirmed that agricultural sources usually contribute the greatest load of nitrogen whereas the same applies for point urban sources and phosphate loads. In general, household sources of nutrients tend to be the

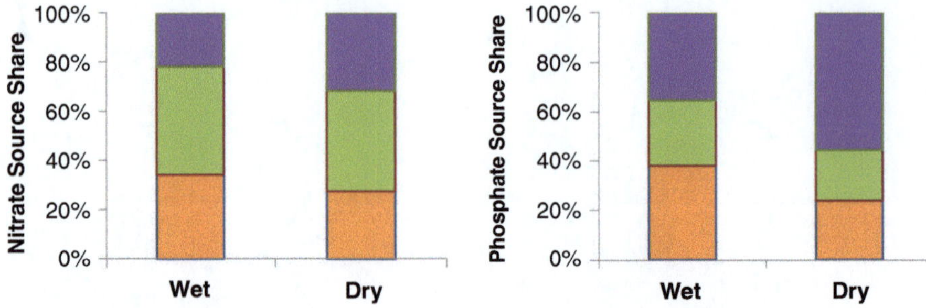

Fig. 10 Mean share (%) for diffuse and point sources for Nitrates (*left*) and Phosphates (*right*) under wet and dry hydrological in the Llobregat River basin within the years 2000–2006

least important in nitrogen and phosphorus export to streams and rivers [28]. However, percentages of nitrogen diffuse sources (resulting from extrapolation to the European scale) range between 0% and 25% [28], probably underestimating the influence of diffuse emissions of nitrate flux in the Llobregat River basin.

Mean nutrient shares varied when total load apportionment percentages were compared between periods of wet and dry conditions (Fig. 10). The portion of nitrate and phosphate flux derived from point source pollution was smaller under wet conditions. However, the point source share for the phosphate model for both hydrological conditions surpassed the percentages found for nitrate loads. The main reason for this was that a high percentage of the measured water flow in streams and rivers under dry climatic conditions is composed of point source effluents. Discharging wastewater directly into rivers with low discharge poses a risk to the ecosystem since dilution does not occur or it is minimal [31].

Phosphate total loads generated in the basin (Fig. 11) varied between years under wet and dry hydrological conditions. Greater loads were predicted during the most humid period. This difference can be attributed to both the influence of the runoff delivery variable in the transport of nutrient flux to the stream network and to the slightly greater capacity of streams to remove phosphate under dry hydrological conditions.

Land-to-water delivery variables, in this case runoff, describe properties of the landscape related to climatic, natural or anthropogenic factors affecting terrestrial nutrient transport [14]. The delivery specification in SPARROW assumes steady-state conditions and thus the transport of nutrients reflects the net effect of the processes involved [14]. The runoff values in the SPARROW model of the Llobregat River basin were expressed as differences from their mean value over all reaches to improve the interpretability of the source coefficient and to provide some stability in its value across alternative specifications of the land-to-water delivery factor.

Diffuse nutrient loads derived from agricultural and natural land increased during years with higher runoff values and decreased when runoff was lower. Overall, climate-associated factors are generally considered the main determinants of nutrient export, particularly in forested land [29]. In an evaluation of regional SPARROW models in the United States, it was recognized that climatic variables played a major role in the delivery of nutrients to streams [16]. In sub-watersheds

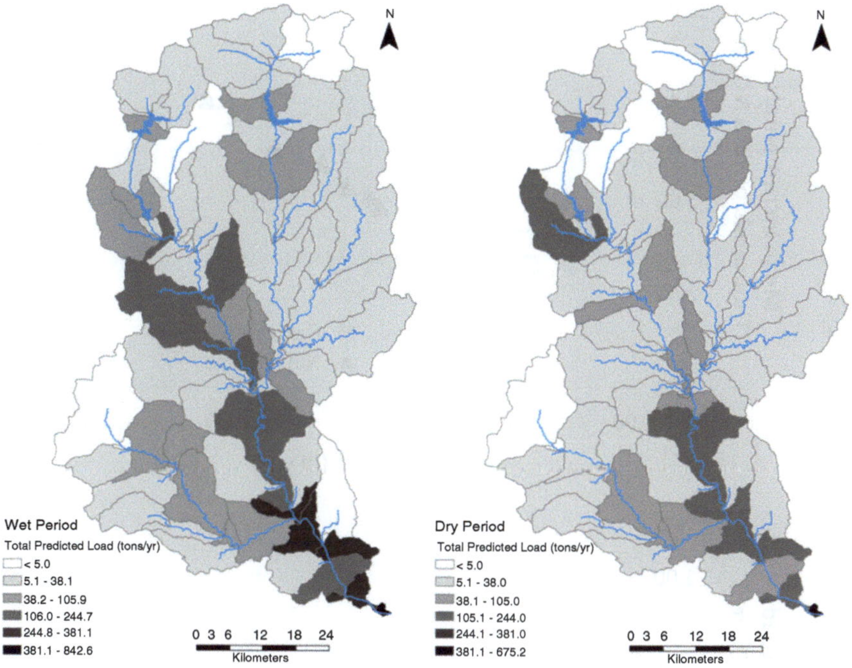

Fig. 11 Phosphate total predicted load variation between wet (*left*) and dry (*right*) periods

with lower runoff (or precipitation), mobilization of nutrients is minimal or smaller than would have been expected if higher precipitation and runoff occurred (Fig. 12).

3.3 *In-Stream Nutrient Loss*

The empirically driven power law relationship to estimate annual net in-stream retention used in SPARROW showed a decline in uptake velocity (v_f) with increasing concentration (Fig. 13). The decreasing values of the uptake velocity along the concentration gradient confirmed a loss in uptake efficiency in streams of all sizes in the Llobregat River basin. However, the slopes were much steeper than those found in literature, both in the nitrate and phosphate models. The regression lines generated with the bibliographical data of uptake velocity and nutrient concentration (Fig. 13) followed the expected power law function with negative exponent attributable to the EL dynamics. The most likely explanation for the differences between the literature and model uptake velocity values is the differences in levels of impairment and in hydrological characteristics in the systems in each data set. The Llobregat River basin is characterized, in most reaches, by higher water flow and in-stream nutrient concentrations.

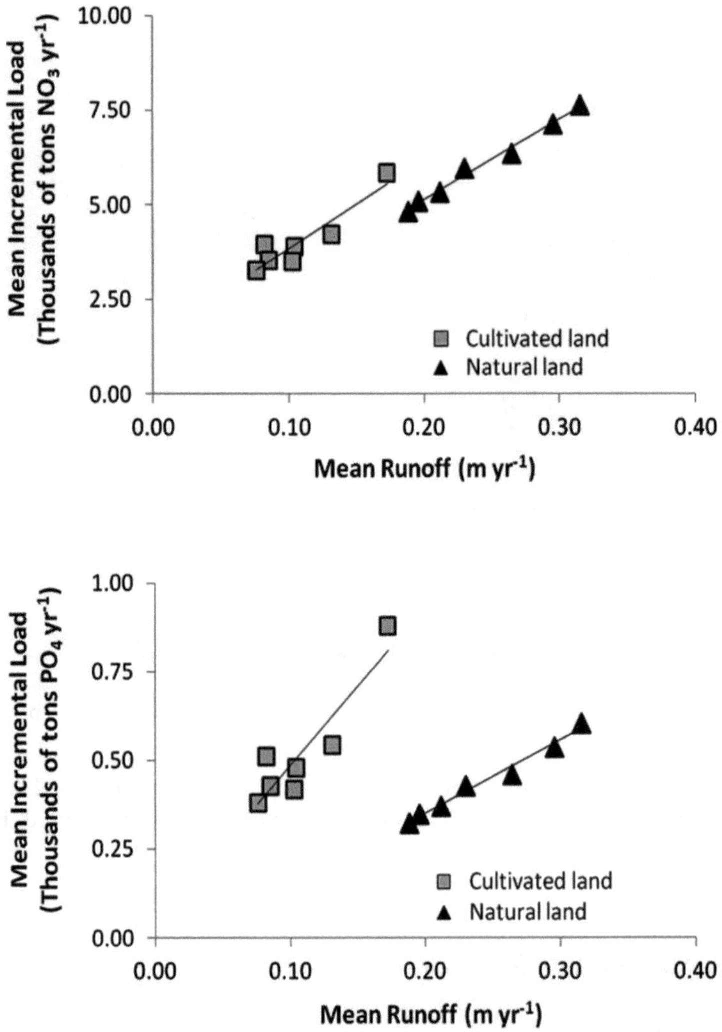

Fig. 12 Relationship between mean incremental load for nitrates (*top*) and phosphates (*bottom*) and mean runoff generated in the basin within 2000–2006

There could be, in fact, several responses (slopes) for distinct types of streams, which are generated according to their actual pollution level and the geomorphologic (or other) changes they underwent. A recently published work [32] concluded that the regression slopes for nutrient uptake might have deviated from the expected behavior if higher nitrate concentration values had been used in the short-term addition approach. Another recent study [33] evaluated the response of streams, characterized by decreasing v_{fden} with increasing concentration, by implementing the river network model based on denitrification measurements from headwater streams found in [23]. The model was set to determine v_{fden} for each segment according to: $v_{\mathrm{fden}} = c[\mathrm{NO_3}]^d$,

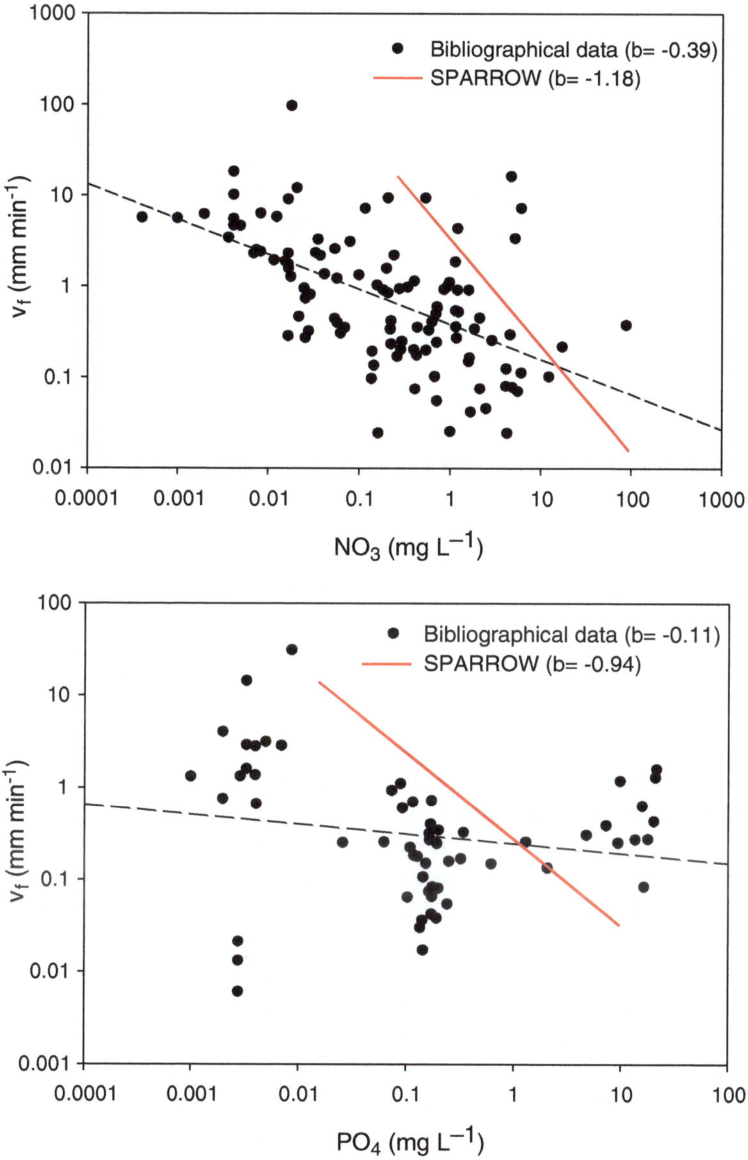

Fig. 13 Uptake velocity vs. concentration for nitrate (*top*) and phosphate (*bottom*); *black dots*: literature values, *red line*: fitted by SPARROW

where parameters *c* and *d* were derived empirically for each catchment involved in the study [33]. The evaluation showed an overall decline in denitrification efficiency (v_{fden}) with increasing nitrate concentration. On the other hand, the strength of this relationship tended to vary significantly when the response of each of the eight

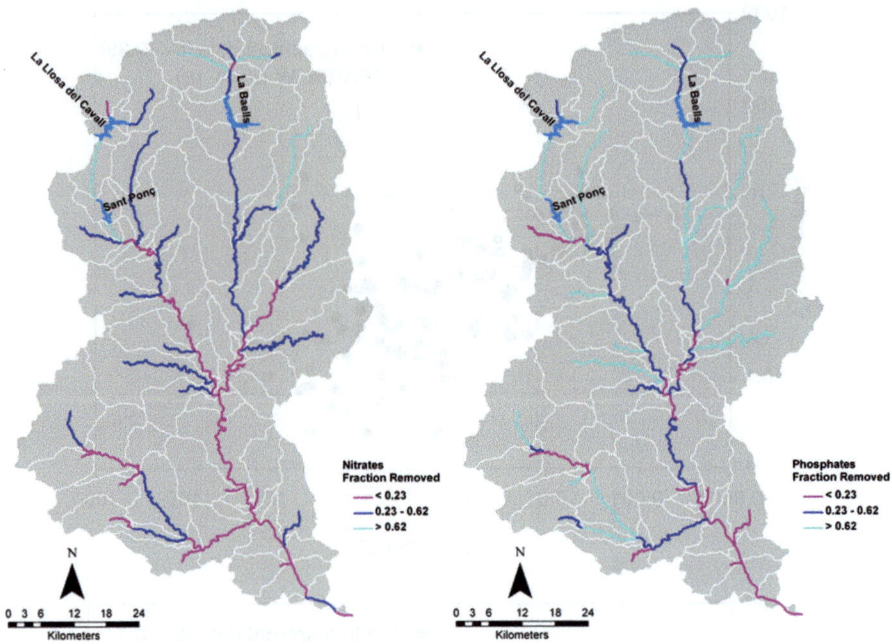

Fig. 14 Mean in-stream nutrient removed fraction obtained from SPARROW reach decay specifications (nitrate model, *left*; phosphate model, *right*)

catchments in the study was considered individually (i.e., nutrient concentration was a primary driver of v_f in some catchments, but not in others).

The fraction values of flux removed (Fig. 14) were extracted from the results under the "*reach decay fraction*" specification in SPARROW (see Sect. 2.2). This is calculated based on the mean available nutrient concentration (C) and hydrological characteristics (represented by the hydraulic load, H_L) in each of the 73 reaches categorized as streams during the 7 calendar years. This very same fraction is assumed to be applied not only to the incremental load generated within the sub-watershed in evaluation, but also to the nutrient load imported from the upstream reach. Overall, the portion of phosphate flux removed in streams in the Llobregat River basin is greater than nitrate removal. The mean annual percentage of nutrient removal was 23% for nitrates and 40% for phosphates. River nitrogen retention in European watersheds is estimated around 10–20% [6].

In-stream removal is generally dictated by the available nutrient concentration (in-stream decay specifications, Sect. 2.2), where the removed fraction is inversely related to increasing concentration (due to decreasing uptake velocity values). In the Llobregat River basin, percent in-stream nitrate removal is greater in the upper part, with fractions of mean removal higher than 0.6 (i.e., 60%); whereas in the lower section, particularly downstream the Abrera monitoring sections, at some reaches, the removal fraction is practically negligible. In the phosphate model, high removal fractions were also observed in reaches of the Anoia River (Fig. 14, right panel).

Low phosphate in-stream retention appears to be related to the presence of urban and industrial nuclei (i.e., the presence of point source discharges). Nitrate low in-stream retention could be related to cultivated land and urban and industrial areas. Point sources effect on the river decreased within the past decades with the implementation of WWTPs. However, in systems such as the Llobregat where the natural flow of the river is low, urban and industrial wastewater discharged into streams and rivers can account for a large part of the streamflow and dilution of nutrients is minimal or negligible. Streams that receive wastewater inputs maintain a determined capacity to retain and transform nutrients, although this self-purification ability is usually lower than that of pristine streams [31]. Overall, the generation of greater loads and the higher level of impairment in the lower section of the basin could explain the decreased capacity of self-purification predicted for the reaches in this area.

In general, variations in the nutrient retention capacity within the river networks are related to hydrological and geomorphological properties that determine the time a solute remains in contact with reactive surfaces [34]. Biological activity is also likely to shape nutrient retention and transport to downstream waterbodies; for instance, a recent study found that biological activity in streams under the influence of agricultural activities was higher relative to pristine streams [35], although nutrient retention efficiency tends to be lower at nutrient enriched streams. In order to be able to describe

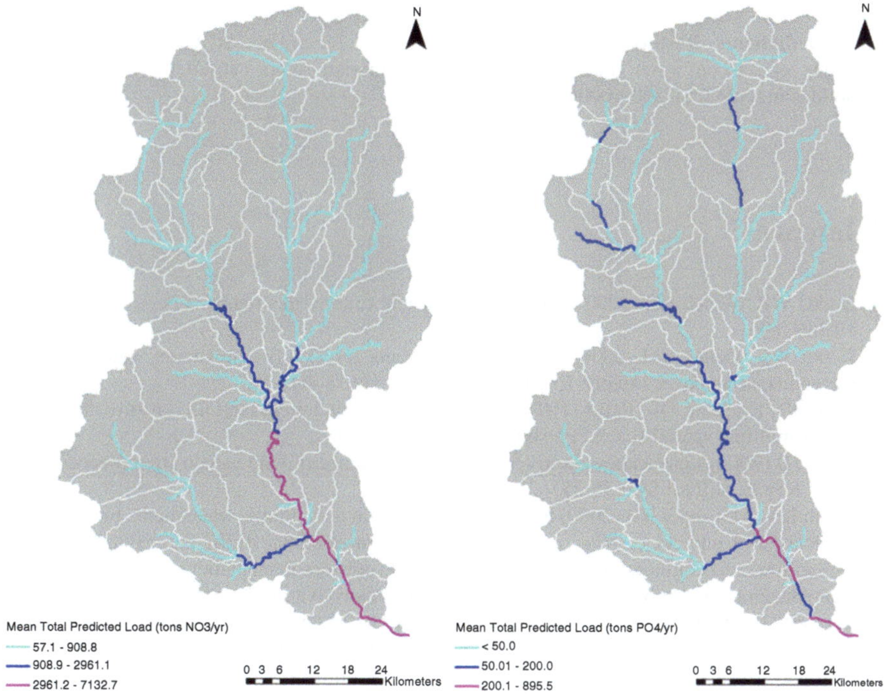

Fig. 15 Mean total predicted load generated after land-to-water delivery and in-stream decay processes

the effective nutrient retention capabilities of river networks, we need to better understand the factors influencing lumped concepts and variables, such as the nutrient spiraling metric v_f at larger scales (other than exclusively studying nutrient retention at the reach level). Reaches in the lower section of the Llobregat River basin do not only receive chronic nutrient inputs but have been also modified by natural and anthropogenic processes and activities, which further influence the self-purification capacity of streams and rivers (Fig. 15).

4 Conclusions

Nutrient source apportionment mirrors the spatial distribution of predominant activities in sub-watersheds of the Llobregat River basin. Although forests and grasslands are the predominant land cover types in the basin, activities related to the presence of agricultural land and human agglomerations cause impairment in water quality and alter the natural flow of rivers and streams. Increased exported nutrient load and decreased in-stream nutrient removal coincide in a downstream direction, probably due to the significant chemical and geomorphological changes particularly suffered in the lower Llobregat River basin.

Acknowledgments Results in this chapter were funded by project SCARCE (assessing and predicting effects on water quantity and quality in Iberian rivers caused by global change) Consolider-Ingenio CSD2009-00065.

References

1. Dodds WK, Welch EB (2000) Establishing nutrient criteria in streams. J N Am Benthol Soc 19:186–196
2. Dodds WK, Oakes RM (2006) Controls on nutrients across a prairie stream watershed: land use and riparian cover effects. Environ Manage 37(5):634–646
3. Allan JD, Erickson DL, Fay J (1997) The influence of catchment land use on stream integrity across multiple spatial scales. Freshwat Biol 37:149–161
4. Allan JD (2004) Landscapes and riverscapes: the influence of land use on stream ecosystems. Annu Rev Ecol Evol Syst 35:257–284
5. Smith RA, Alexander RB (2000) Sources of nutrients in the nation's watersheds. In: Managing nutrients and pathogens from animal agriculture, proceedings from the natural resource, agriculture, and engineering service conference for nutrient management consultants, extension educators, and producer advisors, Camp Hill, Pennsylvania, 28–30, March 2000
6. Grizzetti B, Bouraoui F, de Marsily G, Bidoglio G (2005) A statistical method for source apportionment of riverine nitrogen loads. J Hydrol 304:302–315
7. Gasith A, Resh VH (1999) Streams in mediterranean climate regions: abiotic influences and biotic responses to predictable seasonal events. Annu Rev Ecol Syst 30:51
8. Catalan Water Agency (ACA) (2008) Water in Catalonia. Diagnosis and proposed actions. Significant water management issues raised within the compilation of the River Basin District

Management Plan for Catalonia. Pursuant to Article 14(b) of the Hydrological Planning Regulation (Decree 380/2006)
9. Muñoz I, Prat N (1991) Cambios en la calidad del agua de los ríos Llobregat y Cardener en los últimos 10 años. Tecnol Agua 91:17–23
10. Prat N, Rieradevall M (2006) 25-years of biomonitoring in two mediterranean streams (Llobregat and Besòs basins, NE Spain). Limnetica 25(1–2):541–550 (The ecology of the Iberian inland waters: Homage to Ramon Margalef)
11. Muñoz I, López-Doval JC, Ricart M, Villagrasa M, Brix R, Geiszinger A, Ginebreda A, Guasch H, López MJ, de Alda A, Romaní AM, Sabater S, Barceló D (2009) Bridging levels of pharmaceuticals in river water with biological community structure in the Llobregat river basin (Northeast Spain). Environ Toxicol Chem 28:2706–2714
12. López-Doval JC, Ricart M, Guasch H, Romaní AM, Sabater S, Muñoz I (2010) Does grazing pressure change modify diuron toxicity in a biofilm community? Arch Environ Contam Toxicol 58:955–962
13. Schoumans OF, Silgram M, Groenendijk P, Bouraoui F, Andersen HE, Kronvang B, Behrendt H, Arheimer B, Johnsson H, Panagopoulos Y, Mimikou M, Lo Porto A, Reisser H, Le Gall G, Barr A, Anthony SG (2009) Description of nine nutrient loss models: capabilities and suitability based on their characteristics. J Environ Monit 11(3):506–514
14. Schwarz GE, Hoos AB, Alexander RB, Smith RA (2006) The SPARROW surface water-quality model-theory, applications and user documentation, U.S. Geological Survey, Techniques and Methods, 6-B3, 248 p. and CD-ROM
15. Bouraoui F, Grizzetti B (2008) An integrated modelling framework to estimate the fate of nutrients: application to the Loire (France). Ecol Model 212:450–459
16. Preston SD, Alexander RB, Woodside MD (2011) Regional assessments of the Nation's water quality—Improved understanding of stream nutrient sources through enhanced modeling capabilities, 6 p. Available at http://pubs.usgs.gov/fs/2011/3114/
17. Alexander RB, Elliott AH, Shankar U, McBride GB (2002) Estimating the sources and transport of nutrients in the Waikato River Basin, New Zealand. Water Resour Res 38(12)
18. Alexander RB, Smith RA, Schwarz GE (2000) Effect of stream channel size on the delivery of nitrogen to the Gulf of Mexico. Nature 403:758–761
19. Ensign SH, Doyle MW (2006) Nutrient spiraling in streams and river networks. J Geophys Res 111:G04009
20. Wollheim WM, Vörösmarty CJ, Peterson BJ, Seitzinger SP, Hopkinson CS (2006) Relationship between river size and nutrient removal. Geophys Res Lett 33:L06410
21. Stream Solute Workshop (1990) Concepts and methods for assessing solute dynamics in stream ecosystems. J N Am Benthol Soc 9:95–119
22. O'Brien JM, Dodds WK, Wilson KC, Murdock JN, Eichmiller J (2007) The saturation of N cycling in Central Plains streams: 15 N experiments across a broad gradient of nitrate concentrations. Biogeochemistry 84:31–49
23. Mulholland PJ, Helton AM, Poole GC, Hall RO Jr, Hamilton SK, Peterson BJ, Tank JL, Ashkenas LR, Cooper LW, Dahm CN, Dodds WK, Findlay SE, Gregory SV, Grimm NB, Johnson SL, McDowell WH, Meyer JL, Valett HM, Webster JR, Arango CP, Beaulieu JJ, Bernot MJ, Burgin AJ, Crenshaw CL, Johnson LT, Niederlehner BR, O'Brien JM, Potter JD, Sheibley RW, Sobota DJ, Thomas SM (2008) Stream denitrification across biomes and its response to anthropogenic nitrate loading. Nature 452:202–205
24. Hall RO, Tank JL, Sobota DJ, Mulholland PJ, O'Brien JM, Dodds WK, Webster JR, Valett HM, Poole GC, Peterson BJ, Meyer JL, McDowell WH, Johnson SL, Hamilton SK, Grimm NB, Gregory SV, Dahm CN, Cooper LW, Ashkenas LR, Thomas SM, Sheibley RW, Potter JD, Niederlehner BR, Johnson LT, Helton AM, Crenshaw CM, Burgin AJ, Bernot MJ, Beaulieu JJ, Arango CP (2009) Nitrate removal in stream ecosystems measured by 15 N addition experiments: total uptake. Limnol Oceanogr 54:653–665
25. Mulholland PJ, Hall RO, Sobota DJ, Dodds WK, Findlay SE, Grimm NB, Hamilton SK, McDowell WH, O'Brien JM, Tank JL, Ashkenas LR, Cooper LW, Dahm CN, Gregory SV,

Johnson SL, Meyer JL, Peterson BJ, Poole GC, Valett HM, Webster JR, Arango CP, Beaulieu JJ, Bernot MJ, Burgin AJ, Crenshaw CL, Helton AM, Johnson LT, Niederlehner BR, Potter JD, Sheibley RW, Thomas SM (2009) Nitrate removal in stream ecosystems measured by 15 N addition experiments: denitrification. Limnol Oceanogr 54:666–680

26. Aguilera R, Marcé R, Sabater S (Submitted) Nutrient retention in impaired rivers: evaluating the Efficiency Loss hypothesis at the watershed scale. J Geophys Res
27. Food and Agriculture Organization of the United Nations (FAO) (2010) The Wealth of waste: the economics of wastewater use in agriculture, FAO Water Report 35
28. Grizzetti B, Bouraoui F (2006) Assessment of nitrogen and phosphorus environmental pressures at European scale. EUR 22526 EN
29. Beaulac MN, Reckhow KH (1982) An examination of land use-nutrient export relationships. Water Resour Bull 18:1013–1024
30. Preston SD, Brakebill JW (1999) Applications of spatially referenced regression modeling for the evaluation of total nitrogen loading in the Chesapeake Bay watershed: U.S. Geological Survey Water-Resources Investigations Report 99: 4054 12 p.
31. Martí E, Aumatell J, Godé L, Poch M, Sabater F (2004) Nutrient retention efficiency in streams receiving inputs from wastewater treatment plants. J Environ Qual 33:285–293
32. O'Brien JM, Dodds WK (2010) Saturation of NO3 uptake in prairie streams as a function of acute and chronic N exposure. J N Am Benthol Soc 29(2):627–635
33. Helton AM, Poole GC, Meyer JL, Wollheim WM, Peterson BJ, Mulholland PJ, Bernhardt ES, Stanford JA, Arango C, Ashkenas LR, Cooper LW, Dodds WK, Gregory SV, Hall RO Jr, Hamilton SK, Johnson SL, McDowell WH, Potter JD, Tank JL, Thomas SM, Valett HM, Webster JR, Zeglin L (2011) Thinking outside the channel: modeling nitrogen cycling in networked river ecosystems. Front Ecol Environ 9(4):229–238
34. Bukaveckas PA (2007) Effects of channel restoration on water velocity, transient storage, and nutrient uptake in a channelized stream. Environ Sci Technol 41:1570–1576
35. Bernot MJ, Tank JL, Royer TV, David MB (2006) Nutrient uptake in streams draining agricultural catchments of the midwestern United States. Freshwat Biol 51:499–509
36. Marcé R, Honey-Rosés J, Manzano A, Moragas L, Catllar B, Sabater S (2012) The Llobregat River Basin: A Paradigm of Impaired Rivers Under Climate Change Threats. Hdb Env Chem, DOI 10.1007/698_2012_152

The Effect of Multiple Stressors on Biological Communities in the Llobregat

I. Muñoz, E. García-Berthou, and S. Sabater

Abstract This chapter reviews the diversity and composition of algal, invertebrate and fish communities in the Llobregat basin and analyses the factors affecting their distribution. Phytoplankton develops only during short periods and in particular areas where the hydraulic residence time is high. Diatoms are by far the most diverse and abundant in autotrophic benthic biofilms. The invertebrate community is highly rich taxonomically, whereas the native fish richness is relatively low, with about eight freshwater species and some others of marine origin. Some of the fish native species are threatened and two have been extirpated. Many exotic fish species have been introduced into the basin and are suspected to have contributed to the decline of native fish. Salt inputs, organic pollution and hydrological conditions lead to further alterations in water quality in the middle part of the river and downstream, thus reducing ecological quality. Multiple pressures on fluvial biodiversity severely compromise the integrity of this river's ecosystem. To recover the structure and functioning of the biological communities, it is important to maintain the integrity of headwater ecosystems to ensure a species pool for colonisation downstream and reduce water abstraction, regulation and nutrient and pollutant inputs.

Keywords Algae, Diversity, Fish, Invertebrate, Physical and chemical stressors

I. Muñoz (✉)
Department of Ecology, University of Barcelona, Av. Diagonal 643, 08028 Barcelona, Spain
e-mail: imunoz@ub.edu

E. García-Berthou
Institute of Aquatic Ecology, University of Girona, Campus Montilivi, 17071 Girona, Spain

S. Sabater
Institute of Aquatic Ecology, University of Girona, Campus Montilivi, 17071 Girona, Spain

Catalan Institute of Water Research (ICRA), Edifici H2O, Emili Grahit 101, 17003 Girona, Spain

S. Sabater et al. (eds.), *The Llobregat: The Story of a Polluted Mediterranean River*, Hdb Env Chem (2012) 21: 93–116, DOI 10.1007/698_2012_148,

Contents

Abbreviations

AMA	Aminopeptidase activity
APCs	Alkylphenolic compounds
DW	Dry weight
EQS	Environmental quality standard
NP1EC, NP2EC	Nonylphenol (mono)di-ether carboxylate
WWTP	Wastewater treatment plants

1 Introduction

Substantial research on river ecology has been undertaken to elucidate the relative contributions of physical, chemical and biological factors to the structure and function of stream biota (e.g. [1–4]). Natural factors of variation may co-occur with others due to human activities, with different origins and modes of action (e.g. nutrient enrichment, toxicant occurrence, hydraulic infrastructures and introduction of non-native species, as well as changes in riparian vegetation and habitat). The studies on the effects of disturbances have provided important information on the integrated assessment of ecological conditions of streams and rivers based on communities [5].

The organisms inhabiting Mediterranean areas present life cycle adaptations and specific traits that make them resistant and resilient to the effects of environmental changes, such as droughts and periodic floods [6, 7]. Mediterranean regions are considered a global biodiversity hotspot [8], with regard to both the terrestrial vegetation and the diversity of aquatic Mediterranean ecosystems [9]. Unfortunately, Mediterranean freshwater systems are one of the most stressed systems due to the rising number of and long-lasting human activities. Historical establishment of human settlements in these warm climatic areas has put widespread pressures on natural water resources and few systems currently remain in a near-pristine state.

Waters in the Llobregat River are used for irrigation, industry (tannery, textile, chemical, pulp and paper) and as drinking water for the densely populated watershed. Such pressures on water resources and their scarcity triggered the monitoring of the river many years ago, including some components of its biological communities. The monitoring of benthic communities started at the end of the 1970s and beginning of the 1980s on the urging of the regional Administration (Diputació de Barcelona). These early studies demonstrated that the water quality of the watershed had already deteriorated in the upper–middle part of the catchment due to evaporite-bearing geological formations, mining and industrial activities related to potash exploitation, as well as subsequent increasing sodium and potassium chloride concentrations in the water. It was also obvious that water from the Anoia and Rubí tributaries determined the few and highly tolerant species occurring in the benthos downstream ([10] for invertebrates; [11]) for diatoms. The common carp *Cyprinus carpio* was the only fish detected in the lower part of the river [12]. In the 1990s, some major infrastructural works, including wastewater treatment plants (WWTP) and a brine collector that collected and transported mining wastewater to the sea, helped improve ecological quality [13, 14]. Nevertheless, the middle and lower part of the main river and some tributaries continued to show a poor quality due to the densely populated area and the increase in industrial facilities. Furthermore, the lower summer discharge intensified the effects of organic matter inputs into the river coming from WWTP.

This chapter reviews the composition of algal, invertebrate and fish communities along the river and links them with the main factors affecting their distribution. The chapter is organised in two sections. The first one illustrates the spatial and temporal distribution of the three main biological elements in the river (benthic diatoms, invertebrates and fish communities). In the second section, special attention is given to the effects of multiple pressures on biological communities by focusing on three examples: hydrological changes and pollution effects on fish, the effect of chemical substances on benthic communities and the response of cyanobacteria to multiple stressors occurring in the river.

2 Spatial Patterns of Biodiversity Along the River

2.1 Algae: Planktonic and Benthic Communities

Phytoplankton development in the Llobregat only occurs in some standing water areas in the main river course in addition to the vicinity of the river mouth. In contrast to temperate large rivers, phytoplankton is only present for short periods of the year, which is characteristic of medium-sized Mediterranean rivers [15, 16], where the hydraulic residence time is too short to allow for the development of a permanent phytoplankton community.

Data on the occurrence of planktonic organisms in the Llobregat are scarce. A sampling collection performed in spring 1999 confirmed that true phytoplankton existed only for short periods and in particular areas. Pools created by weirs (Fig. 1, aimed to derive water for irrigation or hydraulic energy; see [86]) are scattered along the main river course and allow the development of plankton during spring, late winter and summer. The most abundant phytoplankton group (Table 1) observed in the sampling collection was centric diatoms (800–20,000 cells/mL), followed by chlorococcales (green algae) and a minor presence of planktonic cyanobacteria. This community is characteristic of spring phytoplankton growths in eutrophic waters. Drifting algae (benthic diatoms) were very common in the samples, originating in the sediment and resuspended in the shallow water column. Their occurrence increased with higher water flows, while the relative importance of true phytoplankton organisms becomes irrelevant.

Macrophytes develop in some areas of the river, though in small patches. Algae are the most common autotrophs in the benthic compartment, with diatoms by far

Fig. 1 (**a**) Large reservoirs such as the Llosa del Cavall, with a dam 122 m high, have profoundly altered the Llobregat; (**b**) the chub *Squalius laietanus* is a cyprinid endemic of the Ebro, Catalonia and a few southeastern French basins, which has disappeared from most of the Llobregat; (**c**) small weir (Llobregat, Pont de Vilomara) to derive water for irrigation and hydraulic production. A flow reduction is observed downstream; (**d**) reaches such as this one in the Cardener have been profoundly altered in hydromorphology

Table 1 Organisms present in water samples collected at several large pools scattered along the Llobregat River. Abundance is given as cells/mL. The distance to the source of the river is indicated for each location

	Castellgalí 72 km	El Cairat 88 km	Abrera 102 km
Diatoms			
Cyclotella meneghiniana	800	12,267	20,240
Cyclotella comensis	7,920	9,733	3,360
Fragilaria construens	160		
Melosira varians		267	
Navicula sp.	640		
Nitzschia gracilis		133	
Synedra ulna	80		
Green algae			
Chlamydomonas	320	4,133	1,440
Kirchneriella sp.			320
Micractinium pusillum		2,133	640
Monoraphidium komarkovae		133	1,120
Pediastrum duplex		4,267	
Scenedesmus spinosus			1,920
Euglenophyta			
Euglena sp.	240		
Cryptophyta			
Cryptomonas		667	160
Cyanobacteria			
Oscillatoria nigro-viridis	800		
Spirulina sp.	800		
Total abundance	11,760	33,733	29,200

the most diverse and abundant in Llobregat autotrophic biofilms [17]. The composition of the community changes from the headwaters to the middle and lower stretches of the river. The most characteristic species in the headwaters are *Ceratoneis* (*Hannaea*) *arcus*, *Diatoma hiemale v. mesodon*, *Meridion circulare*, *Encyonema minutum*, *Diatoma elongatum*, *Denticula tenuis v. crassula*, *Gomphonema olivaceum* and *Achnanthidium minutissimum*. These are cosmopolitan species adapted to high water velocity, low temperatures and low nutrient concentrations. This community is restricted to the Pyrenean and pre-Pyrenean areas of the watershed, the steep slopes from la Pobla de Lillet to Balsareny.

The physical conditions change quickly downstream to slower-moving waters, with more dissolved elements and relatively high nutrient concentrations. Therefore, a diverse diatom community develops downstream from Balsareny. The genera *Navicula* (*lanceolata*, *gregaria*, *subminuscula*, *capitoradiata*, *cryptocephala*, *tripunctata*, *pelliculosa* and *radiosa v. tenella*) and *Nitzschia* (*dissipata*, *fonticola* and *palea*) contribute with many species. Other diatom taxa abundant in this area are *Gomphonema angustatum*, *Rhoicosphenia abbreviata* and *Melosira varians*.

The composition of these assemblages remains diverse until high nutrient inputs (from urban sewage or agricultural activities) enter the river. The diatom community becomes less diverse and tolerant taxa become abundant. *Nitzschia palea*, *Nitzschia capitellata* and *Nitzschia inconspicua* may reach relative abundances of 30–60% in some samples. Organic pollution has receded in recent times because of the sanitation plans implemented in the Llobregat, but nutrient levels are still very high in the area from Balsareny to Pont de Vilomara. Non-diatom benthic algae in this section of the river include the filamentous taxa *Cladophora glomerata*, *Ulothrix zonata*, *Vaucheria* sp. and *Oedogonium* sp. Cyanobacterial masses (mostly composed of *Oscillatoria* spp.) develop in slow-moving areas of the river (see Sect. 3.3).

The salt inputs from the Cardener tributary lead to further changes in water quality. Water conductivity increases to up to 3,000 $\mu S\ cm^{-1}$ and salinity-tolerant taxa become frequent. The filamentous green alga *Enteromorpha* sp. is indicative of high water salinity in the river. The taxa *Nitzschia frustulum*, *Surirella ovata* and *Mayamaea atomus* are dominant amongst the diatoms. *Amphora veneta*, *A. coffeaeformis*, *Thalassiosira weissflogii*, *Biddulphia levis* and *Amphiprora alata* are present in the lower part of the river and continue to be present in the potamoplankton. The entrance of the tributary Anoia increases water pollution in the lower part of the Llobregat. Some sites of this tributary are extremely polluted and have the most tolerant diatom taxa (80% of the diatom community is made up of *N. inconspicua* in Pobla de Claramunt; [18]). The small and highly polluted tributary, Riera de Rubí, has a highly tolerant diatom community (*Fistulifera* (*Navicula*) *saprophila*, 90% abundance), which reflects its extremely high water pollution. Overall, in an analysis of Catalan watersheds based on the diatom communities, the sites of the lower Anoia and lower Llobregat were amongst the most polluted in the whole data set [19]. The areas showing the lowest diatom quality were Llobregat at Martorell and Anoia at Vilanova del Camí, the species *Fistulifera* (*Navicula*) *saprophila*, *Nitzschia auriariae*, *N. frustulum* and *Nitzschia desertorum* being the most abundant.

2.2 Aquatic Invertebrates: Distribution in Hard and Soft Sediments: Macrofauna and Meiofauna

The longitudinal distribution of the invertebrates in the Llobregat basin results from four main environmental factors: altitude, salinity, pollution and hydrological regime. The highest diversity of the macroinvertebrate community is observed in the Llobregat headwaters, where low water temperature and high oxygen concentration coincide with high habitat quality due to substrate diversity, current velocity gradients and good riparian quality. The high water quality of the headwaters quickly deteriorates in the upper–middle part of the catchment due to salt inputs and industrial and residential activities. The composition of the macroinvertebrate community diverges between tributaries with a continuous flow regime and intermittent rivers or reaches. The occurrence of small dams along the basin strongly alters the hydrological conditions of the downstream reaches and produces daily changes in the water level (Fig. 1). The less tolerant invertebrate species disappear and the ecological quality of the reach pronouncedly decreases. Climatic conditions, mainly heavy rainfall, determine differences in discharge that exacerbate the effects of dams and organic inputs. In general, the biological quality of the river is higher in wetter than drier years. During the drought of 2006–2009, lower biological quality was more frequent [87].

Most of the macroinvertebrate groups have been widely studied. The community of worms (Oligochaeta) was primarily described by Muñoz et al. [20] in the middle part of the river. Individuals from the families Lumbricidae, Nadidae and Tubificidae are the most abundant and six species occur frequently: *Nais* spp. (mainly *N. elinguis*), *Pristina* spp., *Tubifex tubifex*, *Ophidonais serpentine*, *Limnodrilus udekemianus* and *L. hoffmeisteri*. This group is the most abundant in sand and mud sediments [21], pollution favouring the dominance of the Tubicidae family and the *Limnodrilus* spp.

Grazing on biofilms growing on rock and cobble surfaces is common in gastropod snails like *Ancylus fluviatilis*, *Physella acuta*, *Planorbis* sp., *Lymnaea* sp. and *Potamopyrgus antipodarum*. The last species is an invader from New Zealand. *P. acuta* is tolerant to pollution and desiccation and is distributed along the basin. The bivalve *Pisidium* sp. also occurs in the upper and middle parts of the river.

The high insect diversity in Mediterranean water bodies is a result of the interaction between the endemicity and complexity of historical and ecological factors [22]. The most primitive groups (dragonflies, stoneflies and mayflies) have a nymph stage well adapted to aquatic life. Most species of dragonflies and stoneflies act as a key species in the fluvial trophic web due to their large dimensions and predatory behaviour. Dragonflies (Odonata) feed on insects, crustacean, small amphibians and fishes. The indicator species of good ecological river quality are *Calopteryx virgo meridionalis*, *Boyeria irene* and *Cordulegaster boltonii*. Stoneflies (Plecoptera) mainly live in the headwaters of the main course and tributaries, while few species occur downstream of the La Baells reservoir [23]. The headwaters in the pre-Pyrenean mountains are characterised by less tolerant and stenothermic species (Table 2). The carnivorous stoneflies *Perla marginata* and *Dinocras cephalotes* alternate in the sites since they occupy the same ecological niche. The genus with the widest distribution amongst the mayflies (Ephemeroptera) is *Baetis*, which has species tolerant to pollution and salinity (Table 2). *Caenis luctuosa* is also common in the entire basin and tolerant to pollution. Sand and gravels are the habitat for *Ephemera danica*, *Ephoron virgo* and *Potamanthus luteus*. Their presence indicates a good conservation status, and they are able to persist in the mid-lower reaches if pollution is low or moderate.

Trichoptera (caddisflies) distribution along the Mediterranean coast has been described mainly in ecological studies (e.g. [24, 25]). Some Trichoptera species live inside a portable tube or case, that is, covered by detritus, sand or small pieces of wood or leaves. In the Llobregat basin, shredders and woody case caddis (*Limnephilus* spp., *Halesus* spp. and *Potamophylax* sp.) are present in reaches with

Table 2 Longitudinal distribution of some insect species in the Llobregat basin

	Headwaters	Middle reaches	Low reaches	Intermittent tributaries
Plecoptera	*Isoperla* sp. *Protonemura beatensis* *P. risi* *Amphinemura sulcicollis* *Perla marginata* *Dinocras cephalotes*	*Protonemura beatensis* *Leuctra inermis*		*Nemoura cinerea* *Capnia vidua* *C. mitis*
Ephemeroptera	*Baetis alpinus* *Rithrogena diaphana* *Epeorus sylvicola* *Torleya major* *Ephemerella ignita*	*Baetis vardarensis* *B. rhodani* *Ecdyonurus angelieri* *E. insignis* *Ephemerella ignita* *Caenis luctuosa* *Cloeon inscriptum*	*Baetis nigrescens* *B. lutheri* *B. meridionalis* *B. vardarensis* *Caenis luctuosa*	*B. muticus* *Habrophlebia fusca* *H. modesta* *Cloeon inscriptum*
Trichoptera	*Philopotamus montanus* *Sericostoma pyrenaicum* *Hydropsyche siltalai* *H. dinarica* *H. instabilis* *Halesus digitatus* *Limnephilus guadarramicus* *Potamophylax cingulatus* *P. latipennis* *Tinodes assimilis* *T. dives* *T. maclachlani* *Brachycentrus maculatum* *Beraea maurus* *Odontocerum albicorne* *Allogamus auricolis* *Rhyacophila dorsalis* *R. fasciata* *R. meridionalis* *R. gr tristis* *Agapetus fuscipes*	*Odontocerum albicorne* *Mystacides azurea* *Tinodes waeneri* *Polycentropus flavomaculatus* *Cheumatopsyche lepida* *Ecnomus deceptor* *Psychomyia pusilla* *Hydropsyche exocellata* *H. incognita gr pellucidula* *Chimarra marginata* *Hydroptila vectis* *Rhyacophila dorsalis* *R. relicta*	*H. exocellata* *Chimarra marginata* *Hydroptila vectis* *Psychomyia pusilla*	*Mesophylax aspersus* *Limnephilus lunatus*
Diptera Chironomidae	*Diamesa* spp.	*Orthocladius–Cricotopus* group *Potthastia gaedii* group *Cricotopus* (*Isocladius*)	*Eukiefferiella* *Chironomus* spp.	

(continued)

Table 2 (continued)

	Headwaters	Middle reaches	Low reaches	Intermittent tributaries
		Eukiefferiella spp. *Micropsectra* spp.		
Diptera Simulidae	*Simulium variegatum* *S. cryophillum* *S. gaudi* *S. angustipes*	*S. velutinum* *S. bezzii* *S. ornatum* *S. intermedium*		

high riparian structure and heterogeneity. In general, the maximum diversity of caddis is in the Llobregat headwaters (Table 2). Conductivity determines the distribution barrier of most caddis species downstream. The middle reaches are associated with Hydropsychids, Philopotamids, Polycentropodids and Psychomiids (Table 2). The three former families are filterers and elaborate silk nets to capture transported fine detritus material. The high content of suspended solids is also an important factor determining caddis distribution. For example, different species of the net-spinning Hydropsychidae are segregated among the different concentrations of suspended solids because of their feeding and net morphological requirements [26]. High organic matter content characterising the middle and lower stretches of the Llobregat determines the areas with low oxygen concentrations and anaerobic conditions, which is even more extreme during low summer flows. The species most tolerant to this condition is *Hydropsyche exocellata*, present in most sites in the low part of the river. *Mesophylax aspersus* is one of the most abundant and frequent species in the Mediterranean because of its ability to avoid dry periods through behavioural adaptations [27]; it is located in the Llobregat in some intermittent reaches. Caddis larvae are more diverse in autumn, winter and spring than in summer because of the higher emergence of adults between June and September and the life cycle adaptation of some species, which are absent in summer to avoid drought-related problems.

Small streams that become intermittent preferentially host Odonata, Coleoptera and Heteroptera (OCH) rather than Ephemeroptera, Plecoptera and Trichoptera (EPT). OCH share biological traits that allow them to survive in non-flowing streams (only pools) or migrate when the stream dries out [28]. The ratio of OCH to ETP indicates the water permanence of the reach. Some of the most common Heteroptera species in this basin are *Micronecta meridionalis*, *Sigara lateralis* and *Notonecta maculate* [29], while the most common Coleoptera (beetles) species belong to the families Dryopidae and Elmidae [24].

Chironomidae (midges, Diptera) is one of the most diverse and abundant freshwater insect family both in lentic and in lotic habitats. The major groups in the Llobregat reaches are indicated in Table 2 [30]. The dominant midge when pollution increases and oxygen decreases is *Chironomus* spp. This organism has large amounts of haemoglobin in its haemolymph, allowing them higher oxygen affinity, which is very useful in these environments. These highly tolerant species develop large biomasses mainly in soft sediments of highly polluted sites [21]. The Simulidae family (black flies) has also been deeply studied in the Llobregat basin [31]. Black flies live in water with high current velocities; their anchoring systems (tiny hooks and silk holdfasts) enable them to hold on to the substrata and resist being washed downstream, while their transformed mandibles are used as fans to filter water and retain detritus. Only a few species of this group support relatively low oxygen concentrations (2–4 ppm). Blackflies disappear when salinity increases (Table 2), the only exception being *S. equinum*, which tolerates some salinity.

Less attention has been paid to the distribution of meiofauna (invertebrates whose size ranges between 32 and 500 μm). Although the contribution of meiofauna to total stream biomass may be relatively small, it varies greatly due to influences of seasonality and species composition. The nematode community has been studied in the middle [32] and lower parts [21] of the Llobregat. Up to 26 nematode species have been recorded, the most abundant being those from the genera *Tobrilus*, *Eumonhystera* and *Monhystera*. *Monhystera* is associated with low chemical pollution of river sediments, whereas *Tobrilus* is associated with high pollution and is present at the lowest sections of the main river and tributaries.

2.3 Fish Diversity

There is little knowledge about the abundance and distribution of fish assemblages in the Llobregat. Fish populations are not regularly monitored by the Spanish administration and available data correspond mostly to scattered samplings from a few scientific publications. The historical assemblage of the Llobregat (Table 3) consisted of brown trout (*Salmo trutta*) in the headwaters, followed downstream by the barbel *Barbus haasi* and the chub *Squalius laietanus* (Fig. 1), which was recently distinguished from *S. cephalus* [33]. The eel *Anguilla anguilla* was present throughout the basin but has nearly disappeared now mainly because of the existence of dams and weirs. The endangered cyprinodont *Aphanius iberus* is only present in the delta, where it is threatened by the introduced mosquitofish (*Gambusia holbrooki*). There are a few records of *Salaria fluviatilis* and the tench *Tinca tinca* in the Llobregat basin (e.g. [34]), although these species are usually not widespread in the Iberian Peninsula and the former is less easy to detect. Many fish, mostly from other European countries or North America, have been introduced and established in the Llobregat, including bleak, common carp, pumpkinseed sunfish, European catfish, rudd, largemouth bass and pike-perch (Table 3). Some introduced species such as *Ameiurus melas* are not frequent and *Oncorhynchus mykiss* is still stocked by the administration for angling purposes but has been barely established. Some species are translocated, i.e., they are supposed to be native in one part of the Iberian Peninsula but not to the Llobregat basin. This is the case for *Phoxinus* sp., *Gobio lozanoi*, *Barbus graellsii* and *Parachondrostoma miegii*. Some native euryhaline secondary freshwater species, such as *Atherina boyeri*, *Pomatoschistus microps* and *Platichtys flesus*, or mullet species (*Mugil cephalus*, *Chelon labrosus*,

Table 3 Fish from the Llobregat River

Common name	Scientific name	Family	Native status	European conservation status
Twaite shad	*Alosa fallax*	Clupeidae	NE	LC
Eel	*Anguilla anguilla*	Anguillidae	N	CR
Iberian toothcarp	*Aphanius iberus*	Cyprinodontidae	N	EN
Sandsmelt	*Atherina boyeri*	Atherinidae	N	LC
Iberian redfin barbel	*Barbus haasi*	Cyprinidae	N	VU
Thicklip mullet	*Chelon labrosus*	Mugilidae	N	LC
Stickleback	*Gasterosteus aculeatus*	Gasterosteidae	NE	LC
Thinlip mullet	*Liza ramada*	Mugilidae	N	LC
Sharpnose mullet	*Liza saliens*	Mugilidae	N	LC
Striped mullet	*Mugil cephalus*	Mugilidae	N	LC
European flounder	*Platichthys flesus*	Pleuronectidae	N	LC
Common goby	*Pomatoschistus microps*	Gobiidae	N	LC
Freshwater blenny	*Salaria fluviatilis*	Blenniidae	N	LC
Brown trout	*Salmo trutta*	Salmonidae	N	LC
Chub	*Squalius laietanus*	Cyprinidae	N	LC
Black-striped pipefish	*Syngnathus abaster*	Syngnathidae	N	LC
Tench	*Tinca tinca*	Cyprinidae	I?	LC
Pyrenean gudgeon	*Gobio lozanoi*	Cyprinidae	I?	LC
Minnow	*Phoxinus* sp.	Cyprinidae	I?	LC
Bleak	*Alburnus alburnus*	Cyprinidae	I	–
Black bullhead	*Ameiurus melas*	Ictaluridae	I	–
Common carp	*Cyprinus carpio*	Cyprinidae	I	–
Mosquitofish	*Gambusia holbrooki*	Poeciliidae	I	–
Pumpkinseed sunfish	*Lepomis gibbosus*	Centrarchidae	I	–
Largemouth bass	*Micropterus salmoides*	Centrarchidae	I	–
Rainbow trout	*Oncorhynchus mykiss*	Salmonidae	I	–
Ebro nase	*Parachondrostoma miegii*	Cyprinidae	I	–
Roach	*Rutilus rutilus*	Cyprinidae	I	–
Pikeperch	*Sander lucioperca*	Percidae	I	–
Rudd	*Scardinius erythrophthalmus*	Cyprinidae	I	–
European catfish	*Silurus glanis*	Siluridae	I	–

Native status in the basin: *N* native, *NE* extirpated native, *I* introduced. European conservation status follows that of Freyhof and Brooks [84], except for fish not native to the basin. For these, a conservation status was not assigned following the IUCN guidelines [85]

Liza ramada and *L. saliens*) are only present in the delta and the river mouth. The three-spined stickleback (*Gasterosteus aculeatus*) was recorded in the Llobregat delta around 1903, but seems to have been extirpated from the basin [35]. Similarly, Aparicio et al. [34] mentioned that the shad *Alosa* sp. (most likely *Alosa fallax*) was present in the Llobregat in the early 1900s, but has since been locally extirpated. Sturgeons (*Acipenser sturio*) were recorded at the coast of Barcelona (before disappearing from Spain after 1970), but we have found no records in the Llobregat itself [36]. Therefore, the Llobregat originally had a fauna of about eight freshwater fishes, two of which have been extirpated from the basin, plus several other euryhaline species of marine origin. At least 12 exotic fish species have been introduced and established in the basin.

The original zonation of the fish fauna was thus similar to that observed in other Iberian streams, with trout, barbel and eel zones (e.g. [37]). Many European rivers display the classic fish zonation described with the "trout", "grayling", "barbel" and "bream" zones [38]. However, grayling or bream, for instance, are not native to the Iberian Peninsula (e.g. [39]) and Iberian streams, thus displaying different longitudinal variation in their fish assemblages. Fish fauna in the Sant Ponç and La Baells reservoirs is dominated by the introduced species mentioned above, while native species are rare [40]. Fish ecology and biology of some cyprinid species have received some attention, particularly in the Aigua d'Ora stream, a tributary of the Cardener [41] and the Vallvidrera stream, a downstream tributary very close to the city of Barcelona [42–44]. Aparicio and de Sostoa [42] studied the life history of *B. haasi* in the Vallvidrera stream, showing reduced growth rates and longevity compared with other barbel species. They also showed that most individuals in this population were highly sedentary and resided within a range of less than 20 m [43]. Casals [41] studied the growth, reproduction, abundance and production of *B. haasi*, *P. miegii* and *S. laietanus* in the Aigua d'Ora stream. No studies on the ecology of brown trout in the Llobregat headwaters have been published. The genetics of this species is well known, including variable levels of introgression in the Llobregat with individuals from hatchery stockings [45–48].

3 Main Pressures on Biological Communities: Establishing Cause–Effect Relationships

3.1 Effects of Pollution, Hydrological Alteration and Invasive Species on Fish

The pollutant burden on fish from the Llobregat is very well known, particularly regarding organic chemicals, such as pesticides, surfactants and plasticisers [49, 50, 88], as well as heavy metals and metalloids [51], which are important in the lowlands and headwaters [52]. Severe pollution has caused frequent fish kills during the last century and even in recent years (Table 4), but these kills are probably less frequent nowadays because of the development of numerous WWTP by the Catalan Water Agency. Organic pollutants (mainly alkylphenolic compounds, APCs) have recently been shown to have endocrine-disrupting effects in the Llobregat, such as inducing the presence of plasma vitellogenin in male carp (*C. carpio*) and the occurrence of intersex individuals [49, 52–54]. Overall, fish abundance is very low in the lowermost reaches of the Llobregat (electrofishing passes often yield no captured fish), probably due to the chronic effects of pollution.

Table 4 Some fish kills recorded in the Llobregat River basin. "?" = Unknown data

Date	Locality	Number of dead fish and main species	Suspected cause	Reference access
November 1933	Sallent	?	Chemical pollution after drains	1
11–12 December 1957	?	?	Chemical pollution after drains resuming drought	2
January 1967	Manresa	8,000 kg	Oil spill	3
7 July 1998	Guardiola de Berguedà and Cercs	>24,000 fish; *S. trutta*, *B. haasi*, *C. carpio*	Drought in the river due to water abstraction to a canal for hydroelectrical generation	4, 5
June 1999	Sant Vicenç de Castellet	Some hundreds; *Barbus* sp.	Hydrogen peroxide and sodium carbonate	6
October 2001	Sant Boi de Llobregat	400 fish	Chemical spill	7
January 2002	Martorell	Some hundreds; *C. carpio*	Surfactant	8
April 2002	Berga	A few thousands; *S. trutta*	Discharge of hypolymnetic water from La Baells reservoir	9
8 May 2002	Monistrol de Montserrat	Some thousands; *Barbus* sp., *C. carpio*	Dye spill	10
November 2006	Anoia stream, between Sant Pere de Riudebitlles and Torrelavit	About 600 fish; *Barbus* sp., *C. carpio*	Chemical spill	11

1. http://hemeroteca.lavanguardia.com/preview/1933/11/30/pagina-17/33181124/pdf.html
2. http://hemeroteca.lavanguardia.com/preview/1958/01/04/pagina-35/32748495/pdf.html
3. http://hemeroteca.lavanguardia.com/preview/1967/01/28/pagina-23/34340604/pdf.html
4. http://www.324.cat/noticia/51683/altres/Es-demanen-2-anys-de-preso-per-la-mort-de-mes-de-10000-peixos-al-riu-Llobregat-el-1998
5. http://www.elpais.com/articulo/sociedad/CATALUNA/BENJAC_/EMPRESA/Expediente/empresa/muerte/miles/peces/rio/Llobregat/elpepisoc/19980709elpepisoc_15/Tes
6. http://www.elpais.com/articulo/cataluna/ESPANA/CATALUNA/JUNTA_DE_AGUAS_DE_CATALUNA/Aparecen/centenares/peces/muertos/cauce/rio/Llobregat/elpepiautcat/19990611elpcat_25/Tes
7. http://www.elpais.com/articulo/cataluna/Alerta/Sant/Boi/aparicion/densa/espuma/Llobregat/elpepiautcat/20011101elpcat_1/Tes
8. http://www.elpais.com/articulo/cataluna/vertido/detergente/causa/mortandad/peces/Llobregat/elpepiautcat/20020116elpcat_18/Tes
9. http://www.vilaweb.cat/noticia/500664/20020425/central-dendesa-provoca-mortaldat-peixos-llobregat.html
10. http://www.lamalla.cat/medi_ambient/danys_medi/article?id=51696
11. http://www.324.cat/noticia/16432/altres/Medi-Ambient-investiga-la-mort-duns-600-peixos-en-un-tram-del-riu-Anoia

In contrast, the effects of the massive regulation and water abstraction in the Llobregat have largely not been described, although they are thought to be enormous. The Llobregat is a highly fragmented river, with a few very large dams and dozens of many small weirs for hydroelectric, irrigation or supply purposes (Fig. 1, [86]). This regulation is likely to affect fish populations by impeding migration and colonisation, changing the flow and thermal regimes, and degrading habitats. An example of the effects of this disruption is that the eel, which has historically been present throughout most of the basin, is largely absent in most of it and confined to the delta and river mouth. Another species that was widespread in the basin and now has almost disappeared is the chub. This species is a large-sized cyprinid, which prefers pools and the water column as a microhabitat and is probably affected by water regulation and abstraction and changes in spawning grounds. The spawning grounds of many native fish such as brown trout, redfin barbel and chub, which have litophilic reproduction that require coarse substrata for spawning, are most probably altered and restricted to areas below large dams, such as La Baells, Sant Ponç or La Llosa del Cavall. However, this has not been formally studied.

Invasive species are another strong pressure on native fish assemblages in the Llobregat. As in other Iberian river basins [55], the Llobregat currently has a higher number of introduced species than native ones, with the introduced species dominating many stretches. The ecological impact of these invasive species has not been studied in the Llobregat, but implications for a few species are known from other rivers in the region. For instance, the European catfish, introduced to the Llobregat in the last few years through the La Baells reservoir [56], has been shown to affect native fish and waterbirds in the Ter and Ebro river basins [40], while mosquitofish is known to impact the endemic, endangered cyprinodont *A. iberus* [57, 58]. The illegal stocking of exotic species is not well controlled and still frequent in Spain [56], often introducing other unnoticed species, such as fish parasites [59] and zebra mussels (*Dreissena polymorpha*). In October 2011, zebra mussels were reported in the La Baells reservoir and their expansion will certainly have enormous economic impacts throughout the basin, which supplies water to the Barcelona region.

3.2 Links Between Environmental Factors and Structure and Function of Benthic Communities: The Role of Priority and Emerging Compounds

The biological community living in fluvial systems reflects the historical and current effects of the combined impacts of chemical, physical and biological stressors [60]. Interactions between stressors may be accumulative, but synergistic interactions between natural stressors and toxicants are common phenomena in river ecosystems, mainly in industrialised countries [61]. These interactions make it difficult to understand the effects of mixtures and establish definitive relationships between disturbances and ecosystem integrity [62].

Over 43 million organic and inorganic compounds are commercially available [63], which can be released into fluvial systems and remain in sediment. Information about their occurrence and effects on organisms and ecosystems is not yet available, and risk assessments have not been conducted. A few of them (e.g. 33 compounds are classified as priority substances by the Water Framework Directive) have been evaluated, classified and finally included in the priority substance list. Estimation of the overall bulk loads of organic compounds (131 compounds) in the Llobregat shows the following rank order: pharmaceuticals > alkylphenols > pesticides > illicit drugs > > estrogens [64].

Recent works in the Llobregat [65–67] describe the experimental field approach as a tool for understanding the multiple toxic effects of priority and emerging compounds on benthic communities. This approach is based on using multivariate techniques to assess disturbances and determine potential relationships between chemical stressors and the functional and structural composition of benthic communities. Out of the 22 pesticides from seven chemical families in water and sediment, mainly triazines affect the composition of diatom biofilms [66], while conductivity and nutrients are the main factors determining macroinvertebrate distribution. Regarding functional biofilm characteristics, the percentage of variance explained by pesticides and physical and chemical variables differs among the biofilm metrics. Chlorophyll-*a* and photosynthetic efficiency are influenced mainly by the presence of pesticides, while bacterial extracellular enzymatic activities associated with the heterotrophic biofilm compartment are affected mainly by water temperature and sulphate content. Fauna and biofilm differ in their response to pesticides, indicating that the influence of the detected compounds depends on their target organisms. Of the 20 chemicals determined, only 15% are insecticides and the rest herbicides, which nearly always have higher concentrations. Muñoz et al. [65] determined a potential causal association between the concentration of some pharmaceuticals and the abundance and biomass of several benthic invertebrates. Their multivariate analysis showed that the concentration of some anti-inflammatories (indomethacin and ibuprofen) and β-blockers (propranolol) correlated with higher densities and biomass of midges and worms. These results have been confirmed in laboratory conditions [68] and sediment exposure experiments [69]. A higher density of pharmaceutical concentrations downstream increases the toxicological risk on communities, as was observed by Ginebreda et al. [70], who studied hazard quotient indices and their relationship with diversity indices. Areas of no risk were only observed in a few sites of the middle part of the river and in upstream tributaries.

Brix et al. [67] described that APCs had a certain role in the distribution of the benthic community in the Llobregat. APCs have shown a downward trend in the Llobregat over the last decade [71]. Although the maximum allowable concentrations defined in the European Union's environmental quality standard (EQS) for nonylphenol and octylphenol have not been exceeded, there is still a potential risk of oestrogenic activity [67]. The hydrophobic metabolites of APCs may associate

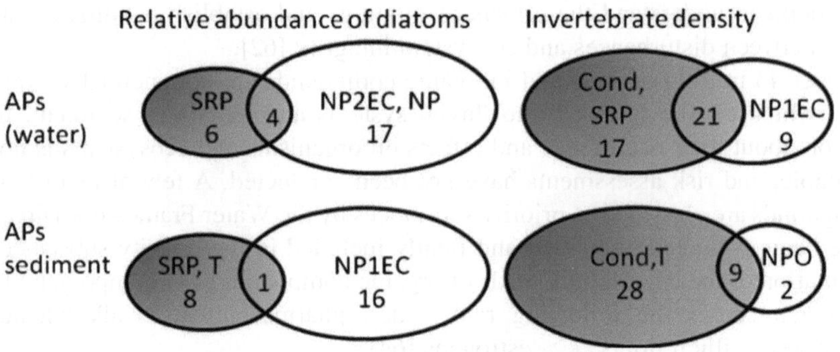

Fig. 2 Effects of APCs on benthic communities in the Llobregat. Percentages of partial variance explained by each group and covariance (shared) between groups of variables from the redundancy analyses are shown. Only significant factors are indicated. *NPO* sum of nonylphenol ethoxylates, *SRP* soluble reactive phosphorus, *Cond* conductivity, *T* temperature

with organic matter in sediments, suspended particulate matter and lipid content of organisms, thus becoming more hazardous to the organisms in this habitat and providing a potential path for bioaccumulation and transference through the food web. Both diatom and macroinvertebrate communities show significant sensitivity to APCs (Fig. 2). While the nonylphenol di-ether carboxylate (NP2EC) has been linked with diatoms in water, NP1EC is a significant stressor in sediment. Other environmental variables such as conductivity and soluble reactive phosphorus also contribute partly to the variance of the invertebrate density distribution along the river.

In the previous referred studies, the relationships found can be tentatively interpreted as cause–effect. In the case of pharmaceuticals, complementary experiments have confirmed some links. In other cases, the proportion of unexplained variance is high for both communities, showing that other stressors could also be important for species distribution along the pollution gradient in the Llobregat.

3.3 Dynamics of Cyanobacterial Mats and Geosmin Production: An Example of a Community Response to Multiple Stressors

Geosmin is a bicyclic terpenoid by-product (4,8a-dimethyl-decahydronaphthalene-4a-ol) produced by cyanobacteria [72] and actinomycetes [73], and consequently, characteristic of wet soil and standing waters. Cyanobacteria are the main producers of geosmin in standing or slow-flowing waters, but its production only occurs under particular conditions. When present, geosmin produces an earthy-musty odour that can be perceived at very low concentrations (ca. 4 ng L^{-1}) by humans consuming the water. Its elimination requires additional water treatment before consumption, and therefore, significantly increases the cost of water purification. Recorded

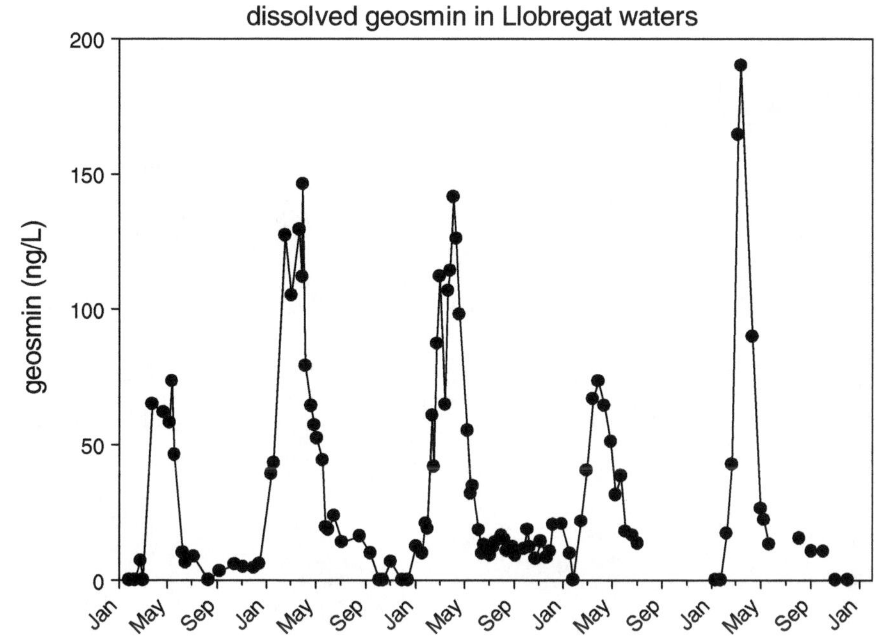

Fig. 3 Historical record of geosmin dissolved in water in the period 1998–2003. Data were obtained by Aigües Ter-Llobregat (ATLL) at the Abrera drinking water plant

geosmin concentrations in the Llobregat have reached up to 190 ng L^{-1} (data provided by Aigües Ter-LLobregat; Fig. 3) and follow a remarkable seasonal pattern, the highest values being recorded in late winter and spring. Geosmin is bound to cell structures, such as chloroplast lamellae and lipophilic cell materials [74], and therefore only leaves the cell during cell lysis. Treatment with chlorine, copper sulphate or potassium permanganate during water purification ruptures existing cells and increases geosmin concentration in the water [75].

Mass production of cyanobacteria coincides with peaks of the odorous metabolite geosmin in the Llobregat, suggesting that the ecological mechanisms behind its production are linked with cyanobacterial growth. Cyanobacterial toxins and odour/taste metabolites may be regulated by complex environmental factors affecting their physiological state and growth stage [76]. In physiological terms, geosmin synthesis might be interpreted as a mechanism for dissipating excess carbon during growth. Synthesis of geosmin in culture has been linked with changes in cell growth caused by nutrient deficiency [77], as well as during the lag-growth phase when the population is not at its optimum.

Benthic cyanobacterial mass growth in the Llobregat starts in early January and lasts until the end of May. Benthic cyanobacteria develop largely in littoral areas or immediately downstream of dams, where waters are shallow and slow moving (Fig. 4). A few species dominate; the filamentous cyanobacteria *Oscillatoria limosa* and *Oscillatoria tenuis* account for more than 90% of the abundance in the mat,

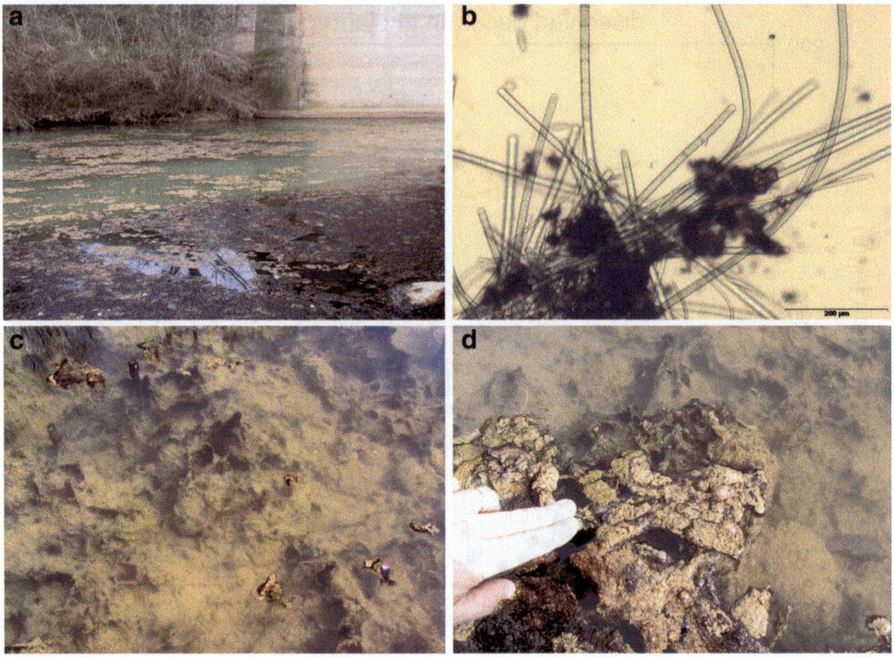

Fig. 4 Mass growth of benthic cyanobacteria in the Llobregat. From *left* to *right* and from *top* to *bottom*, (**a**) aspect of the river with cyanobacterial masses; (**b**) *Oscillatoria*-forming species under a light microscope; (**c**) attached masses; (**d**) drifting masses

while other non-cyanobacteria such as *Vaucheria* sp. and a few diatoms form part of the cyanobacterial mat. The occurrence of these masses is seasonal. Prior to cyanobacterial development, diatom communities predominate in the river producing brown mats that usually completely cover the riverbed. During the growth of benthic cyanobacteria, the brown mats remain in the riffle zones. Cyanobacterial masses progressively grow in thickness and extension, covering up to 70% of the total riverbed surface area. Their growth starts in shallow areas of the river (where current velocity is not much higher than 1 cm s^{-1}) and progressively extends towards the riffle zones (avoiding, however, the areas with currents that are too fast). Significant fractions of the attached mats become unattached and free floating, and consequently, drift downstream to colonise further areas below. Cyanobacterial mass growth finishes when waters become warmer and flow increases (usually towards the end of May). At that moment, a community of green algae dominated by *Cladophora* replaces the cyanobacterial mats. The periods when cyanobacteria prevail show high light availability and low water flow (0.25–0.27 m^3 s^{-1}), as well as a moderate water temperature and high phosphorus and lower dissolved inorganic nitrogen levels (lower N/P ratios) [78].

Chlorophyll-*a* concentration in the attached and free-floating cyanobacterial mats ranges from 200 to 500 mg Chl m^{-2} for most of the period. Chlorophyll concentration positively correlates with water nutrient content. The masses have lower geosmin concentration in the attached mats (0.55 ± 0.97 ng geosmin mg DW^{-1}) than in the drifting free-floating mats (5.25 ± 4.96 ng geosmin mg DW^{-1}), its concentration correlating with cell density [78]. These drifting masses have a very high density of meiofaunal organisms [32]. Chironomidae, Tardigrada, Oligochaeta and particularly Nematoda account for a large number of individuals in these masses.

In the Llobregat, there is a high concentration of nitrate but lower values of ammonia, which could be disadvantageous for cyanobacteria in the light-limited conditions that occur inside the mat. High aminopeptidase activity (AMA) observed in the masses indicates that inorganic nitrogen is obtained from organic sources [79, 80]. Vilalta and Sabater [81] demonstrated that small hypoxic areas within the mats were also areas of high AMA, potentially able to sustain oxygenic nitrogen fixation to complement cell requirements for nitrogen. These areas could have the highest production of geosmin.

The conditions favouring the development of cyanobacterial masses in the Llobregat, as well as their production of geosmin, are low turbulence, full light availability, nutrient-rich conditions and imbalanced N/P availability (low availability of nitrogen, low N/P ratios). These complete a favourable environment for the mass development of benthic cyanobacteria in the Llobregat, which achieve chlorophyll concentrations (200–500 mg Chl m^{-2}) commonly recorded in eutrophic situations [82, 83]. The occurrence of huge cyanobacterial masses and geosmin cannot be understood without the conjoint stressors occurring in the river. Avoiding them would require a more strict control of nutrient inputs derived from diffuse and local sources, as well as hydrological restoration of the river (including removal of unused dams and the maintenance of water in the channel).

4 Concluding Remarks

As detailed in this and other chapters of this book, the Llobregat is a clear example of a Mediterranean river that has been overexploited over the last decades. Physical and chemical pressures constrain community diversity and species distribution along the basin. Algal and invertebrate distributions are mainly determined by water salinity, nutrient concentration and pollution. The result is a general decrease in diversity in the middle and lower parts of the main course and the most important tributaries, where only the most tolerant species are present. The continuous presence of weirs and reservoirs alters the composition of the macroinvertebrate community and limits the fish species abundance and distribution. Moreover, the introduction of tolerant, non-native species adds further pressure on native species and their potential recovery. Dry conditions, common in late spring and summer, exacerbate these problems and some reaches suffer low flow conditions, consequently decreasing the capacity to dilute pollution coming mainly from wastewater plants. Still, some reaches in headwaters maintain biological communities with high diversity, despite changes in the riparian area.

Under the scenario of rising land use and water scarcity in Mediterranean systems, major alterations in fluvial biodiversity compromise ecosystem integrity. In a river so heavily managed such as the Llobregat, maintaining headwater ecosystem integrity is a guarantee for ensuring a species pool for colonisation downstream. Controlling water abstraction and limiting nutrient and pollutant inputs downstream is essential for recovering the structure and functioning of biological communities.

Acknowledgements This research was funded by the Spanish Ministry of Science and Innovation, projects SCARCE (Consolider-Ingenio 2010 program, CSD2009-00065), CGL2009-12877-C02-01 and CGL2011-30474-C02.

References

1. Margalef R (1960) Ideas for a synthetic approach to the ecology of running waters. Int Rev Hydrobiol 45:133–153
2. Hynes HBN (1970) The ecology of running waters. Univ Press, Liverpool
3. Barnes JR, Minshall GW (1983) Stream ecology. Plenum Press, New York
4. Statzner B, Higler B (1986) Stream hydraulics as a major determinant of benthic invertebrate zonation patterns. Freshw Biol 16:127–139
5. Barbour MT, Gerritsen J, Snyder BD, Stribling B (1999) Rapid bioassessment protocols for use in streams and wadeable rivers: periphyton, macroinvertebrates and fish, 2nd edn. EPA 841-B-99-002, US Environmental Protection Agency, Office of Water, Washington
6. Gasith A, Resh VH (1999) Streams in Mediterranean climate regions: abiotic influences and biotic responses to predictable seasonal events. Annu Rev Ecol Syst 30:51–81
7. Bonada N, Doledec S, Statzner B et al (2007) Taxonomic and biological trait differences of stream macroinvertebrate communities between Mediterranean and temperate regions: implications for future climatic scenarios. Glob Chang Biol 33:1658–1671
8. Myers N, Russell A, Mittermeier CG et al (2000) Biodiversity hotspots for conservation priorities. Nature 403:853–858
9. Alvarez-Cobelas M, Rojo C, Angeler DG (2005) Mediterranean limnology: current status, gaps and the future. J Limnol 64:13–29
10. Millet X, Prat N (1984) Las comunidades de macroinvertebrados a lo largo del río Llobregat. Limnetica 1:222–233
11. Tomas X, Sabater S (1985) The diatom flora of the Llobregat river and its relation to water quality. Verh Int Verein Limnol 22:2348–2352
12. Prat N, Puig MA, González G et al (1984) Llobregat. In: Whitton BA (ed) Ecology of European rivers. Blackwell, Oxford
13. Muñoz I, Prat N (1992) Cambios en la calidad del agua de los ríos Llobregat y Cardener en los últimos 10 años. Tecnol Agua 91:17–23
14. Muñoz I, Prat N (1994) A comparison between different biological water quality indexes in the Llobregat basin (NE Spain). Verh Int Verein Limnol 25:1945–1949
15. Sabater S (1990) Phytoplankton composition in a medium-sized Mediterranean river: the Ter (Spain). Limnetica 6:47–56
16. Sabater S, Artigas J, Duran C et al (2008) Longitudinal development of chlorophyll and phytoplankton assemblages in a regulated large river (the Ebro River). Sci Total Environ 404:196–206
17. Sabater S, Sabater F, Tomas X (1987) Water quality and diatom communities in two Catalan rivers (NE Spain). Water Res 21:901–911

18. Ricart M (2007) Effects of toxicants on biofilm communities: field and laboratory approaches. MsC Thesis. University of Girona, Girona
19. Tornés E, Cambra J, Goma J et al (2007) Indicator taxa of benthic diatom communities: a case study in Mediterranean streams. Ann Limnol 43:1–11
20. Muñoz I, Prat X, Millet X et al (1986) Heterogeneidad espacial en la distribución de los macroinvertebrados a lo largo de un transecto en el río Llobregat. Limnetica 2:135–145
21. López-Doval J, Großschartner M, Höss S et al (2010) Invertebrate communities in soft sediments along a pollution gradient in a Mediterranean river (Llobregat, NE Spain). Limnetica 29:311–322
22. Balleto E, Casale A (1989) Mediterranean insect conservation. In: Collins NM, Thomas JA (eds) The conservation of insects and their habitats. Harcourt Brace Jovanovich Publishers, London
23. Puig MA (1983) Efemerópteros y Plecópteros de los ríos catalanes. PhD Dissertation. University of Barcelona, Barcelona
24. Puig MA, Bautista I, Tort MJ et al (1981) Les larves de la riviere Llobregat: distribution longitudinales et relation avec la qualite de l'eau. Ser Entomol 20:305–311
25. González MA, García de Jalón D, Terra L (1987) Faunistic studies on Iberian Trichoptera: a historical survey and present state of knowledge. In: Proceedings of the fifth International Symposium on Trichoptera, pp 85–90
26. Décamps H, Besch KW, Vobis H (1973) Influence des produits toxiques sur la construction du filet des larves d'*Hydropsyche* (Insecta, Trichoptera). C R Acad Sci Paris Sér D 276:375–378
27. Bouvet Y (1974) Ecologie et reproduction chez les tricopteres cavernicoles du groupe de Stenophylax. In: Proceedings of the first International Symposium on Trichoptera, pp 105–109
28. Williams DD (1996) Environmental constraints in temporary waters and their consequences for the insect fauna. J N Am Benthol Soc 15:634–650
29. Murillo J (1985) Algunes captures d'Heteròpters aquàtics a Catalunya i altres localitats de la resta de la península Ibérica. Butll Inst Catalana d'Hist Nat 52:139–148
30. Puntí T, Rieradevall M, Prat N (2009) Environmental factors, spatial variation, and specific requirements of Chironomidae in Mediterranean reference streams. J N Am Benthol Soc 28:247–265
31. Gonzalez G (1990) Sistemática y ecología de los Simulidae (Diptera) de los ríos de Catalunya y otras cuencas hidrográficas españolas. PhD Dissertation. University of Barcelona, Barcelona
32. Gaudes A, Sabater S, Vilalta E et al (2006) The nematode community in river cyanobacterial biofilms. Nematology 8:909–919
33. Doadrio I, Kottelat M, de Sostoa A (2007) *Squalius laietanus*, a new species of cyprinid fish from north-eastern Spain and southern France (Teleostei: Cyprinidae). Ichthyol Explor Freshw 18:247–256
34. Aparicio E, Vargas MJ, Olmo JM, de Sostoa A (2000) Decline of native freshwater fishes in a Mediterranean watershed on the Iberian Peninsula: a quantitative assessment. Environ Biol Fishes 59:11–19
35. Gutiérrez i Perarnau C, Pagès i Puig J (2005) Espinós (*Gasterosteus gymnurus*). Apunts sobre la seva distribució entre el Llobregat i la Tordera. Lauro 28:111–113
36. Almaça C, Elvira B (2000) Past and present distribution of *Acipenser sturio* L., 1758 on the Iberian Peninsula. Bol Inst Esp Oceanogr 16:11–16
37. Vila-Gispert A, García-Berthou E, Moreno-Amich R (2002) Fish zonation in a Mediterranean stream: effects of human disturbances. Aquat Sci 64:163–170
38. Lasne E, Bergerot B, Lek S, Laffaille P (2007) Fish zonation and indicator species for the evaluation of the ecological status of rivers: example of the Loire basin (France). River Res Appl 23:877–890
39. García-Berthou E, Moreno-Amich R (2000) Introduction of exotic fish into a Mediterranean lake over a 90-year period. Arch Hydrobiol 149:271–284
40. Carol J, Benejam L, Benito J, García-Berthou E (2009) Growth and diet of European catfish (*Silurus glanis*) in early and late invasion stages. Fund Appl Limnol 174:317–328

41. Casals F (2005) Les comunitats íctiques dels rius mediterranis: relació amb les condicions ambientals. PhD Dissertation. University of Barcelona, Barcelona
42. Aparicio E, de Sostoa A (1998) Reproduction and growth of *Barbus haasi* in a small stream in the NE of the Iberian peninsula. Arch Hydrobiol 142:95–110
43. Aparicio E, de Sostoa A (1999) Pattern of movements of adult *Barbus haasi* in a small Mediterranean stream. J Fish Biol 55:1086–1095
44. Maceda-Veiga A, Salvadó H, Vinyoles D, de Sostoa A (2009) Outbreaks of *Ichthyophthirius multifiliis* in Redtail Barbs *Barbus haasi* in a Mediterranean stream during drought. J Aquat Anim Health 21:189–194
45. Sanz Ball-llosera N, García-Marín JL, Pla C (2002) Managing fish populations under mosaic relationships. The case of brown trout (*Salmo trutta*) in peripheral Mediterranean populations. Conserv Genet 3:385–400
46. Aparicio E, García-Berthou E, Araguas RM, Martínez P, García-Marín JL (2005) Body pigmentation pattern to assess introgression by hatchery stocks in native *Salmo trutta* from Mediterranean streams. J Fish Biol 67:931–949
47. Almodóvar A, Nicola GG, Elvira B, García-Marín JL (2006) Introgression variability among Iberian brown trout evolutionary significant units: the influence of local management and environmental features. Freshw Biol 51:1175
48. Araguas RM, Sanz N, Fernández R, Utter FM, Pla C, García-Marín JL (2009) Role of genetic refuges in the restoration of native gene pools of brown trout. Conserv Biol 23:871–878
49. Petrovic M, Solé M, López De Alda MJ, Barceló D (2002) Endocrine disruptors in sewage treatment plants, receiving river waters, and sediments: integration of chemical analysis and biological effects on feral carp. Environ Toxicol Chem 21:2146–2156
50. Céspedes R, Lacorte S, Raldúa D, Ginebreda A, Barceló D, Piña B (2005) Distribution of endocrine disruptors in the Llobregat River basin (Catalonia, NE Spain). Chemosphere 61:1710–1719
51. Peré-Trepat E, Ginebreda A, Tauler R (2007) Comparison of different multiway methods for the analysis of geographical metal distributions in fish, sediments and river waters in Catalonia. Chemometr Intell Lab Syst 88:69–83
52. Peré-Trepat E, Olivella L, Ginebreda A, Caixach J, Tauler R (2006) Chemometrics modelling of organic contaminants in fish and sediment river samples. Sci Total Environ 371:223–237
53. Solé M, Barceló D, Porte C (2002) Seasonal variation of plasmatic and hepatic vitellogenin and EROD activity in carp, *Cyprinus carpio*, in relation to sewage treatment plants. Aquat Tox 60:233–248
54. Solé M, Raldua D, Piferrer F, Barceló D, Porte C (2003) Feminization of wild carp, *Cyprinus carpio*, in a polluted environment: plasma steroid hormones, gonadal morphology and xenobiotic metabolizing system. Comp Biochem Physiol C Toxicol Pharmacol 136:145–156
55. Clavero M, García-Berthou E (2006) Homogenization dynamics and introduction routes of invasive freshwater fish in the Iberian Peninsula. Ecol Appl 16:2313–2324
56. Benejam L, Carol J, Benito J, García-Berthou E (2007) On the spread of the European catfish (*Silurus glanis*) in the Iberian Peninsula: first record in the Llobregat river basin. Limnetica 26:169–171
57. Rincón PA, Correas AM, Morcillo F, Risueño P, Lobón-Cerviá J (2002) Interaction between the introduced eastern mosquitofish and two autochthonous Spanish toothcarps. J Fish Biol 61:1560–1585
58. Alcaraz C, García-Berthou E (2007) Life history variation of invasive mosquitofish (*Gambusia holbrooki*) along a salinity gradient. Biol Conserv 139:83–92
59. García-Berthou E, Boix D, Clavero M (2007) Non-indigenous animal species naturalized in Iberian inland waters. In: Gherardi F (ed) Biological invaders in inland waters: profiles, distribution, and threats. Springer, Dordrecht
60. Ormerod SJ, Dobson M, Hildrew AG et al (2010) Multiple stressors in freshwater ecosystems. Freshw Biol 55:1–4

61. Holmstrup M, Bindesbøl AM, Oostingh GJ et al (2010) Interactions between effects of environmental chemicals and natural stressors: a review. Sci Total Environ 408:3746–3762
62. Allan JD, Castillo MM (2007) Stream ecology: structure and function in running waters. Springer, Dordrecht
63. CAS (2010) Chemical abstract service. American Chemical Society. http://www.cas.org/expertise/cascontent/ataglance/index.html. Accessed 11 Jan 2012
64. Koeck-Schulmeyer M, Ginebreda A, Postigo C et al (2011) Wastewater reuse in Mediterranean semi-arid areas: the impact of discharges of tertiary treated sewage on the load of polar micro pollutants in the Llobregat river (NE Spain). Chemosphere 82:670–678
65. Muñoz I, López-Doval JC, Ricart M et al (2009) Bridging levels of pharmaceuticals in river water with biological community structure in the Llobregat river basin (NE Spain). Environ Toxicol Chem 2:2706–2714
66. Ricart M, Guasch H, Barceló D et al (2010) Primary and complex stressors in polluted mediterranean rivers: pesticide effects on biological communities. J Hydrol 383(1–2):52–61
67. Brix R, López-Doval J, Ricart M et al (2011) Establishing potential links between the presence of alkylphenolic compounds and the benthic community in a European river basin. Environ Sci Pollut Res. doi:10.1007/s11356-011-0527-z
68. López-Doval JC, Kukkonen JVK, Rodrigo P et al (2011) Effects of indomethacin and propranolol on *Chironomus riparius* and *Physella* (*Costatella*) *acuta*. Ecotoxicol Environ Saf. doi:10.1016/j.ecoenv.2011.11.004
69. Tuikka AI, Schmitt C, Hoss S et al (2011) Toxicity assessment of sediments from three European river basins using a sediment contact test battery. Ecotoxicol Environ Saf 74:123–131
70. Ginebreda A, Muñoz I, López de Alda M et al (2010) Environmental risk assessment of pharmaceuticals in rivers: relationships between hazard indexes and aquatic macroinvertebrate diversity indexes in the Llobregat River (NE Spain). Environ Int 36:153–162
71. Gonzalez S, Petrovic M, Barcelo D (2004) Simultaneous extraction and fate of linear alkylbenzene sulfonates, coconut diethanol amides, nonylphenol ethoxylates and their degradation products in wastewater treatment plants, receiving coastal waters and sediments in the Catalonian area (NE Spain). J Chromatogr A 1052:111–120
72. Persson PE (1996) Cyanobacteriaand off-flavours. Phycologia 30:168–171
73. Aoyama K (1990) Studies on the earthy-musty odour in natural water (IV). Mechanism of earthy-musty odour porduction of actinomycetes. J Appl Bact 68:405–410
74. Wu TJ, Juttner F (1988) Effect of environmental factors on geosmin production by *Fischerella muscicola*. Wat Sci Technol 20:143–148
75. Chow CWK, House J, Velzeboer RMA et al (1998) The effects of ferric chloride flocculation on cyanobacterial cells. Wat Res 32:808–814
76. Paerl HW, Millie DF (1996) Physiological ecology of toxic aquatic cyanobacteria. Phycologia 35:160–167
77. Wu JT, Ma PI, Chou TL (1991) Variation of geosmin content in Anabaena cells and its relation to nitrogen utilization. Arch Microbiol 157:66–69
78. Sabater S, Vilalta E, Gaudes A et al (2003) Ecological implications of mass growth of benthic cyanobacteria in rivers. Aquat Microb Ecol 32:175–184
79. Patel AB, Fukami K, Nishijima T (2000) Regulation of seasonal variability of aminopeptidase activities in surface and bottom waters of Uranouchi Inlet, Japan. Aquat Microb Ecol 21:139 149
80. Sala MM, Karner M, Arin L et al (2001) Measurement of ectoenzyme activities as an indication of inorganic nutrients imbalance in microbial communities. Aquat Microb Ecol 23:301–311
81. Vilalta E, Sabater S (2005) Structural heterogeneity in cyanobacterial mats is associated with geosmin production in rivers. Phycologia 44:678–684
82. Dodds WK, Smith VH, Zander B (1997) Developing nutrient targets to control benthic chlorophyll levels in streams: a case study of the Clark Fork River. Wat Res 31:1738–1750

83. Romaní AM, Sabater S (2000) Influence of algal biomass on extracellular enzyme activity in river biofilms. Microb Ecol 40:16–24
84. Freyhof J, Brooks E (2011) European red list of freshwater fishes. Publications Office of the European Union, Luxembourg
85. IUCN Standards and Petitions Subcommittee (2011) Guidelines for using the IUCN red list categories and criteria. Version 9.0. Prepared by the Standards and Petitions Subcommittee. http://www.iucnredlist.org/documents/RedListGuidelines.pdf. Accessed 11 Jan 2012
86. Marcé R, Honey-Rosés J, Manzano A, Moragas L, Catllar C, Sabater S (2012) The Llobregat River Basin: A paradigm of impaired rivers under climate change threats. Hdb Env Chem doi: 10.1007/698_2012_152
87. Guillén D, Ginebreda A, Eljarrat E, Petrovic M, Barcelo D (2012) Occurrence of persistent organic pollutants in the Llobregat River Basin: An overview. Hdb Env Chem doi: 10.1007/698_2012_144
88. Eljarrat E, Barceló D (2012) Occurrence and behavior of brominated flame retardants in the Llobregat River Basin. Hdb Env Chem doi: 10.1007/698_2011_139

Occurrence of Persistent Organic Pollutants in the Llobregat River Basin: An Overview

Daniel Guillén, Antoni Ginebreda, Ethel Eljarrat, Mira Petrovic, and Damià Barceló

Abstract This chapter provides an overview about the occurrence of persistent organic pollutants (POPs) in the Llobregat River basin (NE Spain). This basin historically has received considerable amounts of contaminants from both urban and industrial origin as well as from agricultural holdings. Among these contaminants, POPs are ubiquitous compounds of great concern due to their high persistence in the environment, high bioaccumulative nature and adverse effects (toxicity, carcinogenic and mutagenic effects, etc.) on wildlife and humans. These persistent contaminants include compounds such as chlorinated pesticides, polychlorinated byphenyls (PCBs) and polyaromatic hydrocarbons (PAHs). This overview has been complemented with a recent monitoring study carried out in the framework of the SCARCE project in which the occurrence of 52 POPs in sediment samples from Llobregat River basin has been evaluated. Concentrations were compared with two approaches derived from a consensus sediment quality guideline (SQG) in order to assess the chemical quality of the sediments along the basin. Finally, the toxic potency associated with PAHs exposure is also evaluated.

D. Guillén (✉), A. Ginebreda, and E. Eljarrat
Institute of Environmental Assessment and Water Research (IDAEA-CSIC), C/Jordi Girona, 18-26 08034, Barcelona, Spain
e-mail: dggqam@cid.csic.es

M. Petrovic
Catalan Institute for Water Research (ICRA), c/Emili Grahit 101, 17003 Girona, Spain

Catalan Institution for Research and Advanced Studies (ICREA), Passeig Lluis Companys 23, 80010, Barcelona, Spain

D. Barceló
Institute of Environmental Assessment and Water Research (IDAEA-CSIC), C/Jordi Girona, 18-26 08034, Barcelona, Spain

Catalan Institute for Water Research (ICRA), c/Emili Grahit 101, 17003 Girona, Spain

King Saud University (KSU), P.O. Box 2455, 11451 Riyadh, Saudi Arabia

S. Sabater et al. (eds.), *The Llobregat: The Story of a Polluted Mediterranean River*, Hdb Env Chem (2012) 21: 117–134, DOI 10.1007/698_2012_144,

Keywords Llobregat, Organochlorinated pesticides, Persistent organic pollutants, Polyaromatic hydrocarbons, Polychlorinated byphenyls, Sediment quality

Contents

Abbreviations and Symbols

ACA	Catalan Water Agency (Agència Catalana de l'Aigua)
BFRs	Brominated flame retardants
DDD	Dichlorodiphenyldichloroethane
DDE	Dichlorodiphenyldichloroethylene
DDT	Dichlorodiphenyltrichloroethane
DLC	Dioxin-like compounds
dw	Dry weight
EPA	Environmental Protection Agency
EQC	Equilibrium criterion
EROD	Ethoxyresorufin-*O*-deethylase
EUSES	European Uniform System for the Evaluation of Substances
HCB	Hexachlorobenzene
HCHs	Hexachlorocyclohexanes
IARC	International Agency for Research on Cancer
OCPs	Organochlorinated pesticides
PAH	Polycyclic aromatic hydrocarbons
PCBs	Polychlorinated biphenyls
PCDDs	Polychlorinated dibenzo-*p*-dioxins
PCDFs	Polychlorinated dibenzofurans
PCNs	Polychlorinated naphthalenes
PEC	Probable effect concentration
POPs	Persistent organic pollutants
SQGs	Sediment quality guidelines
TCDD	Tetrachlorodibenzo-*p*-dioxin
TEC	Threshold effect concentration
TEF	Toxicity equivalent factors
TEQ	Toxicity equivalent quantity
TOC	Total organic carbon
US EPA	United States Environmental Protection Agency

WWTP	Wastewater treatment plants
$\sum$PAHs	Sum of total PAHs
$\sum$PCBs	Sum of total PCBs

1 Persistent Organic Pollutants

Persistent organic pollutants (POPs) are able to remain in the environment during years since they show low degradation rates (months or years). Their environmental effects are well known and some POPs present toxic, carcinogenic and also has been proved they can act as endocrine disruptors affecting important biological processes to organisms, for example dichlorodiphenyltrichloroethane (DDT) and hexachlorobenzene (HCB) [1–3]. Both organochlorinated compounds were recognized as endocrine disruptors by the European Commission [4].

PAHs are widely distributed environmental contaminants that have detrimental biological effects, toxicity, mutagenecity and carcinogenicity [5–7]. Due to their ubiquitous occurrence, recalcitrance, bioaccumulation potential and carcinogenic activity, PAHs have gathered significant environmental concern. Both the International Agency for Research on Cancer (IARC) and the United States Environmental Protection Agency (US EPA) have classified many PAHs into carcinogenic compounds [8, 9]. The widespread occurrence of PAHs is essentially due to their formation and release during the incomplete combustion of organic matter (mainly petroleum derivatives) but PAHs can also be generated during natural processes (forest fires and volcanic eruptions). Their environmental fate includes volatilization, photo-oxidation, chemical oxidation, adsorption on soil particles, leaching and microbial degradation [10]. Lower molecular weight PAHs are partially adsorbed and degraded in the atmosphere by reaction with photochemically produced hydroxyl radicals. PAHs persistence increases with increase in the molecular weight. Then high molecular weight PAHs are extremely stable to degradation processes and are known to persist in the environment during long periods of time.

Organochlorinated pesticides (OCPs) were widely used since 1940 for protecting crops against insects and other animals. These ubiquitous pesticides present long-term adverse effects on ecosystems and human health. Many OCPs, including aldrin, endrin, DDT, dieldrin, hexachlorocyclohexane (HCHs), heptachlor, chlordane and toxaphene, were classified and regulated as POPs under the Stockholm Convention signed by 125 nations. OCPs environmental occurrence can be attributable to their high persistence, which makes them amenable to persist during years in aquatic environments. However, several studies have evidenced that they are still being used illegally for agricultural purposes.

During last 60 years, developed economies have experienced a rapid urban and industrial development. This growth has been accompanied by the undesirable emission of large amounts of products and unintentionally generated byproducts. PCBs are industrially produced compounds and are recognized to be toxic, bioaccumulative, persistent, carcinogenic and mutagenic compounds [11, 12]. PCBs were widespread used in a variety of industrial applications, mainly as dielectric and cooling fluids in electronic devices (transformers, capacitors, engines, etc.). But PCBs production/usage has been banned or restricted worldwide since the early 1970s, but due to their high persistence they are still detected in environmental matrices such as sediments and biota [13, 14].

1.1 Factors Influencing Environmental Fate of POPs

Environmental fate of chemical compounds is directly related to their physico-chemical properties, such as vapour pressure, aqueous solubility, octanol solubility, Henry's law constant, and partition coefficients. These properties determine their environmental behaviour (persistence, bioaccumulation, adsorption, mobilization, etc.) and justify how chemicals are distributed throughout the environmental compartments (water, air, soil or sediment and biota).

POPs are chemicals involved in several fate and transport phenomena. They show high persistence in the environment but due to their semivolatility they have a tendency to partition between air, water and organic phases, which permits the relatively easy movement from phase to phase. Their ability to be present in and move between various environmental phases requires the development of multi-compartmental approaches for assessing their environmental occurrence. Numerous studies have demonstrated that POPs have the ability to circulate globally, as they can travel long distances on air and water currents before being re-deposited in remote areas far from their original source of application [15]. Because transport dynamics can be affected by a variety of factors (volatility, degradability, deposition velocity of each compound, and their respective temperature dependence), it is not obvious how to predict long-range transport of POPs.

Currently, there is an increasing interest in developing multi-media models for predicting the environmental fate of chemicals. Among the several models available ChemCAN, the equilibrium criterion (EQC) model [16], CalTOX [17], and SimpleBox [18], which is a component of the European Uniform System for the Evaluation of Substances (EUSES) [19] are worth to be mentioned. In addition to the intrinsic chemical properties, multi-media models require a great number of data inputs to represent a particular scenario and to provide more realistic and accurate predictions; some of these factors are shown in Table 1.

2 POPs Occurrence in the Llobregat River Basin

The Llobregat River is a basic source of water for Catalonian population, and in last years there has been a special interest to monitor the Llobregat River basin in order to evaluate its chemical quality. The Catalan Water Agency (Agència Catalana de l'Aigua, ACA) as the responsible authority of the Llobregat River basin has created a control network with numerous monitoring points in order to ensure the good

Table 1 Main factors affecting environmental fate and transport of contaminants

Application mode/ emission characteristics	Diffuse	Chemicals released to the environment from a not located emission point (e.g. traffic origin) or emitted directly to the air (e.g. mill smokestacks) where chemicals can be easily dispersed by air currents – *Polyaromatic hydrocarbons (PAHs)* Chemicals applied over extensive areas. They are spread out directly to the environment (e.g. pesticides application in crops) – *Cyclodienes: aldrin, endrin, dieldrin, isodrin, heptachlor, etc.* – *(α,β,γ,δ)-Hexachlorocyclohexane* – *Diphenyl aliphatics: DDT and derivatives*
	Point	Emitted from localized sources: industrial and urban effluents, wastewater treatment plants (WWTP), etc. – *Hexachlorobenzene, pentachlorobenzene, pentachlorophenol, etc.*
	End of useful life	Chemical additives add some property to the final product (textile, electronics, paper, leather, lubricants, etc.) and remain on it. Additives such as plasticizers, waterproof aids, flame retardants, biocides, paintings or coatings. When products reach to the end of their useful life (product becomes waste), these additives enter in the environment – *Polychlorinated biphenyls (PCBs), brominated flame retardants, perfluorooctanesulfonic acid, etc.*
Environmental factors	Matrix composition	– *pH, $CaCO_3$, cationic exchange capacity (CEC)* – *Nutrient status (competitive species in soil and solution)* – *Organic matter content (total carbon and organic carbon content)* – *Redox potential* – *Soil texture, porosity, bulk density*
	Biological processes	– *Biodegradability and its influence on organism's biological processes* – *Transformation products*
	Climate and hydrologic conditions	– *Temperature* – *Humidity* – *Wind speed* – *Solar radiation (photodegradation)* – *Rainfall* – *Water flow* – *Water column depth*
Others	– *Quantity of use* – *Commercial transport* – *Formulation (mixtures format or isolated compound)* – *Industrial or domestic use* – *Waste disposal, waste treatment, recycling*	

In italic some examples of POPs for different application mode and emission characteristics chemical and ecological status in the Llobregat River basin. This information is now available for public access [20].

2.1 POPs in Surface Water

Analysis made by ACA monitoring network reveals low concentrations of POPs in surface waters from Llobregat River [20]. It can be attributed to the hydrophobic nature of POPs that tends to accumulate them in sediments, together with their ability to volatilize from the water masses.

Concentrations detected for PAHs and organochlorinated compounds (DDT, aldrin, endrin, isodrin, hexachlorobenzene and PCBs) were almost always below the detection limits [20]. Only the more polar chlorinated pesticides such as endosulfan and lindane (log K_{OW} 3.83 and 3.72, respectively [21]) showed remarkable levels in surface waters. In Fig. 1 are shown the highest concentrations of POPs in surface waters reported by ACA control network. Monitoring performed last 3 years indicate that POPs are not usually detected in surface waters.

It is generally accepted that POPs fate is directly related to their hydrophobic behaviour and once they enter into water systems, they tend to remain adsorbed onto particles with subsequent sedimentation. As a result, bottom sediment constitutes a sink for hydrophobic compounds. Sediment is thus the matrix of choice for assessing POPs occurrence in aquatic environments.

2.2 POPs in Sediments

Industrial effluents containing high content of organochlorinated chemicals have been historically discharged in Llobregat River. These compounds were widely used for industrial purposes such as bleaching processes (pulp and paper, textiles, etc.) and for preserving products (wood, textiles, cordage fibres, leather, etc.) against the micro-organisms action. Several studies have reported the presence of chlorinated compounds in coastal sediments collected in the Llobregat River mouth (NW Mediterranean Sea) [22–24]. These chlorinated compounds include families such as polychlorinated dibenzo-p-dioxines (PCDDs), polychlorinated dibenzofurans (PCDFs) and polychlorinated byphenyls (PCBs). These findings indicate that the sea coast is highly influenced by the load of pollutants transported by the Llobregat River. Therefore, coastal sediments represent an important sink for hydrophobic compounds which can persist during long periods of time.

Eljarrat et al. determined the presence of 16 PAHs (EPA priority pollutants) in marine sediment samples ($n = 7$), showing a mean value of $\sum$PAHs of 3,659 ng/g dry weight (dw) [22]. This study indicates (Fig. 2) a major contribution of four-ring PAHs (mainly fluoranthene and pyrene, with mean values of 744 and 600 ng/g dw, respectively). Generally, PAHs with four and more rings present a higher lipophilicity and are found predominantly in particulate phase, while PAHs with lower number of rings are found mainly in the gas phase [25]. The concentrations of

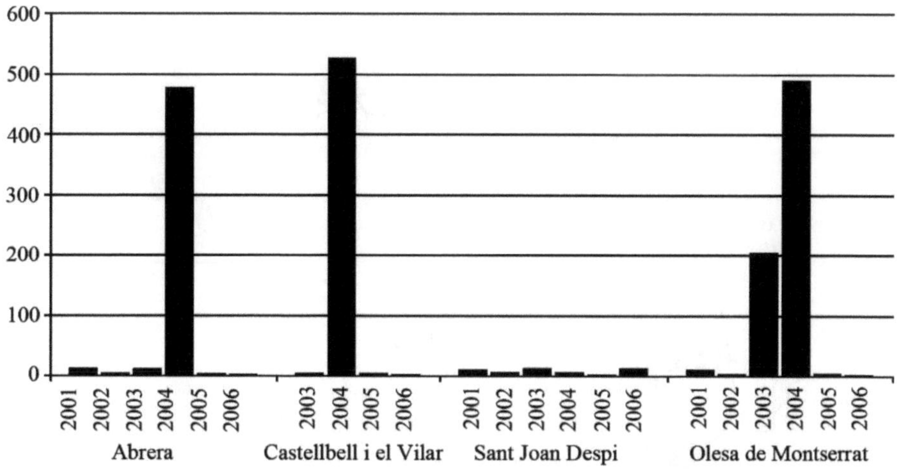

Fig. 1 Concentrations (ng/L) of chlorinated pesticides detected in surface water samples in the Llobregat River basin [6]

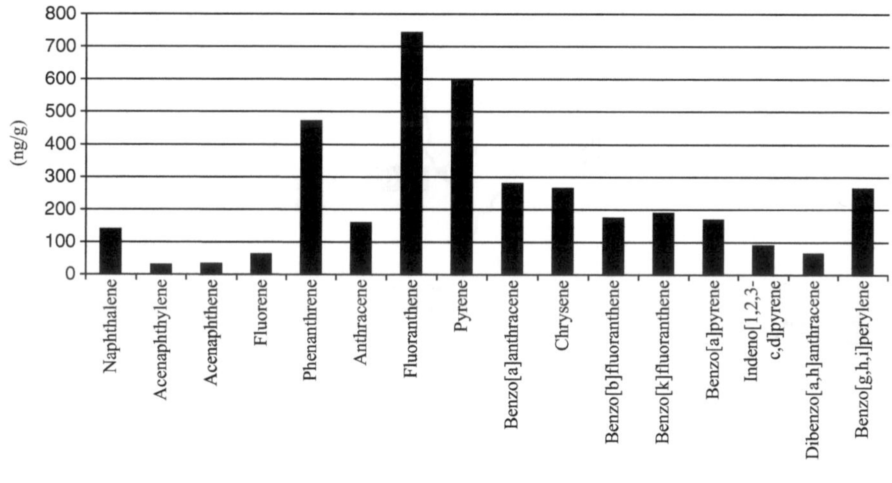

Fig. 2 Concentrations of 16 EPA PAHs in marine sediments at the Llobregat mouth [22]

PAHs in sediments studied were less than the effects range low (ERL) established at 4 μg/g dw for the sum of the 16 priority PAHs [26].

The occurrence of a great number of POPs in the Llobregat River basin has been evaluated in the framework of the SCARCE project (Consolider-Ingenio 2010, CSD2009-00065). Bottom sediment in seven sampling sites was sampled in September 2010 from the main stream of the Llobregat River (sample names: LLO1 to LLO7), and in the two main tributaries, i.e., four samples in the Cardener River (CAR1 to CAR4) and three in the Anoia River (ANO1 and ANO3). In total, 14 samples were collected for characterizing the whole basin. The locations for each sampling point are illustrated in Fig. 3.

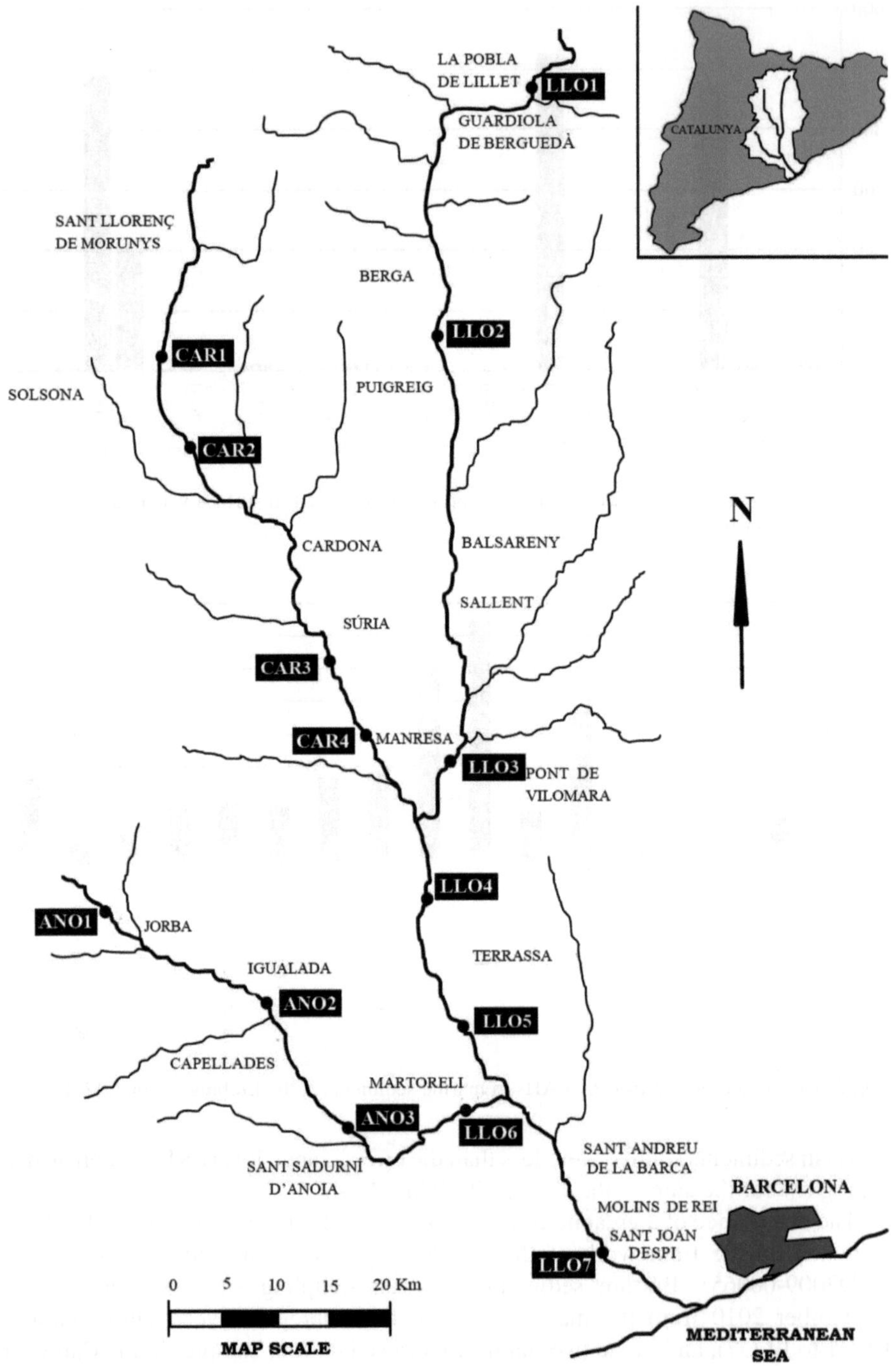

Fig. 3 Location of the sampling sites at the Llobregat River basin

The total organic carbon (TOC) content of the 14 sediment samples comprised between 0.26% and 3.43% by weight, reflecting the variety of sediment composition. For most of the sediment samples, TOC measures show typical values, except for ANO2 (3.43%) that is located in an important industrial zone in the middle course of the Anoia River.

A multiresidue method was applied to determine a total of 52 POPs (listed in Table 2). These POPs comprise the EPA priority PAHs and 36 chlorinated compounds (pesticides, PCBs, HCB and trichlorobenzenes) from agricultural and industrial origin. Among the 52 POPs included in this study, 27 were positively detected, i.e., the 16 EPA PAHs, DDT (and their transformation products) and some congeners of PCB, ranging from penta- to hepta-congeners (penta-PCB-101, hexa-PCB-138, penta-PCB-118, hexa-PCB-149, hexa-PCB-153, hepta-PCB-170 and hepta-PCB-180). Table 3 indicates the concentrations detected for each sampling point.

The results reveal a widespread contamination by PAHs in the whole river basin (Fig. 4). The highest concentrations of PAHs were found near urban and industrialized areas due to the increasing traffic pressure and the scarce dispersion in the atmosphere. ANO2 and LLO4 were the two locations presenting the highest levels of PAHs pollution with $\sum$PAHs 1211 and 784 ng/g dw, respectively. The high pollution level found at ANO2 is attributed to the influence of the WWTP located in Igualada, next to the sampling point. This point also receives the PAHs contribution from the industrial activity of this city (tanneries, pulp and paper, textile, etc.). On the other hand, LLO4 shows high levels of PAHs likely due to the influence of the WWTP located in Manresa, few kilometres upstream to the sampling point. LLO2 sample was collected from an area apart from big cities, but it presented a considerable PAHs pollution ($\sum$PAHs 290 ng/g dw). This concentration could be associated with the emissions from a nearby power plant currently in use.

The PAHs profile in sediments is dominated by hydrocarbons with a high number of aromatic rings. PAHs with lower number of rings rarely remain deposited in sediments, since they are easily involved in exchange processes (resuspension in water, volatilization before adsorption in solid matter, etc.) [21]. Most of the analysed samples containing heavy PAHs (4–6 rings) are in greater proportion than light PAHs (2–3 rings) (Fig. 5).

Four analysed samples have levels of DDT (4,4′ isomer) in the range of 7.50–70.83 ng/g dw, and these concentrations are higher than those of DDT transformation products: dichlorodiphenyldichloroethylene (DDE) and dichlorodiphenyldichloroethane (DDD). It suggests that this pesticide is currently still being used illicitly for protecting crops in the studied basin.

PCBs were also detected in several samples of the Llobregat River. The major occurrence of PCBs was observed in samples collected in the last stretch of the river, where there is an important industrial activity. $\sum$PCBs concentrations were in the range of 2.88–17.2 ng/g dw, and mainly were detected the highly chlorinated PCBs (with 5–7 chlorine atoms).

Table 2 List of POPs included in the multiresidue analysis (between brackets indicated the number of rings for PAHs and the chlorine atoms for PCBs)

Polyaromatic hydrocarbons	Organochlorinated compounds	Polychlorinated byphenyls
Naphthalene [2]	–(1,2,3–1,2,4)-Trichlorobenzene	PCB18 [2]
Acenaphthylene [3]	–Hexachlorociclohexanes (α,β,γ,δ)	PCB28 [3]
Acenaphthene [3]	–Hexaclorobenzene (HCB)	PCB31 [3]
Fluorene [3]	*–Cyclodienes:*	PCB44 [4]
Phenanthrene [3]	Aldrin, isodrin, dieldrin	PCB52 [4]
Anthracene [3]	(α, β)-endosulfan	PCB101 [5]
Fluoranthene [4]	endosulfan sulphate, heptachlor	PCB118 [5]
Pyrene [4]	and (*exo-*,*endo-*) heptachlor	PCB138 [6]
Benzo(*a*)anthracene [4]	*–Diphenyl ethanes/ethenes:*	PCB149 [6]
Chrysene [4]	Dicofol, *p,p*′-DDT, *p,p*′-DDE	PCB153 [6]
Benzo(*b*)fluoranthene [5]	*p,p*′-DDD, *o,p*′-DDT, *o,p*′-DDE,	PCB170 [7]
Benzo(*k*)fluoranthene [5]	*o,p*′-DDD	PCB180 [7]
Benzo(*a*)pyrene [5]		PCB194 [8]
Indeno(1,2,3-cd)pyrene [6]		
Dibenzo(*a,h*)anthracene [5]		
Benzo(g,h,i)perylene [6]		

3 Assessment and Evaluation of Sediment Quality

In order to assess the risk associated with chemical exposure in the environment monitoring programmes must be complemented with information associated with their exposure effects over wildlife. However, toxicity tests with sediments and sediment-dwelling organisms are not available for a broad spectrum of substances; therefore, toxicological data referred to this matrix are limited.

Several numerical sediment quality guidelines (SQGs) have been developed by various federal, state, and provincial environmental agencies for both freshwater and marine ecosystems. A consensus-based SQGs [27] were developed for several chemicals (i.e., metals, PAHs, PCBs and pesticides) in freshwater sediments in order to provide an accurate basis for evaluating the chemical quality of this matrix. In this study, we used two approaches: threshold effect concentration (TEC) and probable effect concentration (PEC). Figure 6 represents the sum of concentration of several PAHs (including naphthalene, fluorene, phenanthrene, anthracene, fluoranthene, pyrene, benzo(*a*)anthracene, chrysene and benzo(*a*)pyrene), PCBs (detected congeners), DDTs (sum DDT and derivatives) besides both quality standards approaches, TEC and PEC.

All the sediment samples from Llobregat River basin presented levels of POPs lower than corresponding PEC values established in SQGs. Eight sediment samples out of 14 exceeded the TEC_{PAHs} value (ANO1, ANO2, CAR2, CAR4, LLO2, LLO3, LLO4 and LLO7) whereas four exceeded the TEC_{DDTs} (CAR2, CAR3, CAR4, LLO2).

In summary, SQGs can be used to identify hot spots with respect to sediment contamination, to determine the potential for and spatial extent of injury to sediment-dwelling organisms, to evaluate the need for sediment remediation and to provide support for the deployment of monitoring programmes to further assess the extent of contamination and the effects of contaminated sediments on sediment-dwelling organisms.

Table 3 Concentrations (ng/g) of POPs in the 14 samples analysed along the Llobregat River basin

	ANO1	ANO2	ANO3	CAR1	CAR2	CAR3	CAR4	LLO1	LLO2	LLO3	LLO4	LLO5	LLO6	LLO7
Naphthalene	85.75	bld	bld	bld	6.09	bld	26.99	bld	54.13	12.49	bld	bld	bld	bld
Acenaphthylene	bld	1.74	bld	19.5	bld	bld	4.44	bld	bld	1.91	bld	bld	bld	19.29
Acenaphthene	bld	14.08	bld	bld	bld	bld	6.24	bld	4.68	bld	bld	bld	bld	bld
Fluorene	3.74	8.92	bld	bld	3.4	bld	15.15	10.42	11.94	4.26	9.45	4.43	bld	bld
Phenanthrene	5.47	121.46	bld	5.07	7.55	bld	47.65	8.32	37.39	9.16	165.09	4.91	bld	25.46
Anthracene	2.4	22.52	bld	bld	2.06	bld	17.09	bld	4.65	2.31	22.57	1.92	bld	bld
Fluoranthene	11.65	385.36	4.45	bld	21.66	bld	bld	7.65	66.02	18.39	210.6	10.01	7.7	54.72
Pyrene	9.55	161.49	5.59	bld	11.83	bld	79.7	5.9	24.85	9.49	135.68	6.14	5.74	17.88
Benzo(*a*)anthracene	2.66	110.11	2.85	bld	5.31	0.93	39.49	bld	15.33	4.59	66.22	4.03	3.37	13.3
Chrysene	7.28	113.54	3.62	3.79	6.21	0.43	54.19	7.29	19.42	6.26	71.34	6.96	4.41	20.68
Benzo(*b*)fluoranthene	3.14	91.56	4.15	bld	3.75	bld	24.53	2.19	11.71	3.9	28.61	2.52	bld	5.58
Benzo(*k*)fluoranthene	bld	31.99	4.76	bld	3.47	bld	18.46	1.45	12.38	3.96	22.09	2.11	bld	4.47
Benzo(*a*)pyrene	3.72	66.21	3.46	1.76	4.7	1.48	19.58	bld	10.08	3.39	26.21	3.57	1.97	6.87
Indeno(1,2,3-cd)pyrene	bld	37.30	2.49	bld	3.21	1.02	11.66	bld	7.18	2.93	11.41	bld7	bld	bld
Dibenzo(*a,h*)anthracene	bld	13.17	1.65	bld	1.61	bld	3.8	bld	2.66	1.72	4.38	bld	bld	bld
Benzo(g,h,i)perylene	2.08	31.91	2.27	blq	3.22	bld	13.41	1.73	8.11	2.72	9.91	bld	bld	1.84
∑PAHs	*137.4*	*1211*	*35.29*	*30.12*	*84.07*	*3.86*	*382.4*	*44.95*	*290.5*	*87.48*	*783.6*	*46.6*	*23.19*	*170.1*
4,4′-DDE	bld	bld	bld	bld	3.43	bld	bld	bld	bld	bld	bld	bld	bld	21.26
2,4′-DDD	bld	0.62	bld	bld	bld	bld	bld	bld	bld	bld	bld	bld	bld	bld
4,4′-DDD	bld	2.43	bld	bld	bld	bld	bld	bld	bld	bld	bld	bld	bld	bld
4,4′-DDT	bld	BLD	bld	bld	26.7	7.50	70.83	bld	27.31	bld	bld	bld	bld	bld
PCB101	2.77	bld	bld	bld	bld	bld	bld	bld	bld	bld	bld	bld	bld	bld
PCB138	5.33	bld	bld	bld	bld	bld	bld	bld	bld	bld	bld	3.19	3.24	bld
PCB118	bld	8.35	bld	bld	bld	bld	3.12	bld	bld	bld	bld	bld	bld	bld
PCB149	5.79	bld	bld	bld	bld	bld	bld	bld	bld	bld	bld	bld	bld	bld
PCB153	3.28	bld	bld	bld	bld	bld	bld	bld	bld	bld	bld	bld	bld	bld
PCB170	bld	5.91	bld	bld	bld	bld	bld	bld	bld	bld	bld	bld	bld	bld
PCB180	bld	0.52	bld	bld	bld	bld	bld	bld	bld	bld	bld	bld	bld	2.68
ΣOCCs	*17.17*	*17.83*	*0*	*0*	*30.13*	*7.5*	*73.95*	*0*	*27.31*	*0*	*0*	*3.19*	*3.24*	*23.94*

bld below limit of detection

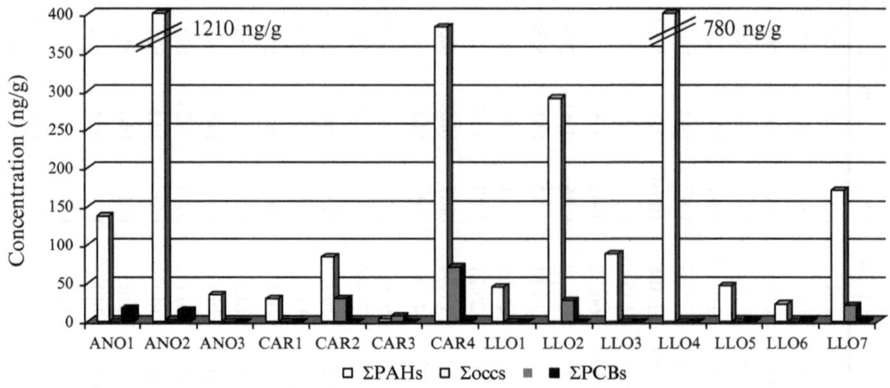

Fig. 4 Total concentrations of PAHs, OCCs and PCBs in sediment samples

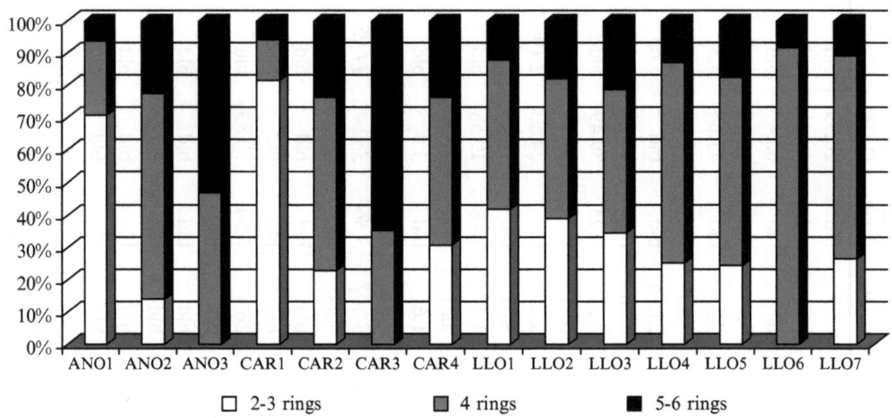

Fig. 5 Ring number composition for PAHs in sediment samples

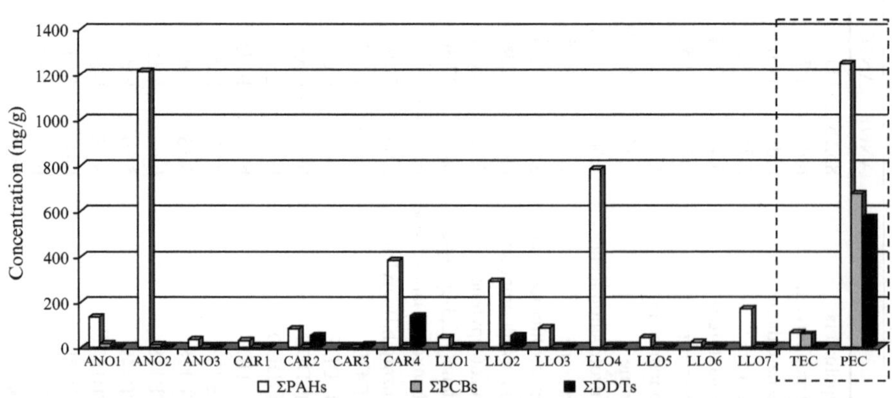

Fig. 6 Sum of PAHs, PCBs, DDTs and the respective quality standard approaches, TEC and PEC

3.1 Toxicity Potency of PAHs in Sediment Samples

A great deal of emphasis has been placed on evaluating the toxic potency of dioxin in different environmental samples. The commonly named dioxin-like compounds (DLCs), such as PCDDs, PCDFs, PCBs and polychlorinated naphthalenes (PCNs), were studied to determine their relative toxic potency. Some studies have indicated that another group of contaminants, the PAHs, can dominate estimates of toxic equivalent quantity (TEQ) for samples containing PAHs and DLCs. The knowledge of the relative contribution of each contaminant to the total dioxin-like activity associated with environmental samples could aid in identifying the most important contributory pollutants [28].

The concentration of each PAH species detected in Llobregat sediment samples was converted to the TEQ concentration based on the toxicity equivalent factors (TEFs) with respect to 2,3,7,8-tetrachlorodibenzo-*p*-dioxin (2,3,7,8-TCDD). Thus, TEQ values express PAHs concentrations as a single number, equivalent to that a toxicity derived exclusively from 2,3,7,8-TCDD. Several substantially different TEF schemes for PAHs are available. TEF values derived vary depending upon the type of bioassays systems, end points and calculation methods for relative potencies. Three TEF_{PAHs} approaches (Table 4) have been used in this study to assess the PAHs toxicity. Willett et al. [29] determined the induction potency of PAHs with respect to 2,3,7,8-TCDD in rat hepatoma H4IIE cells. Clemons et al. [30] examined the ability of PAHs to induce AhR-mediated luciferase activity in mouse hepatoma cells and Klimm et al. [31] determined TEFs by comparing the induction of ethoxyresorufin-*O*-deethylase (EROD) activity by PAHs standards with those of a 2,3,7,8-TCDD standard.

The TEQ concentration of PAH compound "*i*" in the samples was calculated as follows:

$$\mathrm{TEQ}_i = C_{\mathrm{PAH}i} \times \mathrm{TEF}_{\mathrm{PAH}i},$$

where $C_{\mathrm{PAH}i}$ is the concentration of PAHs found in the analysed samples.

Table 4 Three TEF_{PAH} approaches used for calculate TEQ associated with PAHs

Compound	$TEFs_{PAHs}$ [29]	$TEFs_{PAHs}$ [30]	$TEFs_{PAHs}$ [31]
Anthracene	–	0.0001	–
Benzo(*a*)anthracene	0.000025	0.00001	0.000027
Benzo(*a*)pyrene	0.000354	0.00001	0.0003
Benzo(*b*)fluoranthene	0.00253	–	0.00038
Benzo(*k*)fluoranthene	0.00478	0.05	0.00029
Chrysene	0.0002	0.01	–
Dibenzo(*a*,*h*)anthracene	0.00203	0.05	0.000078
Indeno(1,2,3-cd)pyrene	0.0011	–	0.000086
2,3,7,8-TCDD	1	1	1

Table 5 TEQ_{PAH} (ng/g) values obtained applying the different TEF values proposed in the literature

Compound	TEQ_{PAHs} (TEFs from [29])	TEQ_{PAHs} (TEFs from [30])	TEQ_{PAHs} (TEFs from [31])
ANO1	0.0108	0.0731	0.0024
ANO2	0.5013	3.3977	0.0711
ANO3	0.0414	0.3568	0.0044
CAR1	0.0014	0.0379	5.3E-04
CAR2	0.0359	0.3164	0.0044
CAR3	0.0018	0.0043	5.6E-04
CAR4	0.1896	1.6572	0.0229
LLO1	0.0139	0.1454	0.0013
LLO2	0.1099	0.9469	0.0123
LLO3	0.0381	0.3469	0.0042
LLO4	0.2246	2.0401	0.0283
LLO5	0.0192	0.1754	0.0027
LLO6	0.0017	0.0442	6.8E-04
LLO7	0.0424	0.4305	0.0058

Table 5 shows the results of TEQ_{PAHs} obtained for 14 sediment samples from the Llobregat River, by using the three TEF_{PAHs} foregoing approaches proposed in the literature [29–31]. TEQ_{PAHs} differ significantly depending on the TEF_{PAHs} method used for calculations: the lowest values were obtained applying TEFs established by Klimm et al. [31], followed by those of Willett et al. [29], and finally, the highest values were obtained employing Clemons et al. approach [30].

Quality objectives for TEQs have been formulated in order to assess the quality of freshwater and coastal sediments, resulting in a safe sediment value of 20 pg TEQ/g dw [32]. Calculated $TEQs_{PAHs}$ values for all samples are below the suggested safe sediment value, but for a full risk evaluation concentrations for dioxin-like compounds (e.g., PCBs, PCDDs, and PCDFs) would be required.

The work showed that (a) exposure assessment based on total concentrations of PAHs in Llobregat River basin produced different outcomes from those that were based on the TEQ of individual PAHs, (b) the outcomes are influenced by the scale of TEFs used for the TEQ calculation and (c) it is recommended to use different available TEFs in order to obtain a more comprehensive picture about the risk of exposure of pollutants at a certain site.

4 Conclusions

Monitoring programmes suppose a powerful strategy in order to assess the chemical quality of the aquatic environments. POPs are cause of special concern since they are very ubiquitous in the environment and may pose a risk for wildlife and thus for human health. Even though analysis methods for routine POPs monitoring were available several years ago, they are still key contaminants deserving maximum interest from an environmental point of view. Therefore, analytical methodological improvements must be oriented to enhance sample throughput, to minimize sample manipulation, decrease solvent consumption and increase overall method efficiency in terms of selectivity and sensitivity.

ACA studies have evidenced that water bodies of Llobregat River do not present significant concentration of dissolved POPs. In contrast, other studies reveal that sediments constitute a sink for POPs where owing to their hydrophobic characteristics they may remain adsorbed in organic matter during years. Several studies have reported the presence of brominated flame retardants (BFRs), chlorinated pesticides, PCBs and PAHs in both coastal (next to Llobregat mouth) and bottom river sediment samples.

This overview has been complemented with the results from a recent monitoring campaign carried out in the framework of the SCARCE project. In this study, 27 out of the 52 POPs analysed were detected in sediment samples. Results evidence a PAHs widespread contamination and in many sampling points chlorinated compounds from both agricultural and industrial origin were also detected. Efforts to combine analytical measurements with other disciplines such as ecotoxicology, ecology and hydrology are currently ongoing in the framework of the SCARCE project in order to better understand the fate and ecological effects caused by POPS and other contaminants on the aquatic ecosystem of the Llobregat River basin.

References

1. Hardell L, van Bavel B, Lindstrom G, Carlberg M, Dreifaldt AC, Wijkstrom H, Starkhammar H, Eriksson M, Hallquist A, Kolmert T (2003) Increased concentrations of polychlorinated biphenyls, hexachlorobenzene, and chlordanes in mothers of men with testicular cancer. Environ Health Perspect 111:930–934
2. Zhan W, Xu Y, Li AH, Zhang J, Schramm KW, Kettrup A (2000) Endocrine disruption by hexachlorobenzene in Crucian carp (Carassius auratus gibelio). Bull Environ Contam Toxicol 65:560–566
3. Beard J, Australian Rural Hlth Res C (2006) DDT and human health. Science of the Total Environment 355:78–89
4. Commission of the European Communities, "Community Strategy for Endocrine Disrupters", SEC(2007) 1635
5. International Agency for Research on Cancer (IARC) (1983) Benzo[a]pyrene, polynuclear aromatic compounds, Part 1, chemical, environmental and experimental data, vol 32, Monographs on the evaluation of the carcinogenic risk of chemicals to humans. IARC, Lyon, pp 211–224
6. Kapustka LA (2004) Establishing Eco-SSLs for PAHs: lessons revealed from a review of literature on exposure and effects to terrestrial receptors. Hum Ecol Risk Assess Int J 10:185–205
7. Ross JA, Nesnow S (1999) Polycyclic aromatic hydrocarbons: correlations between DNA adducts and ras oncogene mutations. Mutat Res Fund Mol Mech Mutagen 424:155–166
8. National Toxicology Program (NTP) (2005) Report on Carcinogens, eleventh ed. Public Health Service, US Department of Health and Human Services, Washington, DC

9. IARC (1983) IARC monographs on the evaluation of carcinogenic risk of chemicals to humans, Polynuclear aromatic compounds, Part 1, chemical, environmental and experimental data, vol 32. International Agency for Research on Cancer, Lyon
10. Wild SR, Jones KC (1995) Polynuclear aromatic-hydrocarbons in the United-Kingdom environment - a preliminary source inventory and budget. Environ Pollut 88:91–108
11. Blackburn GM, Kellard B (1986) Chemical carcinogens–Part II. Chem Ind 20:687–695
12. IARC (1987) IARC monographs on the evaluation of the carcinogenic risk of chemicals to humans, Suppl.7, vol 1–42, Overall evaluations of carcinogenicity: an updating of IARC monographs. IARC, Lyon
13. Howell NL, Suarez MP, Rifai HS, Koenig L (2008) Concentrations of polychlorinated biphenyls (PCBs) in water, sediment, and aquatic biota in the Houston Ship Channel, Texas. Chemosphere 70:593–606
14. Covaci A, Gheorghe A, Hulea O, Schepens P (2006) Levels and distribution of organochlorine pesticides, polychlorinated biphenyls and polybrominated diphenyl ethers in sediments and biota from the Danube Delta, Romania. Environ Pollut 140:136–149
15. Wania F, Mackay D (1993) Global fractionation and cold condensation of low volatility organochlorine compounds in polar regions. Ambio 22:10–18
16. Mackay D, Di Guardo A, Paterson S, Cowan CE (1996) Evaluating the environmental fate of a variety of types of chemicals using the EQC model. Environ Toxicol Chem 15:1627–1637
17. McKone TE. 1993. CalTOX, a multimedia total exposure model for hazardous-waste sites; Part 1, Executive summary. UCRL-CR-111456-Pt.1; Other: ON: DE93040847 United States10.2172/139702Other: ON: DE93040847Tue Dec 22 07:26:16 EST 2009INIS; OSTI as DE93040847LLNL; SCA: 540220; 540120; 560300; 990200; PA: EDB-93:144177; INS-93:021795; NTS-94:003565; ERA-18:037186; SN: 93001055714English
18. Brandes LJ, de Hollander H, van de Meent D (1996) SimpleBox 2.0: a nested multimedia fate model for evaluating the environmental fate of chemicals. RIVM Report No 719 101 029; Bilthoven, The Netherlands
19. Institute of Public Health and the Environment, The Netherlands (1996) EUSES—The European Union System for the evaluation of substances. European Chemicals Bureau, Ispra
20. ACA, Agencia catalana de l'aigua (Catalan water agency). Available at: http://aca-web.gencat.cat/aca/appmanager/aca/aca/ (Accessed on Nov 2011)
21. Navarro-Ortega A, Tauler R, Lacorte S, Barceló D (2010) Occurrence and transport of PAHs, pesticides and alkylphenols in sediment samples along the Ebro River Basin. J Hydrol 383:5–17
22. Eljarrat E, Caixach J, Rivera J, de Torres M, Ginebreda A (2001) Toxic potency assessment of non- and mono-ortho PCBs, PCDDs, PCDFs, and PAHs in Northwest Mediterranean Sediments (Catalonia, Spain). Environ Sci Technol 35:3589–3594
23. Eljarrat E, De La Cal A, Larrazabal D, Fabrellas B, Fernandez-Alba AR, Borrull F, Marce RM, Barcelo D (2005) Occurrence of polybrominated diphenylethers, polychlorinated dibenzo-*p*-dioxins, dibenzofurans and biphenyls in coastal sediments from Spain. Environ Pollut 136:493–501
24. Eljarrat E, Caixach J, Rivera J (2001) Evaluation of dioxin contamination in sewage sludge discharges on coastal sediments from catalonia, spain. Water Res 35:2799–2803
25. Benner BA, Gordon GE, Wise SA (1989) Mobile sources of atmospheric polycyclic aromatic hydrocarbons: a roadway tunnel study. Environ Sci Technol 23:1269–1278
26. Lee B, Yi J (1999) A statistical approach for determining the environmental impact of polynuclear aromatic hydrocarbons in an oil spill-contaminated coastal area. Environ Pollut 105:391–396
27. MacDonald DD, Ingersoll CG, Berger TA (2000) Development and evaluation of consensus-based sediment quality guidelines for freshwater ecosystems. Arch Environ Contam Toxicol 39:20–31
28. Eljarrat E, Barceló D (2003) Priority lists for persistent organic pollutants and emerging contaminants based on their relative toxic potency in environmental samples. Trends Analyt Chem 22:655–665

29. Willett KL, Gardinali PR, Sericano JL, Wade TL, Safe SH (1997) Characterization of the H4IIE Rat hepatoma cell bioassay for evaluation of environmental samples containing polynuclear aromatic hydrocarbons (PAHs). Arch Environ Contam Toxicol 32:442–448
30. Clemons JH, Allan LM, Marvin CH, Wu Z, McCarry BE, Bryant DW, Zacharewski TR (1998) Evidence of estrogen- and TCDD-like activities in crude and fractionated extracts of PM10 air particulate material using in vitro gene expression assays. Environ Sci Technol 32:1853–1860
31. Klimm CH, Hofmaier AM, Schramm K-W, Kettrup A (1999) Using TEF concept for assessing toxic potency of polycyclic aromatic hydrocarbons in industrial samples. Organohalogen Compd 40:39–42
32. Evers EHG, Laane RWPM, Groeneveld GJJ, Olie K (1996) Levels, temporal trends and risk of dioxins and related compounds in the Dutch aquatic environment. Organohalogen Compd 28:117–122

Occurrence and Behavior of Brominated Flame Retardants in the Llobregat River Basin

E. Eljarrat and D. Barceló

Abstract This chapter summarizes brominated flame retardant research in the Llobregat River basin. The work included the analysis of sediments and biota, with special attention on aspects such as occurrence, bioavailability, and bioaccumulation of these contaminants. The previous studies were focused only on the PBDE contamination issue; however, some emerging BFRs, such as HBB, PBEB, and DBDPE, were also included into the last works. For river sediment samples PBDE levels were between 2 and 10 ng/g dw, whereas coastal sediments showed higher concentrations, with values ranging between 25 and 134 ng/g dw. For both river and coastal sediments, the contribution of deca-BDE-209 on the total PBDE contamination was high, ranging between 50% and 99%. As regards biota samples, and more specifically, feral carps, PBDE levels ranged between 29 and 744 ng/g lw. Differences in PBDE profiles between sediments and fish were noticed, and they can be attributed to differential bioavailability or due to biotransformation processes. Finally, emerging BFRs were also detected in sediment samples, but at lower concentration levels than those observed for PBDEs.

Keywords Bioaccumulation, Bioavailability, Brominated flame retardants, Decabromodiphenylethane, Hexabromobenzene, Pentabromoethylbenzene, Polybrominated diphenyl ether

---

E. Eljarrat (✉)
Department of Environmental Chemistry, IDAEA, CSIC, Jordi Girona 18-26, 08034 Barcelona, Spain
e-mail: eeeqam@cid.csic.es

D. Barceló
Department of Environmental Chemistry, IDAEA, CSIC, Jordi Girona 18-26, 08034 Barcelona, Spain

Catalan Institute for Water Research (ICRA), Parc Científic i Tecnològic de la Universitat de Girona, Pic de Peguera 15, 17003 Girona, Spain

King Saud University, P.O. Box 2455Riyadh 11451, Saudi Arabia

S. Sabater et al. (eds.), *The Llobregat: The Story of a Polluted Mediterranean River*,
Hdb Env Chem (2012) 21: 135–150, DOI 10.1007/698_2011_139,

Contents

Abbreviations and Symbols

BFR	Brominated flame retardant
DBDPE	Decabromodiphenylethane
dw	Dry weight
EROD	7-ethoxyresorufin O-deethylase
GC	Gas chromatography
HBB	Hexabromobenzene
LOD	Limit of detection
lw	Lipid weight
MS	Mass spectrometry
NCI	Negative chemical ionization
PBDE	Polybrominated diphenyl ether
PBEB	Pentabromoethylbenzene
PCB	Polychlorinated biphenyl
PCDD	Polychlorinated dibenzo-*p*-dioxin
PCDF	Polychlorinated dibenzofuran
PLE	Selective pressurized liquid extraction
POP	Persistent organic pollutant
WWTP	Wastewater treatment plant

1 Introduction

Currently, the contamination in aquatic environment includes, together with other persistent organic pollutants (POPs), a variety of brominated flame retardants (BFRs). BFRs are comprised of diverse classes or chemical compounds used in a variety of commercial applications. They are used in plastics, textiles, electronic circuitry, and other materials to prevent fires. The estimated global consumption of BFRs shows that their usage is on the rise [1]. Similarly to other POPs, BFRs appear to be lipophilic and bioaccumulate in biota and humans.

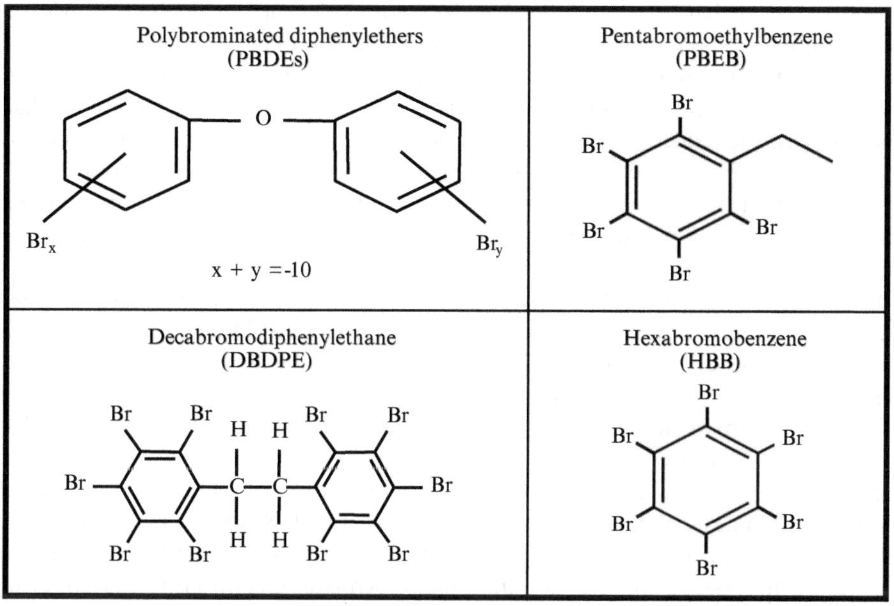

Fig. 1 Chemical structures of PBDEs and emerging BFRs, PBEB, HBB, and DBDPE

The studies performed in the last 15 years have mainly concerned polybrominated diphenyl ethers (PBDEs) (Fig. 1) [2, 3]. PBDEs were typically produced at three different degrees of bromination, i.e., Penta-BDE, Octa-BDE, and Deca-BDE mixtures. Penta-BDE formulation consists of 41–42% tetra-BDEs (mainly BDE-47) and 44–45% penta-BDEs (predominantly BDE-99 and BDE-100), whereas deca-BDE formulation consists mainly of BDE-209 (97–98%), with a small amount of nona-BDES (0.3–3%) [4]. BDE-183 is often taken as indicative of the presence of the octa-BDE formulations, in which hexa-BDE-153 and hexa-BDE-154 were also detected [5].

PBDEs were first discovered in the environment in 1981 [6] when they were found in pike from western Sweden. Subsequent reports, based on analyses of sediments, document the ubiquitous distribution of PBDEs in the environment. Considerable data is available on the levels and composition of PBDEs in the environment, biota, and human samples [7, 8]. Structural similarity to other environmental chemicals with known toxic effects, such as polychlorinated biphenyls (PCBs), polychlorinated dibenzo-*p*-dioxins (PCDDs), and polychlorinated dibenzofurans (PCDFs), could also indicate that PBDEs could be harmful to health. Since 1994, PBDEs are listed as compounds that can affect the regulation of thyroid and steroid hormones. Several studies indicate that commercially obtained tetra- and penta-BDEs are endocrine disrupters, which can exert effects on the thyroid system. The effects of penta-BDEs on thyroxine and the thyroid gland are considered to be principally due to the induction of liver enzymes, although several mechanisms may operate. The liver appears to be sensitive, and for penta-BDEs, a no-observed-adverse-effect level of

1 mg/kg bw/day has been determined, with effects evident at 2 mg/kg bw/day. Meerts et al. [9] have reported estrogenic activities on PBDEs and hydroxylated PBDEs as determined in the human T47D breast tumor cell line stably transfected with estrogen responsive luciferase reporter gene construct. A commercial Penta-PBDE mixture has been reported to reduce circulating thyroxine and to induce rat liver 7-ethoxyresorufin O-deethylase (EROD) activity in the parent animal as well as in the offspring [10].

Due to their toxicological effect, the production and use of PBDE formulations have been banned in Europe, and Penta- and Octa-BDE formulations are now banned in North America. In response to increasing international regulations on BFR formulations, alternative additive flame retardants for achieving commercial product fire safety standards are being developed and used [11]. Some of these non-BDE BFRs are decabromodiphenylethane (DBDPE), hexabromobenzene (HBB), and pentabromoethylbenzene (PBEB) (Fig. 1).

DBDPE is marketed as an alternative for technical Deca-BDE. The molecular structures of the two chemicals resemble each other and they have similar applications, i.e., as additive BFRs in a wide range of polymeric materials. DBDPE became available on the market in the mid-1980s [12]. One reason for the commercialization of DBDPE was that, in contrast to Deca-BDE, DBDPE is less prone to produce dioxins and furans under pyrolysis conditions [13]. HBB is an additive flame retardant, historically used not only in plastics, textiles, and wood but also in paper and electric manufactured goods. Apart from being commercially produced, there are other possible sources for HBB to the environment. HBB was found to be the major product from pyrolysis of Octa- and Deca-BDE technical products [14]. As regards PBEB, their main applications are related to polyester resins, circuit boards, textiles, adhesives, wire and cable coatings, and polyurethane foam. An additional identified source of PBEB emissions was from polymeric BFRs. PBEB was released from the polymer at room temperature and the emissions increased several orders of magnitude when the polymer was exposed to thermal stress, a situation that normally occurs during compounding of thermoplastic polyesters [14].

PBEB, HBB, and DBDPE have been detected in environmental samples from Europe and North America, including sediments [15, 16], sewage sludge [17], air [18], house dust [19], blood, and eggs of gulls [11]. In addition, there is little or no toxicological data available for wildlife or humans for these additive BFRs.

2 Area of Study

The study area is located in the northeast of Spain, along the Llobregat River basin (Fig. 2). The Llobregat River comes from the northwest of Catalonia and flows into the Mediterranean Sea, close to Barcelona city. It is 156.5 km in length and covers a catchment area of approximately 4,948 km^2 of basin surface. The Llobregat River watershed is heavily populated (3,089,465 inhabitants in 1999). Together with its two main tributaries, the Cardener and Anoia Rivers, the Llobregat River is a

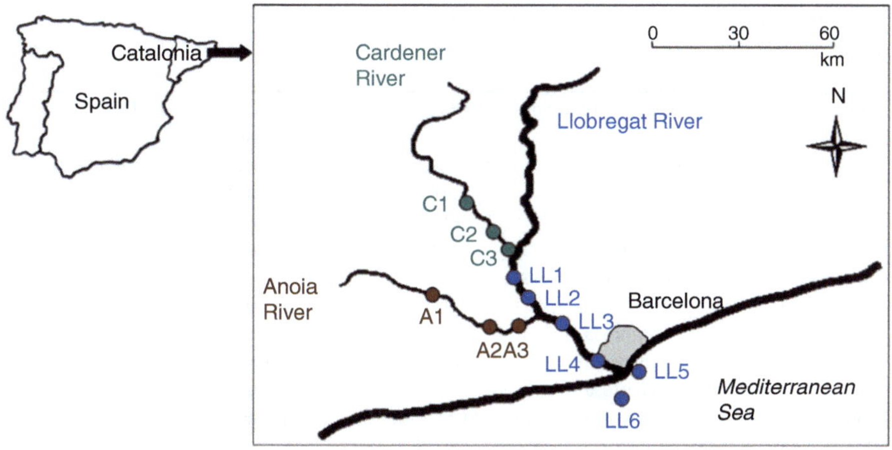

Fig. 2 Geographical location of the area of study and sampling stations along the Llobregat River and their tributaries, the Anoia and Cardener Rivers

paradigm of overexploited Mediterranean Rivers. The river has a mean annual discharge of 693,000,000 m^3, and nearly 30% is used for drinking water. It receives extensive urban and industrial wastewater discharges (137,000,000 m^3/year; 92% comes from the wastewater treatment plants (WWTPs)) that cannot be diluted by its natural flow (0.68–6.5 m^3/s basal flow).

Different sampling stations were selected along the Llobregat River as well as at the Anoia and Cardener Rivers (Table 1). Three sampling sites were selected at the Anoia River, including one sampling site (A1) located 1.5 km upstream of Igualada and two sampling sites (A2 and A3) situated 23 and 27 km downstream, respectively. Igualada is a heavily industrialized town with a very important textile and tannery industry. Regarding the Cardener River, one site (C1) was selected 5 km upstream of Manresa, whereas two sites (C2 and C3) were situated 4 and 8 km, respectively, downstream from Manresa. Manresa is also a heavily industrialized town with an important rubber industry. Finally, four sampling sites were selected from the middle and lower parts of the Llobregat River main channel. These sites were part of a pollution gradient: site LL1 (Castellvell i el Vilar) was the least polluted but received some industrial effluents and surface runoff from agricultural areas; sites LL2 (Martorell) and LL3 (Abrera) were located in a densely inhabited area and received urban and industrial wastewater inputs; site LL4 (Sant Joan Despí) was located in the last section of the river and was the most polluted site. Sampling stations were also selected at the mouth of the Llobregat River in Barcelona (LL5 and LL6).

Different sampling campaigns were carried out from 2000 to 2006 (Table 1). Samples consisted of a series of sediment and feral carp. Surficial sediments (0–2 cm) were collected from the middle river bed using a Van Veen drag at each selected site. Sediments were freeze dried, lyophilized, ground, and homogenized by sieving through a stainless steel 0.2-mm sieve, and stored in sealed containers at

Table 1 Sampling campaigns carried out in the area of study between 2000 and 2006

River	Sample code	Location site	Sampling date	Collected samples	Selected analytes
Anoia	A1	Upstream Igualada	2000, 2005	3 sediments, 6 feral carp	PBDEs, emerging BFRs
	A2	Downstream Igualada	2000, 2005, 2006	4 sediments, 10 feral carp	PBDEs, emerging BFRs
	A3	Downstream Igualada	2000, 2005, 2006	4 sediments, 10 feral carp	PBDEs, emerging BFRs
Cardener	C1	Upstream Manresa	2000	7 feral carp	PBDEs
	C2	Downstream Manresa	2000	5 feral carp	PBDEs
	C3	Downstream Manresa	2000	1 sediment, 5 feral carp	PBDEs
Llobregat	LL1	Castellvell i el Vilar	2005, 2006	3 sediments	PBDEs, emerging BFRs
	LL2	Martorell	2005, 2006	2 sediments	PBDEs, emerging BFRs
	LL3	Abrera	2005, 2006	3 sediments	PBDEs, emerging BFRs
	LL4	Sant Joan Despí	2005, 2006	3 sediments	PBDEs, emerging BFRs
	LL5	Llobregat River mouth	2002	1 sediment	PBDEs
	LL6	Outflow of wastewater	2002	3 sediments	PBDEs

−20°C until analysis. Feral carp (*Cyprinus carpio*) specimens were collected by DC electric pulse. Fish were killed, weighted, and the fork length of each fish was measured. Muscle samples were collected from below the dorsal fin, then lyophilized and homogenized, and preserved frozen at −20°C until analysis.

3 Analytical Methodology

Forty-seven PBDE congeners were included in the analytical work: 3 mono-BDEs (1,2,3), 7 di-BDEs (7,8,10,11,12,13,15), 8 tri-BDEs (17,25,28,30,32,33,35,37), 6 tetra-BDEs (47,49,66,71,75,77), 7 penta-BDEs (85,99,100,116,118,119,126), 5 hexa-BDEs (138,153,154,155,166), 3 hepta-BDEs (181,183,190), 4 octa-BDEs (194,196,197,203), 3 nona-BDEs (206,207,208), and the deca-BDE (209). Additionally, in some samples the analytical methodology also included the determination of the emerging BFRs, PBEB, HBB, and DBDPE.

3.1 Sample Preparation

One gram dry weight (dw) for sediment and muscle samples were spiked with internal standards (tetra-BDE-77, hepta-BDE-181, and ^{13}C-BDE-209) (Wellington Labs., Guelph, Ontario, Canada). Spiked samples were kept overnight to equilibrate. A selective pressurized liquid extraction (SPLE) [20] was carried out using a fully automated ASE 200 system (Dionex, Sunnyvale, CA, USA). Extracts were concentrated to incipient dryness and redissolved with 50 μL of toluene prior to the analysis by gas chromatography (GC) coupled to mass spectrometry (MS), working with negative chemical ionization (NCI).

Determination of lipid content in fish material was carried out by PLE of additional subsample and further gravimetric measurement.

3.2 Instrumental Analysis

GC-NCI-MS analyses were performed on a gas chromatograph Agilent 6890 connected to a mass spectrometer Agilent 5973 Network (Agilent Technologies España, Madrid, Spain). A DB-5ms (15 m $\times$ 0.25 mm, i.e., 0.1 μm film thickness) capillary column was used for the determination of congeners from mono- to hepta-BDEs [21, 22].

Quantification was carried out by internal standard procedure, for mono- to penta-BDEs, PBEB, and HBB using BDE-77 as internal standard, for hexa- to hepta-BDEs and DBDPE using BDE-181 as internal standard, and for deca-BDE-209 using ^{13}C-BDE-209 as internal standard.

Using the described methodology, recoveries ranged from 53% to 84% for sediment, and between 52 and 103% for muscle fish samples. Limits of detection (LODs) were in the range of 6–86 pg/g dw for sediments, and in the range of 77–736 pg/g lipid weight (lw) for fish samples. Relative standard deviations of the method were in the range of 1–13%.

4 Results and Discussion

4.1 Sediment

4.1.1 River Sediments

The first PBDE results obtained at the Llobregat River basin corresponded to the study carried out with samples collected during 2000 [23]. PBDEs were detected in all the sediment samples collected along the Anoia and Cardener Rivers at concentrations ranging from 2 to 10 ng/g dw (Table 2). Of 40 congeners included

Table 2 Results (expressed in ng/g dw) obtained for sediment samples collected along the Llobregat River basin

	Anoia						Cardener	Llobregat					
	2000			2005–2006			2000	2005–2006				2002	
	A1 ($n = 1$)	A2 ($n = 1$)	A3 ($n = 1$)	A1 ($n = 2$)	A2 ($n = 3$)	A3 ($n = 3$)	C3 ($n = 1$)	LL1 ($n = 3$)	LL2 ($n = 2$)	LL3 ($n = 3$)	LL4 ($n = 3$)	LL5 ($n = 1$)	LL6 ($n = 3$)
BDE-47	0.28	0.40	0.23	nd	3.20	8.7	0.26	nq	1.40	nq	2.60	0.07	0.11
BDE-100	0.35	0.39	0.07	nq	nq	nq	0.36	nq	nq	nq	1.00	0.10	0.11
BDE-99	0.78	0.98	0.03	nd	nq	3.3	0.78	nq	nq	nq	10.0	0.14	0.18
BDE-154	nd	0.98	nd	nd	nd	nd	nd	nd	nd	nd	nd	nd	0.07
BDE-153	1.06	1.15	0.77	nq	nd	nd	1.04	9.00	nq	4.4	nq	0.24	0.27
BDE-183	nd	1.23	1.09	nd	nd	nd	0.48	nd	nd	nd	nd	0.40	0.69
BDE-209	nq	4.62	3.96	3.65	15.7	34.7	na	8.47	47.5	13.9	35.3	2.95	76.4
ΣPBDEs[a]	2.47	9.75	6.15	3.65	16.7	38.7	2.92	17.5	48.2	15.4	39.9	3.90	77.8
PBEB	na	na	na	7.70	6.35	7.50	na	5.57	4.70	4.77	8.53	na	na
HBB	na	na	na	nq	nq	0.40	na	nd	2.20	nq	2.40	na	na
DBDPE	na	na	na	12.1	21.5	17.7	na	8.47	17.5	10.7	15.0	na	na

na not determined, *nd* below limit of detection, *nq* below limit of quantification
[a]Total PBDE values were calculated assuming that *nd* = *nq* = 0

in the analytical work, seven different PBDEs were detected, ranging from tetra- to decabrominated compounds: tetra-BDE-47, penta-BDE-99, penta-BDE-100, hexa-BDE-153, hexa-BDE-154, hepta-BDE-183, and deca-BDE-209. For sediments collected at sites A2 and A3, BDE-209 had the highest concentration of the seven PBDEs measured. This compound is generally observed to be the most abundant congener found in sediment samples. The remaining PBDE burden in sediments was primarily comprised of congeners BDE-99, BDE-153, and BDE-183. Contamination levels were higher in sediment samples collected downstream of Igualada (A2 and A3). These higher values were mainly due to BDE-209 levels, which was below the limit of quantification in the sediment sample upstream of Igualada (A1), whereas presented values around 4 ng/g dw in the two samples downstream Igualada (A2 and A3). It should be pointed out that this trend could be attributed to the application of the Deca-BDE mixture in the textile and tannery industries located in the surroundings of this area.

Our PBDE results were slightly lower than those reported in other river sediment studies. PBDEs were determined in sediments along the Spanish River Cinca at concentrations ranging from 2 to 42 ng/g dw [24]. PBDEs were also determined in Swedish river sediments at 8–50 ng/g dw [25]. Similar values were found in Japanese river sediments, with concentration levels between 21 and 59 ng/g dw [26]. PBDEs were also determined in samples from 32 sites along Danube River with PBDE concentrations between 0.06 and 84 ng/g dw [27]. Higher levels up to 1,400 ng/g dw were found in a downstream area of a foam manufacturing plant who is user of PBDEs in United Kingdom [28] and at 120 ng/g dw downstream of an area with textile industries [25].

After the study cited above, the following available data on BFR contamination levels in sediments from the Llobregat River basin corresponded to the study of samples collected between 2005 and 2006 [29]. In this case, PBDEs were detected at concentrations ranging from 8 to 83 ng/g dw (Table 2). Site LL2 was the most contaminated zone followed by site LL4 and A3. These sediment samples were also subjected to a monitoring study of different priority and emerging contaminants, and also LL2, LL4, and A3 sites presented the highest levels of pollution [30]. High levels at site A3 could be attributed to the application of PBDEs in the textile and tannery industries located in the surroundings of this area. As regards, LL2 is under the influence of the effluents of a WWTP located next to the sampling point.

Of 40 PBDEs congeners included in our analytical work, the dominance of deca-BDE-209 in the total PBDEs is unquestionable such as it had been described in literature, which was present at concentrations varying from 3.2 to 82 ng/g dw. In these samples, BDE-209 constituted between 11% and 71% of the total PBDE contamination. The presence of BDE-209 could indicate the use of commercial formulations of Deca-BDE in the studied area.

The study carried out during 2005–2006 included also the determination of some emerging BFRs, such as PBEB, HBB, and DBDPE, in sediment samples. PBEB was detected in 13 of 19 analyzed samples at concentration levels between 3.1 and 10 ng/g dw. HBB was detected just in 3 out of the 19 sediments, at levels ranging from 0.4 to 2.4 ng/g dw. Finally, DBDPE was detected in all samples, at levels between 4.8 and 24 ng/g dw.

Few data are available about these compounds in sediment samples. In the case of PBEB, sediment samples from Michigan lake were analyzed [18], but it was not detected. Thus, the values corresponding to sediments from the Llobregat River basin represented the first data available of PBEB in sediment samples around the world. The current production of this compound is not available, but recent studies reported the presence of relatively high concentrations in Chicago air samples [31].

HBB was widely used in Japan as an additive to paper, plastic, and electronic goods. In sediment samples collected in 1982 and analyzed by the Environmental Agency of Japan, HBB was detected in 3 out of 126 samples (nd – 4.3 ng/g dw) [32]. River sediment samples from Japan collected in 1981–1983 contained HBB at 6–60 ng/g dw, and estuary sediments had lower concentrations (0.5–6 ng/g dw) [33]. HBB is still used in Japan at lower volumes; however, HBB is not reported by the EU industry as a currently produced chemical. Besides, it has to be taken into account that another source of HBB in the environment result from thermal degradation of BDE-209 [31].

DBDPE became commercially important as an alternative to the Deca-BDE formulation in the early 1990s. DBDPE were determined in samples from Western Scheldt River (The Netherlands) at 0.65–9.8 ng/g dw [16]. A higher value was reported by Kierkegaard et al. [34] in sediment from the same river, with a concentration of 24 ng/g dw. Moreover, DBDPE was analyzed in sediments from isolated lakes [35]. Results from the analysis of the lake sediments showed that DBDPE was present in all of the 11 sampled lakes, at levels from 0.23 to 11 ng/g dw. All these values are in accordance with levels found in the Llobregat River. However, higher levels were reported in a study of 15 Chinese river sediment samples, with levels ranging from 19 to 430 ng/g dw [36].

Results of PBDEs and emerging BFRs in the Anoia and Llobregat sediment samples were compared, and a similar distribution was observed in both rivers, being higher the concentration levels of PBDEs (mainly corresponding to BDE-209), followed by DBDPE, PBEB, and finally HBB. As was indicated before, DBDPE applications are similar as for BDE-209 with the advantage of no production of dibenzo-*p*-dioxins and no formation of furans under pyrolysis conditions. Generally, the reported levels of DBDPE were lower than for BDE-209 [34]. In the Llobregat River basin it is observed that in 43% of samples, DBDPE concentrations are higher than BDE-209, with ratios between 1.1 and 3.6. This could be due to the higher input in the use of DBDPE during last years. Although currently DBDPE is not as widely used as BDE-209 because of its higher cost, it is predicted that DBDPE will soon become one of the major flame retardants used by thermoplastics industry. As regards PBEB its contribution represented around the 16% of that of PBDEs, even if its production was low in Europe since 2002. Finally, the contribution of HBB to the total PBDEs corresponded to 2% of that of PBDEs. HBB is not reported by the EU industry as a currently produced chemical. This could be the main reason of its low contribution.

4.1.2 Coastal Sediments

Four coastal sediments collected at the mouth of the Llobregat River in Barcelona were analyzed for determining their PBDE content. PBDE levels ranged from 25 to 134 ng/g dw (Table 2). The congener pattern is dominated by the BDE-209. This congener constituted between 50% and 99% of the total PBDE contamination. This behavior was in accordance with published environmental levels (sediments and sewage sludge), where major congener detected was BDE-209 followed by BDE-47, -99, -100, and -153. Congener patterns observed for samples collected at the mouth of the Llobregat River differed significantly from the Penta-BDE formulation (Fig. 3). Whereas BDE-47, -99, and -100 were detected, high contribution of hexa-BDEs -153 and -154 were also observed. The presence of these hexa-BDEs (-153 and -154) correlated with the Octa-BDE mixture. In this case, we can assume that PBDE congener patterns observed are the result of the use of not only Deca-BDE and Penta-BDE but also Octa-BDE formulations.

Concentration levels in sediments collected at the mouth of the Llobregat River were compared with data about these compounds in coastal sediment samples. Grant et al. [37] determined PBDE levels in marine sediments collected from 41 sites throughout the Strait of Georgia. Total PBDE concentrations ranged from 0.09 to 12.7 ng/g dw, being lower than results obtained in the Llobregat study. The highest concentrations were found near municipal outfalls. The deca-BDE-209 dominated the PBDEs, accounting for 52% of the average total concentration. Ramu et al. [38] determined PBDEs in surface sediments collected from 24 coastal

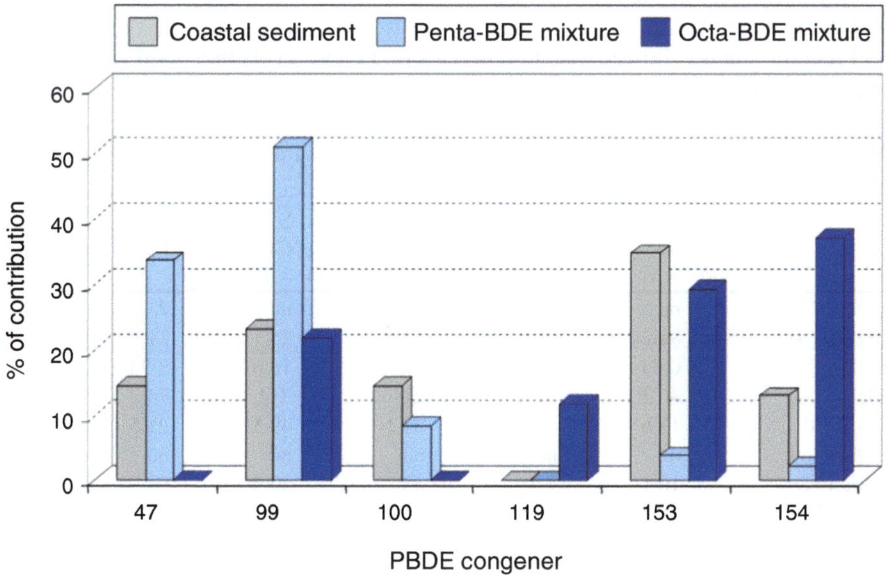

Fig. 3 PBDE congener pattern for the technical mixtures Penta- and Octa-BDE, and for a coastal sediment

locations in Korea. The concentrations of ΣPBDEs (defined as the sum of all targeted PBDE congeners except BDE-209) and BDE-209 ranged from 0.05 to 32 ng/g dw and from 0.40 to 98 ng/g dw, respectively, which are in accordance with levels reported in the Llobregat study. The PBDE congener compositions were also dominated by BDE-209.

4.2 Biota

Data on PBDE contamination levels in biota from the Llobregat River basin corresponded to the study of samples collected during 2000 [23]. PBDEs were detected in the muscle of the different feral carp collected samples at concentrations ranging from 29 to 744 ng/g lw (Table 3). It should be pointed out that the contamination level was similar for both tributaries of the Llobregat River, with slightly higher values found in the Cardener River (mean contamination level of 288 ng/g lw) than in the Anoia River (mean contamination level of 160 ng/g lw). Results were compared with data published in the literature, although studies of PBDE contamination in carp are very scarce. Llobregat results were similar to those reported for carp samples collected in the Des Plaines River, USA [39]. In this work, the whole fish sample was analyzed with levels ranging between 47 and 685 ng/g lw. Lower values were found in carp collected in the Detroit River, USA [39], with values of 29–54 ng/g lw. In contrast, higher levels were found in two carp samples collected at the Hadley lake, USA [40], with contamination between 760 and 2,500 ng/g lw.

Previous studies showed that, at a given site, PBDE levels increased with the age of the fish, indicating bioaccumulation [24]. Fish length is directly related to fish age.

Table 3 Results (expressed in ng/g lw) obtained for feral carp samples collected along the Llobregat River basin

	Anoia			Cardener		
	A1 ($n = 6$)	A2 ($n = 10$)	A3 ($n = 10$)	C1 ($n = 7$)	C2 ($n = 5$)	C3 ($n = 5$)
BDE-25	nd	1.37	3.07	nd	2.19	2.54
BDE-28 + 33	8.62	2.49	11.5	15.1	11.2	21.8
BDE-49	0.64	2.58	12.7	16.6	3.94	11.1
BDE-47	40.2	37.9	211	314	74.2	251
BDE-100	13.6	5.93	31.1	49.5	21.5	39.1
BDE-99	nd	nd	nd	nd	nd	nd
BDE-154	10.1	7.28	33.2	22.0	9.69	18.5
BDE-153	18.8	6.28	15.0	nd	nd	nd
BDE-183	19.4	3.95	14.3	13.9	0.70	nd
ΣPBDEs[a]	82.6	56.6	311	386	111	329

nd below limit of detection

[a]Total PBDE values were calculated assuming that *nd* = 0

Larger fish have normally been exposed for longer periods to POPs, such as PBDEs. In the Llobregat study, there was no clear relationship between PBDE and weight or length at any of the sites, with the exception of site 3 in Cardener River (C3). The regression coefficients at site C3 ($R^2 = 0.94$) indicate a positive relationship. The data in the literature are inconclusive regarding such relationships in carp. Rice et al. [39] concluded that no correlation of PBDE with weight or length was obtained for 24 carp collected from the Detroit and Des Plaines Rivers. In contrast, Loganathan et al. [41] reported increasing PBDE concentrations as a function of age in 8 carp from New York.

Of 40 congeners included in the analytical work, nine different PBDEs were detected, ranging from tri- to heptabrominated compounds: tri-BDE-25 and -28 + 33, tetra-BDE-47 and -49, penta-BDE-100, hexa-BDE-153 and -154, and hepta-BDE-183. However, deca-BDE-209, one of the dominant congeners of the total PBDE contamination in sediments, was not detected in biota samples. Because this compound is very hydrophobic (log $K_{OW} \sim 10$), it has been suggested that BDE-209 has very low bioavailability, although debromination to more bioavailable metabolites has also been suggested to occur in fish tissues. Stapleton et al. [42] studied the debromination of BDE-209 by caged carp following dietary exposure. They concluded that debromination of BDE-209 occurs in carp. Although they did not detect any accumulation of BDE-47 and BDE-99, accumulation of other PBDEs, such as BDE-154, occurs. In the Llobregat work, BDE-154 was only detected in one sediment sample (site A2), whereas this congener was present in the 91% of carp analyzed. The contribution of this congener in the total PBDE contamination in carp from the Anoia and Cardener Rivers ranged from 5% to 22%. This could be attributed partly to the debromination of BDE-209.

Similar relationship may exist between BDE-183 and its debromination product, BDE-154. BDE-183 is another predominant BDE congener in our sediment samples with a contribution in the total BDE burden ranging from 17% to 50% (BDE-209 contribution was not taken into account for these calculations). Stapleton et al. [43] also studied the debromination of BDE-183 in caged carp following dietary exposure. Significant debromination was observed, converting BDE-183 to BDE-154. In Llobregat samples, BDE-183 was only detected in 10 of 43 analyzed carp, indicating that this octa-BDE congener was probably debrominated to hexa-BDE-154 with low presence in sediments, but high contribution in carp samples.

Regarding BDE-99, another important contributing congener in sediment samples (from 19% to 32% of the total), it was not detected in feral carp samples. Some previous studies also showed that BDE-99 does not accumulate in tissues of the common carp [39, 40]. This observation is intriguing considering that this congener typically accounts for 10–60% of the BDE burden in most other biota and is a dominant congener in the commercial mixtures. The most likely reason no appreciable uptake of BDE-99 is observed in carp tissues is due to efficient metabolism. Stapleton et al. [43] studied the debromination of BDE-99 in caged carp following dietary exposure. Significant debromination was observed, converting BDE-99 to BDE-47. In the Llobregat study, the contribution of tetra-BDE-47 in sediment contamination was up to 11%, whereas this congener was the dominant in all the fish samples and contributed 37–90% of the total PBDEs observed. This could be attributed partly to the debromination of BDE-99. The predominance of BDE-47 is consistent with other studies on freshwater fish species [7]. Marked differences in the ratios of BDE-47 to BDE-99 among some examined biological species of the St. Lawrence Estuary food web were observed [5]. BDE-47/99 ratios in some fish species such as rainbow smelts, American eels, American plaice, and smooth flounders were much higher than in other organisms, whereas the BDE-47/100 ratios were similar among all the examined biological species. The depletion in BDE-99 concentrations relative to BDE-47 in some fish samples suggests that some organisms possess the capacity to biotransform this specific congener.

Another interesting difference between PBDE congener distribution in sediments and in carp was observed for hexa-BDE-153, with high contribution in sediments (up to 43%), but low contribution (up to 16%) in carp. Moreover, this congener was only detected in 26% of fish samples. Low levels of BDE-153 were also reported for carp collected from Hadley Lake, USA [40]. Their absence in carp samples could be explained by a low bioavailability potential. However, high fish to sediment ratios have been established for this hexa-BDE congener in fish samples such as barbells [24] or bleaks [44], indicating their bioavailability. Future studies are needed in order to determine the potential transformation or debromination of this congener in carp.

Finally, some lower brominated congeners (tri-BDE-25, 28, and 33, and tetra-BDE-49) were also detected in some carp samples, whereas they were absent in all sediment samples. BDE-25 was detected in 9 of 43 analyzed carp, BDE-28 + 33 were found in 21 of 43 samples and BDE-49 was quantified in 23 specimens. It should be pointed out, however, that the contribution of these congeners in the total PBDE burden was low, ranging from 2% to 17% and 3% to 10% of the total for the Anoia and Cardener samples, respectively. Additional studies are required in order to determine if these congeners come from debromination processes of higher brominated PBDEs or their presence could be attributed to their bioavailability potential.

5 Conclusions and Perspectives

In the light of observed findings, the following conclusions can be drawn. PBDEs were detected in all the sediment samples, showing their ubiquity in river and coastal sediments. All samples were dominated by deca-BDE-209, with the highest values observed in samples collected in the mouth of Llobregat River. Moreover, different congener patterns were established, assuming that different commercial mixtures were used at each site. However, BDE-209 degradation studies are required in order to determine the origin of the contamination. Some studies suggested that the presence of the low brominated congeners (BDE-47, -99, and -100) could be the result of the degradation of BDE-209 and not the result of the use of Penta-BDE formulations.

PBDE congeners were also detected in all the fish samples collected along the Anoia and Cardener Rivers. However, some differences were observed between PBDE profiles in sediments and in feral carp. BFR uptake for carps could be via water, ingested sediment, and food. Due to their feeding ecology, sediments may be an important source of POPs to carp. Then, the reasons for the differences observed in distribution in the sediment samples and the fish samples could be attributed to a different bioaccumulation potential of PBDE congeners or to biotransformation processes, probably via debromination. Previous debromination processes observed in studies with caged carp following dietary exposure of BDE-99, BDE-183, and BDE-209 [42, 43], seems to be also observed in results in field studies with feral carp. Deca-BDE-209 and hepta-BDE-183 seems debrominated to the hexa-BDE-154, and penta-BDE-99 metabolized to tetra-BDE-47. Moreover, another hexabrominated congener, BDE-153, shows different contribution in sediments and fish, probably due to potential transformation or debromination processes of this congener in feral carp. It should be pointed out that these results only attempted to identify qualitatively the different biological processes observed in feral carp. However, future studies involving a great number of samples are needed in order to support the preliminary qualitative data and to determine the quantitative transformation of each BDE congener due to the metabolism of carp.

As regards emerging BFRs, the findings provide evidence that currently, PBEB, HBB, and DBDPE of unrestricted usage are present in the environment. Further measurements are required to investigate the environmental sources and fate of these compounds.

Acknowledgments This research study was funded by the Spanish Ministry of Science and Innovation through the projects CEMAGUA (CGL2007-64551/HID) and SCARCE (Consolider Ingenio 2010 CSD2009-00065).

References

1. Arias PA (2001) Proceedings of 2nd international workshop on brominated flame retardants, Stockholm, Sweden, 17
2. Strandberg B, Dodder NG, Basu I, Hites RA (2001) Environ Sci Technol 35:1078
3. Hale RC, Alaee M, Manchester-Neesvig JB, Stapleton HM, Ikonomou MG (2003) Environ Int 29:771
4. Alaee M, Arias P, Sjödin A, Bergman A (2003) Environ Int 29:683
5. Law RJ, Alaee M, Allchin CR, Boon JP, Lebeuf M, Lepom P, Stern GA (2003) Environ Int 29:757
6. Anderson O, Blomkvist G (1981) Chemosphere 10:1051
7. de Wit C (2002) Chemosphere 46:583
8. Law RJ, Herzke D, Harrad S, Morris S, Bersuder P, Allchin CR (2008) Chemosphere 73:223
9. Meerts IATM, Letcher RJ, Hoving S, Marsh G, Bergman A, Lemmen JG, van der Burg B, Brouwer A (2001) Environ Health Perspect 109:399

10. Legler J, van den Brink CE, Brouwer A, Murk AJ, van der Saag PT, Vethaak AD, van der Burg B (1999) Toxicol Sci 48:55
11. Verreault J, Gebbink WA, Gauthier LT, Gabrielsen GW, Letcher RJ (2007) Environ Sci Technol 41:4925
12. UBA (2001) Erarbeitung von Bewertungsgrundlagen zur Substitution umweltrelevanter Flammschutzmittel. Umweltbundesamt, Germany. http://www.umweltbundesamt.org
13. Kierkegaard A, Björklund J, Friden U (2004) Environ Sci Technol 38:3247
14. de Wit CA, Kierkegaard A, Ricklund N, Sellström U (2011) Emerging brominated flame retardants (BFRs) in the environment. In: Eljarrat E, Barceló D (eds) Brominated Flame Retardants. The Handbook of Environmental Chemistry, 16: 241–286. Springer, Berlin
15. Leonards P, Lopez P, de Boer J (2008) Organohalogen Compd 70:926
16. Lopez P, Pim L, Sicco B, de Boer J (2008) Organohalogen Compd 70:224
17. Eljarrat E, Labandeira A, Martinez A, Fabrellas B, Barcelo D (2005) Organohalogen Compd 67:459
18. Hoh E, Zhu L, Hites R (2005) Environ Sci Technol 39:2472
19. Stapleton HM, Allen JG, Kelly SM, Konstantinov A, Klosterhaus S, Watkins D, McClean MD, Webster TF (2008) Environ Sci Technol 42:6910
20. de la Cal A, Eljarrat E, Barceló D (2003) J Chromatogr 1021:165
21. Eljarrat E, Lacorte S, Barceló D (2002) J Mass Spectrom 37:76
22. Eljarrat E, de la Cal A, Barceló D (2004) Anal Bioanal Chem 378:610
23. Labandeira A, Eljarrat E, Barceló D (2007) Environ Pollut 146:188
24. Eljarrat E, de la Cal A, Raldúa D, Duran C, Barceló D (2004) Environ Sci Technol 38:2603
25. Sellström U, Kierkegaard A, de Wit C, Jansson B (1998) Environ Toxicol Chem 17:1065
26. Watanabe I, Kawano M, Tatsukawa R (1995) Organohalogen Compd 24:337
27. Sawal G, Woitke P, Lepom P (2005) Organohalogen Compd 67:587
28. Allchin CR, Law RJ, Morris S (1999) Environ Pollut 105:197
29. Guerra P, Eljarrat E, Barceló D (2010) J Hydrol 383:39
30. Muñoz I, López-Doval JC, Ricart M, Villagrasa M, Brix R, Geiszinger A, Ginebreda A, Guasch H, López de Alda MJ, Romaní AM, Sabater S, Barceló D (2009) Environ Toxicol Chem 28:2706
31. Gauthier LT, Hebert CE, Weseloh DVC, Letcher RJ (2007) Environ Sci Technol 41:4561
32. Watanabe I, Sakai SI (2003) Environ Int 29:665
33. Watanabe I, Kashimoto T, Tatsukawa R (1986) Bull Environ Contam Toxicol 36:778
34. Kierkegaard A, Sellstrom U, McLachlan MS (2009) J Chromatogr A 1216:364
35. Ricklund N, Kierkegaard A, McLachlan MS (2010) Environ Sci Technol 44:1987
36. Zhang XL, Luo XJ, Chen SJ, Wu JP, Mai BX (2009) Environ Pollut 157:1917
37. Grant PBC, Johannessen SC, MacDonald RW, Yunker MB, Sanborn M, Dangerfield N, Wright C, Ross PS (2011) Environ Toxicol Chem 30:1522
38. Ramu K, Isobe T, Takahashi S, Kim EY, Min BY, We SU, Tanabe S (2010) Chemosphere 79:713
39. Rice C, Chernyak SM, Begnoche L, Quintal R, Hickey J (2002) Chemosphere 49:731
40. Dodder NG, Strandberg B, Hites RA (2002) Environ Sci Technol 36:146
41. Loganathan BG, Kannan K, Watanabe I, Kawano M, Irvine K, Kumar S, Sikka HC (1995) Environ Sci Technol 29:1832
42. Stapleton HM, Alaee M, Letcher RJ, Baker JE (2004) Environ Sci Technol 38:112
43. Stapleton HM, Letcher RJ, Baker JE (2004) Environ Sci Technol 38:1054
44. Eljarrat E, de la Cal A, Raldúa D, Duran C, Barceló D (2005) Environ Pollut 133:501

Inputs of Pharmaceuticals and Endocrine Disrupting Compounds in the Llobregat River Basin

Mira Petrovic and Damia Barceló

Abstract Pharmaceutically active compounds (PhACs) and endocrine disrupting compounds (EDCs) including natural and synthetic hormones and alkylphenolic compounds are two classes of organic contaminants found to be ubiquitous in the Llobregat river basin. This chapter summarizes the findings of several extensive monitoring studies conducted in order to evaluate the main sources and identify compounds present in the river water and sediment. WWTP outlets are identified as major contributors of these substances and although the load of pharmaceuticals and EDCs in WWTP outlets are considerably reduced after the treatment, still considerable concentrations are found in receiving river water. However, direct ecotoxicological effects are very difficult to estimate due to the great number and variety of factors, such as is the presence of other organic and inorganic contaminants and impact of other environmental stressors. However, two studies unequivocally confirmed that EDCs present in the Llobregat river have negative effect on the reproductive physiology of endogenous fish feral carp (*Cyprinus carpio*).

Keywords Endocrine disrupting compounds, Nonylphenol, Pharmaceuticals, Steroid hormones, Wastewater treatment plants

---

M. Petrovic (✉)
Catalan Institute for Water Research (ICRA), H2O Building, Scientific and Technological Park of the University of Girona, 101-E-17003 Girona, Spain

Catalan Institution for Research and Advanced Studies (ICREA), Barcelona, Spain
e-mail: mpetrovic@icra.cat

D. Barceló
Catalan Institute for Water Research (ICRA), H2O Building, Scientific and Technological Park of the University of Girona, 101-E-17003 Girona, Spain

Department of Environmental Chemistry, Institute of Environmental Assessment and Water Research (IDAEA-CSIC), Barcelona, Spain

S. Sabater et al. (eds.), *The Llobregat: The Story of a Polluted Mediterranean River*, Hdb Env Chem (2012) 21: 151–166, DOI 10.1007/698_2011_141,

Contents

Abbreviations

AP_1EOs	Alkylphenol monoethoxylates
APEOs	Alkyphenol ethoxylates
APs	Alkylphenols
DWTP	Drinking water treatment plant
EDCs	Endocrine disrupting compounds
LC–MS	Liquid chromatography-mass spectrometry
LOD	Limit of detection
NOEC	No-observed effect concentration
NP	Nonylphenols
NPECs	Nonylphenol carboxylates
NSAIDs	Nonsteroidal anti-inflammatory drugs
OP	Octylphenol
PhACs	Pharmaceutically active compounds
RYA	Recombinant yeast assay
VTG	Vitellogenin
WWTPs	Wastewater treatment plants

1 Introduction

Both priority and emerging contaminants have been shown to pollute the water of the Llobregat River to significant extent as a result of the intense and ever-increasing human activities (agriculture, industry, urban) taking place in the Barcelona area and upstream [1]. Among the organic contaminants found to be present in the Llobregat River are pharmaceutically active compounds (PhACs) and endocrine disrupting compounds (EDCs) including natural and synthetic hormones and alkylphenolic compounds.

PhACs are a group of chemical substances that have medicinal properties, and they are produced worldwide on a 100,000 t scale. Most of the modern drugs are small organic compounds with a molecular weight below 500 Da, which are moderately water soluble as well as lipophilic, in order to be bioavailable and biologically active.

In the European Union (EU), around 3,000 different PhACs are used in human medicine (i.e., analgesics and anti-inflammatory drugs, β-blockers, lipid regulators, antibiotics). After administration, pharmaceutical can be excreted as an unchanged parent compound, in the form of metabolites or as conjugates of glucuronic and sulfuric acid, primarily via urine and feces, thus their main route into the aquatic environment is disposal via wastewater. The ubiquity of PhACs in sewage waters and in the aquatic environment is related to specific sales and practices in each geographic area and under normal conditions, the inputs of PhACs are generally considered to be constant. However, for some pharmaceuticals (i.e., antibiotics) differences between winter and summer influent loads can be expected, probably because of higher attenuation in summer and more use of pharmaceuticals in winter.

The significance of PhACs as environmental contaminants in waterways is due to several reasons: (1) the continuous introduction via effluents from sewage treatment facilities and from septic tanks; (2) they are developed with the intention of performing a biological effect; (3) they often have the same type of physico-chemical behavior as other harmful xenobiotics (persistence in order to avoid the substance to be inactive before having a curing effect, and lipophilicity in order to be able to pass membranes); (4) some pharmaceutical substances are used by man in rather large quantities (i.e., similar to those of many pesticides).

Numerous studies revealed their presence in wastewaters, as well as surface, ground, and drinking water [2]. This is mainly due to the fact that some compounds are not efficiently removed during wastewater treatment processes, being able to reach surface and groundwater and, subsequently, drinking waters. Although conventional wastewater treatment plants (WWTPs) can effectively reduce carbon and nitrogen content of raw sewage, the removal of PhACs is often insufficient and WWTP effluents are frequently pointed out as the main source of these microcontaminants [3]. Two PhACs: diclofenac and ibuprofen are on the list of substances subject to a review for possible identification as "priority substances" or "priority hazardous substances" and candidates for inclusion in forthcoming update of Directive 2008/105/EC on environmental quality standards in the field of water policy.

Another group of contaminants covered in this chapter is EDCs. A broad spectrum of anthropogenic and naturally occurring chemicals can alter the normal endocrine function and physiological status of wildlife species. Among chemicals that act as EDCs are natural and synthetic hormones (estrogens and progestogens) and some common industrial and household products such as plasticizers, antioxidants, pesticides, and nonionic surfactants and their degradation products (Table 1). Some of the main sources of EDCs include tannery effluents, paper and pulp mill effluents, crude oil derivatives, and sewage effluents. 17β-estradiol is the principal endogenous phenolic steroid estrogen. It is metabolized to estrone and estriol. These metabolites are mostly conjugated with glucuronides and, to a smaller extent, sulfates and excreted in the urine. A minor amount of the estrogens are excreted via feces as unconjugated metabolites [4, 5].

The human daily excretion of estradiol, estrone, and estriol varies from men (1.6, 3.9, 1.5 μg) to women (3.5, 8, 4.8 μg) maintaining similar proportions with estrone being the most abundant estrogen [5]. Pregnant women show a different profile

Table 1 Some categories of substances with reported endocrine disrupting properties

Substance	Examples	Uses	Modes of action
Naturally occurring			
Phytoestrogens	Isoflavones; lignans; coumestans	Present in plant material	Estrogenic and anti-estrogenic
Female sex hormones	17β-estradiol; estrone	Produced naturally in animals (including humans)	Estrogenic
Synthetic			
Polychlorinated organic compounds	Dioxins	Byproducts from incineration and industrial chemical processes	Anti-estrogenic
	Polychlorinated biphenyls (PCBs)	Dielectric fluids	
Organochlorine pesticides	DDT; dieldrin; lindane	Insecticides	Estrogenic and anti-estrogenic
Organotins	Tributyltin	Anti-fouling agent	
Alkylphenols	Nonylphenol	Used in production of NPEOs and polymers, degradation product	Estrogenic
Alkylphenol ethoxylates	Nonylphenol ethoxylate	Surfactants	Estrogenic
Phthalates	Dibutyl phthalate (DBP); butylbenzyl phthalate (BBP)	Plasticisers	Estrogenic
Bi-phenolic compounds	Bisphenol-A	Component in polycarbonate plastics and epoxy resins	Estrogenic
Synthetic steroids	Ethinyl estradiol	Contraceptives	Estrogenic

with higher levels of estradiol and estrone by a factor of 10 and estriol daily excretion at 6,000 μg. Women taking contraceptives based on ethinylestradiol excrete daily 35 μg of this synthetic estrogen [6]. Besides contraception, the uses of estrogens can largely be put into three main groups: the management of the menopausal and postmenopausal syndrome (its widest use); physiological replacement therapy in deficiency states; and the treatment of prostatic cancer and of breast cancer in postmenopausal women. Therefore, the main sources of estrogens to WWTPs are from the natural excretion of estrogens by humans, from hormone and estrogen replacement therapies and the intake of hormone contraceptives containing ethinylestradiol.

In sewage effluents and in receiving surface waters, another group of compounds responsible for estrogenic activity is believed to be alkylphenolic compounds such as alkylphenols (APs) and alkylphenol monoethoxylates (AP_1EOs), both degradation products of alkyphenol ethoxylates (APEOs). APEOs are effective nonionic surfactants, widely used as industrial cleaning agents and wherever their interfacial effects of detergency, (de)foaming, (de)emulsification, dispersion or solubilization can enhance products or process performance. Although parent APEOs are not classified into highly toxic substances (EC_{50}, 48 h, *Daphnia magna* 1.5 mg/L), their environmental acceptability is strongly disputed because of estrogenic metabolic products (APs and short chain ethoxylates (AP_1EOs) generated during wastewater treatment. Because of these findings, APEOs are banned or restricted in Europe. Throughout the northern Europe (Scandinavia, UK, Germany), a voluntary ban on APEO use in household cleaning products began in 1995 and restrictions on industrial cleaning applications in 2000 [7]. This resulted in significant reduction of APEO concentrations found in Catalonia typical levels of nonlyphenol (NP) measured in WWTPs in 1998 and 1999 ranged from 100 to 200 $\mu g\ L^{-1}$ in influents, while 2002–2003 data show almost a tenfold decrease, which suggest a gradual withdraw and replacement of APEOs by Spanish tanneries and textile industry [8].

Due to their estrogenic activity, NP and octylphenol (OP) have been listed as priority hazardous substances in the field of water policy by the European Community Water Framework Directive 2000/60/EC and the final European Union decision No. 2455/2001/EC, while 17a-ethinylestradiol and 17β-estradiol have been proposed as substances subject to a review for possible identification as "priority substances" or "priority hazardous substances."

The risks derived from the presence of these two groups of contaminants in the Llobregat River have been pointed out in several monitoring programs that, integrating both chemical and biological analyses, have identified steroids and alkylphenolic compounds as the contaminants responsible for the induction of estrogenic effects, such as feminization in feral carp [9].

2 Occurrence of Pharmaceuticals in the Llobregat River

2.1 Sources, Levels, and Distribution

The Llobregat River basin has been the object of several studies dealing with the presence of PhACs in surface water. Pharmaceuticals and their toxicity have been studied in the upper part of the Llobregat basin [10] and a short list of them was included in a study covering a wide range of emerging pollutants in the same area [11]. The lower part of the Llobregat River and especially area around drinking water treatment plant (DWTP) Sant Joan Despi that supplies drinking water to a great part of the Barcelona metropolitan area and uses water from the Llobregat River as a source have been extensively studied [11–15].

In all monitoring studies, the PhACs detected in the Llobregat River closely matched those identified by the Spanish National Health System as those most consumed. Generally, the most abundant loads are reported for nonsteroidal anti-inflammatory drugs (NSAIDs), such as ibuprofen, acetaminophen (paracetamol), diclofenac and ketoprofen, with the contamination loads increasing downstream along the river. The highest concentrations were found in two hot spots: the Rubi creek and a channel receiving bypasses from the most contaminated fractions of the river. For example, in the lower part of the Llobregat River and in some tributaries (Anoia River, Rubi creek), ibuprofen is detected at high concentrations (in μg/L) [15]. Although the percentage of elimination of this drug is very high [3], it is still detected in rivers downstream WWTPs due to a very high usage in human medicine. Other pharmaceuticals frequently detected were lipid regulators; gemfibrozil and bezafibrate; antibiotics: sulfamethoxazole and ofloxacin, and β-blocker metoprolol. Although its consumption is not very high, psychiatric drug carbamazepine was among the most ubiquitous compounds (detected in all samples) due to its very poor or no removal in conventional biological treatment processes operating in WWTPs in the area. In most of the samples, the concentrations of individual compounds were in the range of 10–1,000 ng/L, with occasional detection of concentrations above that range corresponding to hot spots and drought periods. Generally, the concentrations vary a lot depending upon the sampling time and hydrological conditions (water flow and consequently dilution factor).

Lopez-Roldan et al. [12] studied the occurrence of 28 multi-class pharmaceuticals and detected the highest concentrations for the β-blockers metoprolol (8,042 ng/L) and sotalol (788 ng/L), the antibiotic ofloxacin (1,904 ng/L), and the lipid regulator gemfibrozil (1,014 ng/L). Two sites (channel receiving waters from the Anoia River and Sant Feliu WWTP and Rubí Creek), out the eight monitored in the lower Llobregat River, showed distinctly higher concentrations of PhACs as shown in Fig. 1; however, their waters are diverted to reach the river at locations close to the mouth and downstream of the inlet of the Sant Joan Despí DWTP in order to protect the quality of the source water.

García-Galán et al. [13] studied the occurrence of sulfonamide antibiotics in the lower part of the Llobregat River. Sulfamethoxazole, sulfapyridine, sulfamethazine, and sulfamethizole were the compounds most frequently detected at max concentrations of 2,482 ng/L for sulfamethazine and 4,297 ng/L for sulfamethoxazole. However, median concentrations were lower than 50 ng/L, except for some outlier values at the sampling site near the mouth of the Llobregat River.

A recent study [14] focused on the lower stretch of the Llobregat River between the towns of Molins de Rei and Sant Joan Despí, the latter being particularly important because the intake of an important waterworks supplying drinking water to the city of Barcelona is located there. Data regarding the occurrence of a wide range of emerging contaminants, including pharmaceuticals, were gathered during a period of water reuse of treated effluent that took place as a consequence of the severe drought that took place along the years 2007–2008. Water from the WWTP tertiary treatment of El Prat de Llobregat was pumped upstream ca. 15.6 km and discharged into the river. As far as pharmaceuticals are concerned, 56 drugs out of 71 monitored were detected at least in one sample in concentrations in the low ng/L range for most of the compounds analyzed. Therapeutic classes which were present at higher concentrations at all sampling points were antihypertensives, antibiotics, diuretics, and anti-inflammatories reaching maximum total concentration of around 500 ng/L upstream of discharge of tertiary effluent and around 1 μg/L downstream, with comparable contributions in terms of loads of both the river upstream basin and the tertiary effluent.

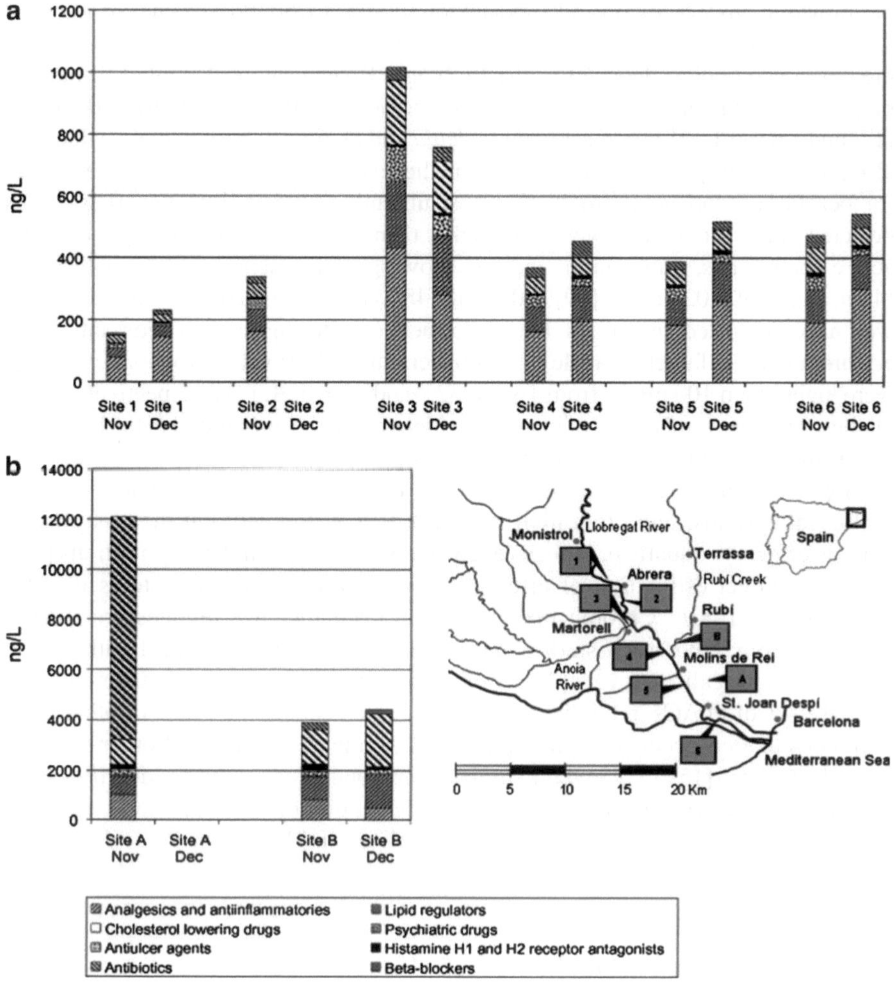

Fig. 1 Cumulative levels of pharmaceuticals, grouped by therapeutic class, detected at sites 1–6 (**a**) and sites A and B (**b**), modified from [12]

2.2 Ecotoxicological Relevance

Concentrations of PhACs detected in surface waters of the Llobregat River basin are at least 1,000 times lower than the levels reported to cause acute toxicity [16]. However, as a wide spectrum of pharmaceuticals has been detected, effects of mixtures should also be taken into account; hence, the overall toxicity could be the result of the sum of individual concentrations, or the interaction of different compounds, occurring effects at the no-observed effect concentration (NOEC) of individual substances. Regarding some examples, diclofenac seems to be the compound having the highest acute toxicity within the class of NSAIDs, since tests performed in phytoplankton and zooplankton showed EC50 (96 h) values of 14.5 mg/L and 22.43 mg/L, respectively [17], nearly two orders of magnitude higher than the highest concentration detected in the Llobregat River (low μg/L range). β-blocker propranolol is the one showing higher acute toxicity, with EC50 (48 h) of 0.8 mg/L [17] and 1.6 mg/L [18] for *Ceridophnia dubia* and *D. magna*, respectively, which are also approximately 1,000 times higher than the levels detected. EC50 values reported for bezafibrate and gemfibrozil, the most ubiquitous blood lipid lowering agents detected in the Llobregat River, were higher than 100 mg/L [16]. Among the neuroactive compounds, fluoxetine is the one showing higher acute toxicity, with EC50 (48 h, algae) of 0.024 mg/L [19] and LC50 (48 h) of 2 mg/L [20], but this compound was never detected in any of the performed monitoring studies. For antibiotics erythromycin, sulfamethoxazole and ofloxacin, the EC50 values for *Vibrio fischeri* were higher than 100 mg/L, from 16.9 to 32.2 and 100 mg/L [21], respectively, being all these levels several orders of magnitude higher than the ones found in the Llobregat River (max concentrations in the low μg/L range).

However, due to specific characteristics of PhACs (i.e., high reactivity with biological systems; and often high stability), it is postulated that they may cause chronic effects to aquatic organisms even at low environmental concentrations [16]. However, direct estimation of effects caused by PhACs on ecosystems is not a straightforward task. Chronic toxicity data for aquatic invertebrates are usually obtained from laboratory studies performed with pharmacological rather than environmental realistic levels [22]. Moreover, the estimation of effects or environmental risks of PhACs in the field should take into account the effect of other toxic pollutants that co-occur with pharmaceuticals, same as the effect of other environmental stressors [15, 18]. Several studies were conducted aimed to identify causal links between the occurrence of PhACs and the quality of ecological systems in the Llobregat River and the main findings are outlined in [38].

3 Occurrence of EDCs

3.1 Levels and Distribution

3.1.1 Hormones

Hormones with the strong estrogenic potential, such as the natural hormone estradiol and the synthetic hormone ethynyl estradiol, were never detected in any of the published studies conducted in the Llobregat River [9, 11, 23, 24]. The hormones most frequently detected along the river are less potent estrogens estriol, estrone, and estrone sulfate, the main metabolites of estradiol. The highest concentrations were found in sewage impacted tributaries Anoia and Cardener Rivers and in one channel receiving water from Anoia River, the Rubí creek and Sant Feliu WWPT to bypass it downstream of the DWTP and near Castellvell, an area influenced by urban activities.

The levels are in the low ng/L level, which for this type of contaminants are concentrations that may pose a risk to aquatic organisms and cause estrogenic effects [25]. The negative findings for strong estrogens (estradiol and ethynyl estradio) do not mean that these compounds were not present at levels that may present a risk, but the analytical capabilities (i.e., limit of detection) in some studies did not permit detection at environmentally relevant concentrations (below ng/L).

For example, one of the most comprehensive studies of estrogens in the Llobregat River was conducted in 2000 [9] when the influents, effluents and sludge from four WWTPs (all having biological treatment), water and sediment from the Anoia and the Cardener River (downstream and upstream of WWTP discharge) and feral carp, collected at the same sampling points, were systematically analyzed over a 7-month period in order to study the temporal variation of endocrine disrupters and their effects on carp. The main metabolites of estradiol, estriol, and estrone, and the natural hormone progesterone, with average concentrations 6.3 ng/L, 8.0 ng/L, and 4.3 ng/L, respectively, were the steroids most frequently found in the river waters analyzed. Estradiol and ethynyl estradiol were not detected in any sample possibly attributable to their relatively high detection limits (LOD in water 5 ng/L). Similar levels of estrone and estrone sulfate in river water (4–22 ng/L and 4–7 ng/L, respectively) were reported by Rodriguez-Mozaz in 2004 [23], while more recent studies show estrone and estrone sulfate concentrations not surpassing 2 ng/L [11]. Moreover, the study of Petrovic et al. [9] reported for the first time the detection of hormones in sediments with several sampling points showing rather high levels of estrogens (see Fig. 2).

3.1.2 Alkyphenolic Compounds and Other EDCs

Several comprehensive monitoring studies focused on alkylphenolic EDCs and reported widespread occurrence of this group of contaminants in the Llobregat River basin [8, 9, 11, 23, 26, 27]. Alkyphenolic compounds such as NP, NPEOs, and nonylphenol carboxylates (NPECs) have been detected in Llobregat River and all tributaries with the highest concentrations in the lower part of the basin and especially at sampling sites downstream of WWPT and in areas with strong presence of tannery and textile industry (such as the Igualada area). The concentrations ranged from a few μg/L to several hundred μg/L for most polar NPECs with a general trend in increase in concentrations from the upper to the lower course of the river.

Petrovic et al. [9] found NPEOs and NP in both water and sediment samples, while more polar acidic byproducts (NPE_1C) were detected only in the water samples. NP levels ranged from 0.5 to 15 μg/L, showing a clear increase in concentrations at sites downstream of WWTPs and at several points exceeding the maximum allowable concentration (MAC) of 2 μg/L established by the EU directive on Priority Substances (Directive 2008/105/EC). The highest concentration (15 μg/L) was found in water of the Anoia River downstream of three WWTPs (Igualada, Piera, and Calaf). These levels exceed or are close to the NOEC for the induction of vitellogenesis in caged trout (5–20 μg/L) and suggest that long-

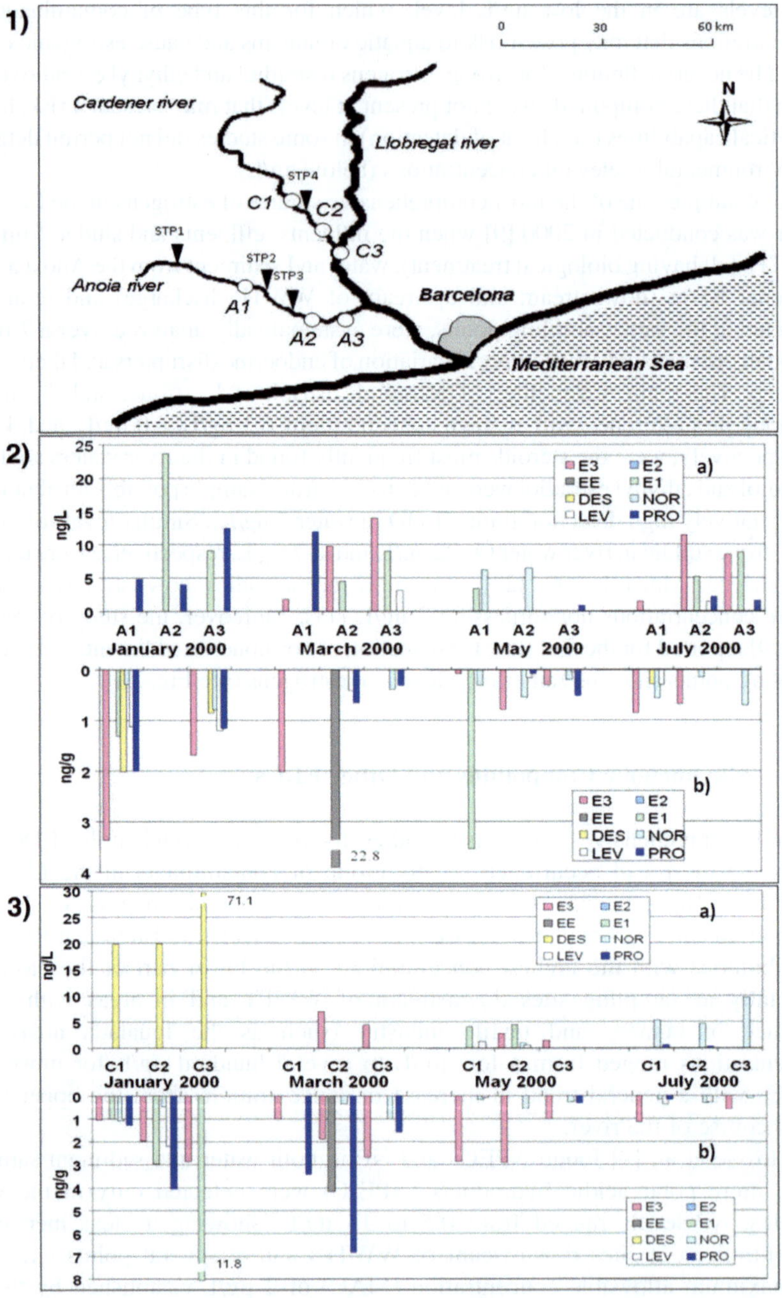

Fig. 2 Seasonal variation of natural and synthetic estrogens and progestogens in water (1) and sediment (2) samples: (**a**) Map of the sampling sites; (**b**) Concentration level in the Anoia River; (**c**) Concentration levels in the Cardener River. Modified from [9]

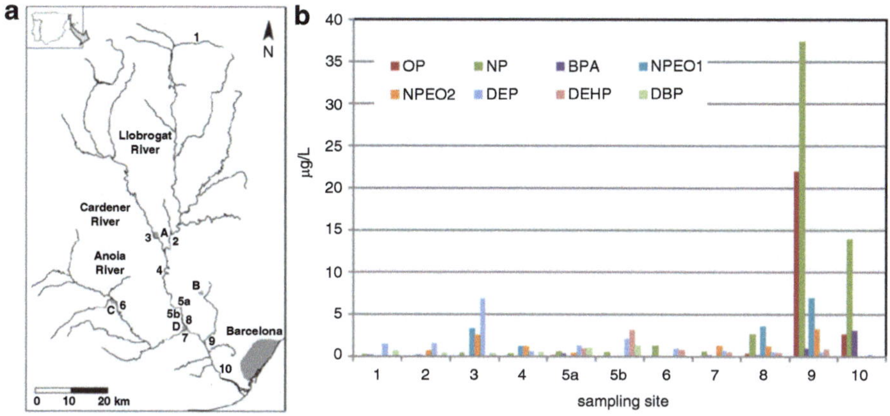

Fig. 3 (**a**) Map of sampling sites in the Llobregat basin. Numbers 1–10 indicate sampling points for river water, letters A–D correspond to STP locations. Gray areas indicate approximate locations of cities near to these STPs: A – Manresa; B – Terrassa, C – Igualada; D – Martorell. The position of the city of Barcelona is given to facilitate localization of the area in the map. (**b**) Concentrations of different EDCs analyzed by LC–MS. Modified from [8]

term exposure may exert an estrogenic effect on fish populations [28]. The concentrations of NPEOs in the Anoia River were from 11 μg/L (upstream) to 41 μg/L (downstream of discharges). NPEOs were far less abundant in the Cardener River, with concentrations ranging from 2.2 to 8.4 μg/L. Furthermore, the study points out that sediment appears to act as a sink for NPEOs and their nonpolar degradation products. At sites situated downstream of the WWTP effluent discharge sediment concentrations of NPEOs and NP were up to 818 and 655 μg/kg in the Anoia River and 150 and 380 μg/kg in the Cardener River, respectively. All these values are in the upper range of values reported for other river basins in Europe, USA, and Japan [29].

Another study [26] that covered more extended area of the river basin and focused only on alkyphenolic compounds and some other EDCs such as bisphenol-A and phathalates (but not hormones), the increasing contaminant burden of the Llobregat and Anoia Rivers was also detected (Fig. 3). The maximum contamination levels (21.9 μg/L of OP and 37.3 μg/l of NP) corresponded to a small stream, Riera de Rubí, into which many different industrial and domestic wastewaters discharge. The authors combined liquid chromatography-mass spectrometry (LC–MS) analysis with the functional recombinant yeast assay (RYA) to evaluate the potential of samples from the Llobregat River for endocrine disruption. In parallel, theoretical estrogenic activities of samples were calculated from the chemical data by multiplying the concentration of each individual EDC and the corresponding relative estrogenicity factor [25]. The results of such a calculation compared to the actual RYA values are shown in Fig. 4 showing that the predicted and actual estrogenicity values followed very similar patterns, both in respect to the increasing estrogenic activity at the lower course of the Llobregat. However, the total estrogenicity calculated from

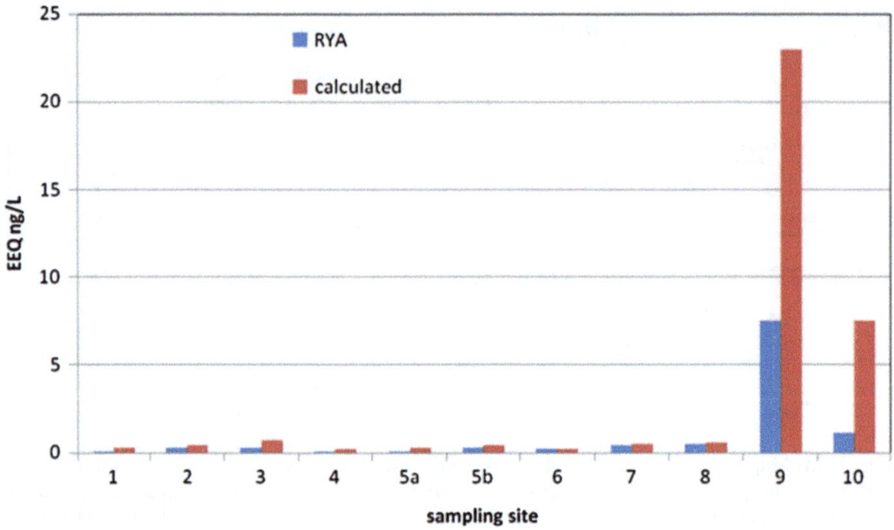

Fig. 4 Total estrogenicity values in estradiol equivalents (EEQ in ng/L) for river water samples corresponding to actual values from RYA vs. estrogenicity values calculated from LC-ESI-MS analysis. (For sampling sites, see Fig. 3.) Modified from [8]

chemical composition of the samples overestimated the actual RYA values by a factor of approximately 3, on average. This overestimation of the predicted estrogenicity values can be explained in several plausible ways. In case of alkyphenolic compounds, the estradiol equivalences were obtained for the para-isomers, which are the most estrogenic alkylphenol isomers, while the actual samples contained most probably technical mixtures of alkylphenol isomers, which are undistinguishable by LC-MS. Therefore, the presence of nonactive isomers would result in an overestimation of the total estrogenicity of the sample. Another explanation for lower measured estrogenicity is due to an anti-estrogenic effect of carboxylated derivatives (NPECs) that may mask the apparent estrogenicity of nonylphenol [30].

3.2 Effects

The effect of xenobiotic and natural EDCs in rivers receiving WWTP effluents has been demonstrated in numerous field studies using endogenous fish species or caged fish [31–33]. Typically, indicators of feminization in male fish included induction of vitellogenin (VTG, an oviparous female egg yolk precursor) and intersex gonads (containing both male and female tissue). Therefore, VTG induction has been selected as a very reliable biomarker of estrogenic exposure [34, 35]. In female fish, the synthesis of VTG is regulated by estradiol circulating levels in the plasma. In males, as a consequence of exposure to substances that mimic natural estradiol, VTG can also be synthesized as a redundant protein, which has been associated with pathological changes in gonadal material and with intersex gonads formation [33, 36].

Two studies unequivocally confirmed that EDCs present in the Llobregat River have negative effect on the reproductive physiology of endogenous fish feral carp (*Cyprinus carpio*). In both studies, feral carp were collected at the same sampling points as river water and sediment in order to study the effects of WWTP effluents on the reproductive physiology of feral carp, in situ using plasma VTG concentration as a biomarker of exposure to estrogenic compounds and indices of fish sexual disruption, and to follow the effects over the annual reproductive cycle in both males and females. In the first study, performed in 1998 [27], a VTG increase was encountered in both studied rivers (Anoia and Cardener), both upstream and downstream of the WWTPs, while the second study in 2000 that involved more extensive monitoring [9] intersex males carps with high VTG content were present after the main WWTP of the Anoia River (site A2) (Fig. 5). Intersexuality was evaluated macroscopically as the co-existence of ovarian and testicular tissue in the same animal (gross abnormalities). In some male fish, there were fully developed eggs. In contrast, other males from the same site did not express VTG. It is most likely that intersex occurs after exposure to estrogenic compounds during the early life stages when sexual determination takes place. In the natural environment, where carp larvae live sedentary on the rivers bottom for 2–3 days before migrating to the upper water, pollutants linked to the sediment (see Fig. 4) may affect fish during their most critical period.

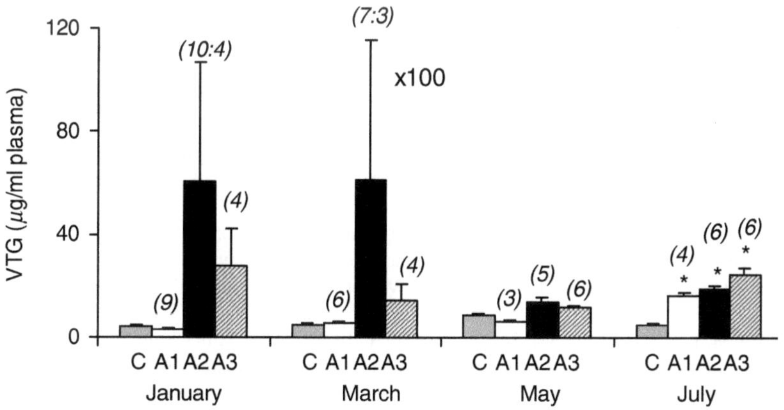

Fig. 5 Vitellogenin levels, in μg/mL plasma, of male carp (*Cyprinus carpio*) in Anoia River. In brackets the number of fish analyzed and the intersex out of the total is indicated for each site. *Asterisk* denotes significance ($p < 0.05$) after *t*-student analysis vs. each control (C). Modified from [9]

4 Conclusions

The wide spectrum of pharmaceuticals and EDCs detected in the Llobregat River indicates that WWTP outlets are major contributors of these substances in the aquatic environment. Although the loads of pharmaceuticals and EDCs in WWTP outlets are considerably reduced after the treatment, still considerable concentrations are found in receiving river water. The highest contamination is detected in smaller tributary rivers, mainly due to low dilution of discharged effluents. In some tributaries, such as the Anoia River or Rubi creek, WWTP effluents may represent a significant percentage of the total flow of the river. Moreover, the situation could be enhanced in drought periods leading to adverse ecotoxicological effects. However, direct cause-effect relationships are very difficult to establish due to the great number and variety of factors, such as the presence of other organic and inorganic contaminants and impact of other environmental stressors. Therefore, more extensive monitoring focused on combining chemical and biological data is needed and is currently undertaken within the frame of Consolider SCARCE project [37]. This new data will help to understand the fate and impact of pharmaceuticals and EDCs, as well as other contaminant classes, in the Llobregat River basin and to bridge chemical contamination with biological effects.

References

1. González SA, López-Roldán R, Cortina JL (2012) Environ Pollut 161:83
2. Petrovic M, Barcelo D (eds) (2009) Analysis, fate and removal of pharmaceuticals in the water cycle. Elsevier, Amsterdam
3. Gros M, Petrović M, Ginebreda A, Barceló D (2010) Environ Int 36:15
4. Schubert W, Cullberg G, Edgar B, Hedner T (1994) Maturitas 20:155
5. Kuhl H (1990) Maturitas 12:171
6. Ying G-G, Kookana RS, Ru Y-J (2002) Environ Int 28:545
7. Knepper TP, Eichhorn P, Bonnington LS (2003) Aerobic degradation of surfactants. In: Knepper TP, Barceló D, de Voogt P (eds) Analysis and fate of surfactants in the aquatic environment. Elsevier, Amsterdam, p 525
8. Gonzalez S, Petrovic M, Barcelo D (2004) J Chromatogr A 1052:111
9. Petrovic M, Solé M, López de Alda MJ, Barceló D (2002) Environ Toxicol Chem 21:2146
10. Farré M, Ferrer I, Ginebreda A, Figueras M, Olivella L, Tirapu L, Vilanova M, Barceló D (2001) J Chromatogr A 938:187
11. Kuster M, López de Alda MJ, Hernando MD, Petrovic M, Martín-Alonso J, Barceló D (2008) Analysis and occurrence of pharmaceuticals, estrogens, progestogens and polar pesticides in sewage treatment plant effluents, river water and drinking water in the Llobregat River basin (Barcelona, Spain). J Hydrol 358:112–123
12. López-Roldán R, de Alda ML, Gros M, Petrovic M, Martín-Alonso J, Barceló D (2010) Chemosphere 80:1337
13. García-Galán MJ, Villagrasa M, Díaz-Cruz MS, Barceló D (2010) Anal Bioanal Chem 397:1325

14. Köck-Schulmeyer M, Ginebreda A, Postigo C, López-Serna R, Pérez S, Brix R, Llorca M, Lopez de Alda M, Petrović M, Munné A, Tirapu L, Barceló D (2011) Chemosphere 82:670
15. Muñoz I, López-Doval JC, Ricart M, Villagrasa M, Brix R, Geiszinger A, Ginebreda A, Guasch H, López de Alda MJ, Romaní AM, Sabater S, Barceló D (2009) Environ Toxicol Chem 28:2706
16. Fent K, Weston AA, Caminada D (2006) Aquat Toxicol 76:122
17. Ferrari B, Mons R, Vollat B, Fraysse B, Paxeus N, Giudice RL, Pollio A, Garric J (2004) Environ Toxicol Chem 23:1344
18. Ginebreda A, Munoz I, de Alda ML, Brix R, Lopez-Doval J, Barcelo D (2010) Environ Int 36:153
19. Brooks BW, Foran CM, Richards SM, Weston J, Turner PK, Stanley JK, Solomon KR, Slattery M, LaPoint TW (2003) Toxicol Lett 142:169
20. Kümmerer K (2004) Pharmaceuticals in the environment, 2nd edn. Springer, New York
21. Isidori M, Lavorgna M, Nardelli A, Pascarella L, Parella A (2005) Sci Total Environ 346:87
22. Crane M, Watts C, Boucard T (2006) Sci Total Environ 367:23
23. Brix R, Postigo C, González S, Villagrasa M, Navarro A, Kuster M, de Alda MJL, Barceló D (2010) Anal Bioanal Chem 396:1301
24. Rodriguez-Mozaz S, López de Alda MJ, Barceló D (2004) J Chromatogr A 1045:85
25. Petrovic M, Eljarrat E, Lopez de Alda MJ, Barceló D (2004) Anal Bioanal Chem 378:549
26. Céspedes R, Lacorte S, Raldúa D, Ginebreda A, Barceló D, Piña B (2005) Chemosphere 61:1710
27. Sole M, Lopez de Alda MJ, Castillo M, Porte C, Ladegaard-Pedersen K, Barcelo D (2000) Environ Sci Technol 34:5076
28. Jobling S, Sheahan D, Osborne JA, Matthiessen P, Sumpter JP (1996) Environ Toxicol Chem 15:194–202
29. Knepper TP, Petrovic M, de Voogt P (2003) Analysis and fate of surfactants in the aquatic environment. Elsevier, Amsterdam, pp 675–693
30. García-Reyero N, Requena V, Petrovic M, Fischer B, Hansen PD, Díaz A, Ventura F, Barcelo D, Piña B (2004) Environ Toxicol Chem 23:705
31. Bjerregaard LB, Lindholst C, Korsgaard B, Bjerregaard P (2008) Ecotoxicology 17:252
32. Burki R, Vermeirssen ELM, Ko¨rner O, Joris C, Burkhardt-Holm P, Segner H (2006) Environ Toxicol Chem 25:2077
33. Jobling S, Beresford N, Nolan M, Tyler C (2002) Biol Reprod 66:272
34. Sumpter HP, Jobling S (1995) Environ Health Prespect 103:173
35. Heppell SA, Denslow ND, Folmar LC, Sullivan CV (1995) Environ Helth Prespect 103:9
36. Gimeno S, Komen H, Jobling S, Sumpter JP, Bowmer T (1998) Aquat Toxicol 43:93
37. SCARCE project – Assessing and predicting effects on water quantity and quality in Iberian rivers caused by global change, Consolider-Ingenio 2010 CSD2009-00065, Spanish Ministry of Science and Innovation. www.idaea.csic.es/scarceconsolider
38. Muñoz I, García-Berthou E, Sabater S (2012) The Effect of Multiple Stressors on Biological Communities in the Llobregat. Hdb Env Chem, DOI: 10.1007/698_2012_148

Occurrence and Fate of Sulfonamide Antibiotics in Surface Waters: Climatic Effects on Their Presence in the Mediterranean Region and Aquatic Ecosystem Vulnerability

María Jesús García-Galán, M. Silvia Díaz-Cruz, and Damià Barceló

Abstract Surface water bodies are constantly exposed to pollutant inputs of different origin. Wastewater effluents discharge directly on the receiving natural streams, and are among the main entrance pathways for sulfonamides. Strong contrast between seasons, with the consequent fluctuations in the flow rates, and heavy contamination pressures from extensive urban, industrial, and agricultural activities are characteristics of water courses located in the Mediterranean area. The low base flows of Mediterranean rivers makes their hydrology cycle heavily dependent on wastewater inputs, and therefore removal efficiencies of wastewater treatment plants are key to the health of the aquatic ecosystem.

Keywords Environmental risk assessment, Mediterranean region, Removal efficiency, Sulfonamide, Surface waters, Wastewater treatment plant

Contents

M.J. García-Galán (✉) and M.S. Díaz-Cruz
Department of Environmental Chemistry, IDAEA-CSIC, Jordi Girona 18-26, 08034 Barcelona, Spain
e-mail: mggqam@cid.csic.es

D. Barceló
Department of Environmental Chemistry, IDAEA-CSIC, Jordi Girona 18-26, 08034 Barcelona, Spain

Catalan Institute for Water Res (ICRA), Parc Científic i Tecnològic de la Universitat de Girona, C/Emili Grahit, 101 Edifici H2O, 17003 Girona, Spain

King Saud University, P.O. Box 2455, Riyadh 11451, Saudi Arabia

S. Sabater et al. (eds.), *The Llobregat: The Story of a Polluted Mediterranean River*, Hdb Env Chem (2012) 21: 167–192, DOI 10.1007/698_2011_140,

Abbreviations

AcSMZ	N^4-acetylsulfamethazine
ARGs	Antibiotic resistance genes
CAFO	Confined animal-feeding operation
CAS	Conventional activated sludge
EC_{50}	Median effective concentration
EMEA	European medicine agency
ERA	Environmental risk assessment
EU	European Union
FEDESA	European federation of animal health
HQ	Hazard quotient
LC_{50}	Median letal concentration
MBR	Membrane bioreactor
ME	Measured environmental concentration
NOEC	Non observed effect concentration
PEC	Predicted environmental concentration
PhP	Pharmaceuticals
PNEC	Predicted no-effect concentration
RE%	Removal efficiency
SA	Sulfonamide
SDM	Sulfadimethoxine
SDZ	Sulfadiazine
SMP	Sulfamethoxypyridazine
SMR	Sulfamerazine
SMX	Sulfamethoxazole
SMZ	Sulfamethazine
SPY	Sulfapyridine
STZ	Sulfathiazole
US	United States
WFD	Water frame directive
WHO	World Health Organization
WWTP	Wastewater treatment plant

1 Introduction

As a consequence of the increasing human population density and more intensive animal farming techniques, fresh water systems have become highly susceptible to be at risk of potential contamination by different pharmaceutical products (PhPs) from both human and veterinary use. Awareness of the presence of PhPs in wastewaters and aquatic ecosystems is growing as investigations regarding new pollutants increase and analytical techniques for detecting these chemicals improve. At present, approximately 3,000 different pharmaceutical ingredients are used in the European Union (EU), including antibiotics, β-blockers, lipid regulators, antidepressants, etc. [1]. Estimations of the potential environmental impact of PhPs are usually based on the quantities produced and consumed, their potency and also on their tendency to bioaccumulate in the environment. The risk posed by antibiotics could be explained in terms of any of these premises. In first place, their role is superlative in modern agriculture and livestock, and this fact is reflected in their high consumption rates. Although information on their usage is not available to the general public either in the United States (US) or in the European Union (EU), estimations indicate sales over the 16,000 t in US in 2001, of which 9,300 t are used in animal-feeding operations [2]. According to the European Federation of Animal Health (FEDESA), the annual consumption of antibiotics in the EU in 1999 was in total 13,288 t with 29% for veterinary medicine, 6% as antibiotic feed additives, and 65% in human medicine. In addition, prescription drugs are generally sold in quantities one order of magnitude lower than nonprescription drugs [3]. Regarding their potency, these substances are designed to cause a biological effect in the target organism or patient at relatively low concentrations. Once discharged in the environment, they may have numerous unexpected effects on nontarget, or as yet unknown, receptors. It has been demonstrated in different studies that the environmental presence of antimicrobials leads to the development of antibiotic resistance in bacteria, threat that has been recognized by, among others, the World Health Organization (WHO) and is a well-documented fact nowadays. They can also be toxic to different nontarget organisms, including beneficial bacteria in both natural and urban environments; for instance, wastewater treatment processes may be disrupted [4, 5] or degrading microbiota from different ecosystems can be negatively affected [6]. Finally, antimicrobial resilience and persistence in the environment has been demonstrated [7, 8]. This is a direct consequence of their physicochemical properties such as polarity or liposolubility (they can go through biological membranes), which makes them very persistent compounds in order to stay active and therefore very prone to bioaccumulate

2 Environmental Presence of Sulfonamide Antibiotics: Sources and Occurrence

Sulfonamides (SAs) are one of the most widely used antibiotics in human and especially in animal husbandry and fish farming [9, 10]. They are usually applied in combination with diaminopyrimidines such as trimethoprim due to the enhancement of their activity [11]. In EU, SAs are the second most widely used veterinary antibiotics, representing 21% of the sales in the United Kingdom in 2000, and 11–23% in several other European countries. In US, SAs account for the 2.3% of the total amount of antibiotics used [a2]. SAs are widely used because they are inexpensive, effective against a broad spectrum of common bacterial infections, and have high effectiveness in growth promotion in veterinary applications, although this last use has been banned in the EU since 2006 for all antibiotics [12]. The increase in the number of confined animal-feeding operations (CAFOs), which often lack proper waste management practices, has led to a higher use of these antibiotics and, therefore, to a greater occurrence of these substances in the environment. Following treatment, livestock will excrete 50–90% of the administered dose, the parent drug making up for 9–30%. These amounts of the unchanged substance vary depending on the form of the drug and the animal age and species [13, 14]. Animal excreta are considered one of the major sources of environmental contamination by SAs; residues of these antimicrobials have been detected in manure from medicated animals, which is frequently applied as nutrient amendment in agriculture as it is regarded as a very valuable fertilizer containing essential nutrients for plant growth such as nitrogen, phosphorous, organic carbon or potassium [15–19]. The extensive use of manure in crop fields is among the major routes by which veterinary antibiotics enter the environment [19–21] and, eventually, the different water systems. The consequent diffuse pollution is difficult to prevent and deal with due to the large areas of application. Once on the topsoil and due to their weak sorption to soil tendency and high solubility, the excreted residues of SAs become very mobile and may reach surface waters during runoff episodes and even percolate and contaminate the aquifers [15, 22]. This possibility has already been proved in several publications, showing the presence of SAs at different concentrations in groundwater from various sites close to animal farming facilities [23–30]. On the other hand, although veterinary antibiotics such as SAs only reach wastewater treatment plants (WWTPs) to a limited extent, they have been frequently detected in influent and most importantly, in effluent wastewaters [31–34] due to their generally low biodegradation and elimination efficiency during sewage treatment. As these effluents commonly discharge into natural water courses, in the last decade a growing awareness in the scientific field has been manifested regarding the danger posed by the WWTPs inputs to river ecosystems. River basins and catchment areas can therefore be considered highly vulnerable systems regarding SAs contamination. It should also be considered the frequent application of biosolids from WWTPs as organic amendments in agriculture, opening a different entrance pathway into the environment for these substances [35]. Other secondary input pathways are waste effluents of the manufacturing processes or hospitals, the disposal of unused or expired drug products (solid waste or "flushing"), accidental spills during manufacturing or distribution and leakage from septic systems and agricultural waste-storage facilities [36–38]. Antibiotics that reach landfill sites as solid waste are subjected to biologic degradation processes, but some may persist and leach into surrounding groundwater or reach river courses after flood episodes [39–41]. Another critical scenario is that of aquaculture and antibiotics direct addition to receiving waters, formulated as feed additives, with 70–80% of the administered amount entering the environment [42].

2.1 Presence of Sulfonamides in Wastewater Treatment Plants

Given the relevance of WWTPs discharges as indirect entrance pathway for SAs and many other pollutants onto surface waters, a first step to evaluate the health of a river ecosystem would be to determine the loads of pollutants in these WWTPs effluents. Degradation and vulnerability of river systems are directly dependent on the removal efficiencies (RE%) of the WWTPs regarding these contaminants; however, data on the RE% of these compounds during wastewater treatment is still scarce. In general, Spanish WWTPs apply primary and secondary biologic treatments, the latter usually based on conventional activated sludge (CAS). Tertiary treatments such as ozonation, which have demonstrated to be highly efficient in the removal of different PhPs including SAs, are seldom applied [43–45]. Table 1 summarizes some of the RE% values found recently in the literature. Recently, frequencies of detection and RE%s were reported for the seven main WWTPs located along the Ebro River Basin [31]. SAs of human application such as sulfamethoxazole (SMX), sulfapyridine (SPY) and sulfadiazine (SDZ, also used in veterinary therapies) were the most frequently detected (>85%) and at the highest concentrations (650 ng L^{-1} for SMX and 227 ng L^{-1} for SPY) in both influent and effluent samples. RE% values obtained were hard to interpret, as SAs were not regularly present in all the WWTPs, and values ranged from negative removals to 100% elimination. SDZ was in average the SA eliminated most efficiently in these seven WWTPs, whereas SPY showed intermediate to high RE% values. SMX showed both RE% higher than 50% but also negative values in many WWTPs. These higher concentrations detected in the effluents are usually attributed to the presence of SA conjugates and metabolites, which usually are not comprised within the scope of the different studies; these conjugates can be transformed back during treatment into the original compound, as demonstrated recently [46] and could therefore explain higher concentrations of SAs in effluents than in influent waters [47, 48]. Alternative secondary treatments, such as membrane bioreactors (MBRs), have been investigated in recent years to obtain an improvement in the RE% values. However, this treatment technology has proved not to be especially good, in particular for SMX and SPY, the two most relevant SAs in terms of frequencies of detection and concentration. Recent works demonstrated that although elimination rates for SMX were higher in the MBRs than in CAS, removal was only partial as nearly half of the SMX input could still be detected in their respective effluents [49–52]. On the contrary, MBRs worked more efficiently than CAS for other SAs, such as SDZ, which was completely removed after MBR treatment, whereas it was removed only 49% during the CAS treatment. Regarding acetylated metabolites, N^4-acetylsulfamethazine (AcSMZ) was 100% removed after MBR treatment, and only in a 54% after the CAS treatment [51]. Tertiary treatments such as ozonation and nanofiltration have demonstrated high efficiencies in SAs removal [44, 53–56], but still its application in WWTPs is scarce and the fate of the transformation products generated unknown [57].

Table 1 Removal efficiencies (RE%) for sulfonamides upon different wastewater treatments

	Removal efficiency (%)											
WWTP treatment	References	SMX	AcSMX	SDZ	AcSDZ	SMZ	AcSMZ	SDM	SMT	STZ	SMP	SPY
CAS	Clara et al. (2005)	66[a]										
CAS	Carballa et al. (2004)	57–67										
CAS	[28]	18–100										
CAS	Brown (2006)	20										
CAS+filtration+chlorination	Peng (2006)	–		97								
CAS	Choi et al. (2007)	93				99		93		98		
MBR + filtration	Senta et al. (2011)	95–100		90–100		95–100				85–100		90–100
CAS	[37]	60[a]	81–96									72[a]
Ozonation	[44]			95		99			90	99		
MBR	Radjenovic et al. (2009)	78.3–80.8										
CAS		73.8										
MBR	[52]	60.5										
CAS		55.6										
Sand filtration		26.9										14.6
Ozonation	[55]	87.4										93.9
CAS		61.5										4
CAS	[48]	30–92		43–98								
CAS + UV or chlorination	Xu et al. (2007)	35–65		50		50						
CAS	[31]	88[a]		100[a]		100[a]		100[a]	100[a]	65–100	100[b]	96[a]
MBR	[51]	52–55		100		35–50	100	100		30–50	100	20–40
CAS		42		50		55	52			55	35	
CAS	[74]	65	84	93	87	82	100[a]	74	69	58	87	72
CAS	[38]	48		37		22						
Ozonation	[45]	98				98		98				
MBR+ozonation	[54]	90–99	90–99	90–99						90–99		90–99
CAS + chlorination	Yang et al. (2005)	81				80		57				

WWTP treatment	Removal efficiency (%) AcSPY	SSX	SSD	SNT	SZI	SQX	SDX	SuSTZ	SBZ
CAS									
CAS									
CAS									
CAS									
CAS+filtration+chlorination					98				
CAS									
MBR + filtration									
CAS									
Ozonation									
MBR									
CAS									
MBR									
CAS									
Sand filtration									
Ozonation									
CAS									
CAS									
CAS + UV or chlorination									
CAS									
MBR		neg	100[a]	100	38[a]	69[a]	100[a]	100	33–100
CAS		70							
CAS									
CAS	45	70		100	99	52	100[b]		100
Ozonation									
MBR+ozonation									
CAS + chlorination									

Clara et al. (2005) Water Res 39:4797–4807; Carballa et al. (2005) Water Sci Technol 52(8):29–35; Brown et al. (2006) Sci Total Environ 366:772–783; Peng et al. (2011) J Environ Monitor 13:446–454; Choi et al. (2007) Chemosphere 66: 977–984; Senta et al. (2011) J Hazard Mat 192:319–328; Radjenovic et al. (2009) Water Res 43:831–841; Xu et al. (2007) Water Res 41:4526–4534; Yang et al. (2005) J Chrom A 1097:45–53

[a]Only value

[b]Range from negative RE%

2.2 Presence of Sulfonamides in Surface Waters

The first reported case of surface water contamination by SAs was in England in 1982, when Watts et al. detected at least one compound from SAs family in river water at concentrations of 1 μg L^{-1} [58]. Nowadays, in Europe the EU Water Framework Directive (WFD) specifies the need to monitor PPs (SAs among them) in surface waters as an informative step to protect and improve the quality of the European water resources [59]. Given that SAs have been frequently detected in WWTP effluents, several studies have aimed to highlight the state and vulnerability of the receiving freshwaters downstream of urban areas and WWTP facilities, focusing especially in the presence in these water matrices of SAs of human consumption, which are the most commonly detected in the wastewater effluents. The low natural biodegradation of SAs [60], and low tendency to adsorb to solid matrices (from the river bed) [61, 62] together with the SAs inputs, both agricultural and urban, that the river may receive all along the basin would lead to a marked concentration gradient from the source to the mouth of the water course. When interpreting the obtained data, seasonal changes should also be taken into account. Generally, the highest concentrations of human SAs (SMX, SPY) are expected during the dry seasons, as the dilution exerted by the receiving streams is lower. For instance, Kim and Carlson [63] detected SMX at a maximum average concentration of 230 ng L^{-1} during the winter and of 320 ng L^{-1} during the summer, in the dry season, in Cache La Poudre River, in northern Colorado. During the rainy season, whereas concentrations of SAs from human use would be more diluted, runoff from irrigated rural areas may increase the concentrations in freshwater of veterinary SAs, denoting its runoff origin from crop lands after heavy rain periods. For instance, in the study by Kim et al., runoff from irrigated rural areas increased the concentrations in freshwater of sulfamerazine (SMR) and sulfadimethoxine (SDM), veterinary SAs (40 and −60 ng L^{-1}), respectively. Cold conditions can also contribute to higher concentrations due to reduced biodegradation of these contaminants in water. Other studies on the distribution of SAs in surface waters yielded similar outcomes, with higher levels of human SAs (SMX,SPY) detected during the dry periods and higher levels of veterinary SAs, such as SDM, sulfamethazine (SMZ), or SDZ during high flow conditions [37, 64, 65]. In some occasions, the release of untreated wastewaters due to strong rainfall events can also contribute to higher concentrations of human SAs than expected [66]. In Europe, the impact of urban inputs was also demonstrated during two sampling campaigns carried out along the Ebro River Basin (Spain) in 2007–2008 [31]. Samples corresponding to tributaries of the main water course presented the highest total concentration of SAs due to their lower flows and dilution exerted on the effluents loads. In 2008, strong rainfall and subsequent runoff events from agricultural land accounted for the highest total SAs concentrations detected in two sampling points in the Ebro River located upstream of two WWTPs (without urban influence). SMX was again the SA most frequently detected in the different surface water samples investigated, being present in the 100% of the samples during the dryer period (2007), with an average concentration of 89.8 ng L^{-1}, and in the 69% of the samples during a higher waterfall period (2008), with an average concentration of 25.5 ng L^{-1}. SPY was also detected in the 100% of the samples during the dry period, at an average concentration of 11 ng L^{-1}, and in the 62% of the samples during the rainy season, with a lower average concentration of 2.7 ng L^{-1} . Another 16 SAs and one acetylated metabolite were detected at concentrations ranging from 0.1 to 127 ng L^{-1}. SMX was also present in freshwater from the Douro River in Portugal, with a maximum concentration of 53.3 ng L^{-1} and an occurrence of 33% [67]. SMX was detected in the Seine River in all the samples investigated over a period of 6 months in 2006, with average concentrations between 37 and 140 ng L^{-1} [66]. In this study, the concentration of SMX seemed to increase after heavy rain episodes, which was attributed to the release of untreated wastewaters and not to surface runoff in agriculture areas, as SMX is mainly used in human medicine. Lower concentrations of SMX were detected by the same author in the Oise River, Marne River, and again Seine River (12–26 ng L^{-1}) [68]. SMX was detected also different sampling sites along the Elbe River in Germany and the Czech Republic during 1999 and 2000 at concentrations in the range of 30–70 ng L^{-1} [69]. The presence of SAs not only in river water samples but also in their sediments [7, 63, 65, 70–72], despite their low distribution coefficients (K_d), highlights the river systems vulnerability against these antimicrobials. Furthermore, the presence of SAs metabolites such as their acetylated or glucuronidated moieties has been already demonstrated and the neglection of these compounds would mean to underestimate the real SAs concentration in the water matrix under study, and also the potential adverse effects derivated from the ecosystems exposure to these substances. For instance, the acetylated form of SMX has been detected in natural streams at higher frequencies and concentrations than its parent molecule [73]. A recent study has also demonstrated that N4-acetyl-SPY is more toxic than its parent compound, SPY, to aquatic bacteria [74].

3 Sulfonamide Presence in the Mediterranean Region: The Case of the Llobregat River Basin

The semiarid conditions present in the Mediterranean region aggravate the adverse ecological effects derived from the presence of SAs and other PPs in natural water courses [75]. The hydrology of streams and rivers of these regions are characterized by high seasonal variability with periods of low or intermittent flow disrupted by acute floods [76, 77]. The increasing population density has resulted in not only a higher water demand for irrigation or human consumption, but also in the intensification of wastewater inputs on the receiving streams, which usually present low natural base flows due to the aforementioned long draught periods. These inputs are among the major stressors of receiving streams and rivers, as they contain an excess of nutrients together with a wide range of emerging contaminants. The Llobregat River is an illustrative example of the hydrological pattern of Mediterranean rivers, with low winter and summer discharges and periodic floods in spring and autumn. It is located in the northeast of Catalonia (Spain) and flows into the Mediterranean Sea south of the city of Barcelona. Along its 156 km, it covers a catchment area of about 4,957 km^2, which is densely populated (3,089,465 inhabitants, data from 1999), especially in its middle and lower sections. Together with its two main tributaries, the Cardener River and the Anoia River, the Llobregat is subjected to heavy anthropogenic pressure, receiving extensive industrial and urban discharges from more than 50 WWTPs (137 Hm^3 $year^{-1}$; 92% from WWTPs) [78]. These inputs are only partially diluted by its natural flow (0.68–6.5 m^3 s^{-1} basal flow). Furthermore, 30% of the annual discharge of the river (693 Hm^3) is used for drinking water supply, including the city of Barcelona. The average monthly flow registered in 2000–2008 period showed peaks of 100 m^3 s^{-1} together with minimum values of 1 m^3 s^{-1} (www.gencat.cat/aca). The Llobregat has therefore been chosen in several studies as the typical case study of the problematic of a Mediterranean overexploited river. Recently, within the framework of the European project MODELKEY, several works have been devoted to study the presence of emerging contaminants [79–81]. During three sampling campaigns carried out along the Llobregat River and one of its main tributaries, the Anoia River, different types of emerging contaminants, including PhPs and SAs, were monitored [82]. Samples were taken in June and November of 2005 and May of 2006, covering spring and autumn periods (maximum flow periods). In the case of SAs, the highest concentrations were detected in the low course of the river and near its mouth (Fig. 1). Due to the cumulative effect along the basin mentioned in Sect. 2.2, SAs followed a pollution gradient and these high

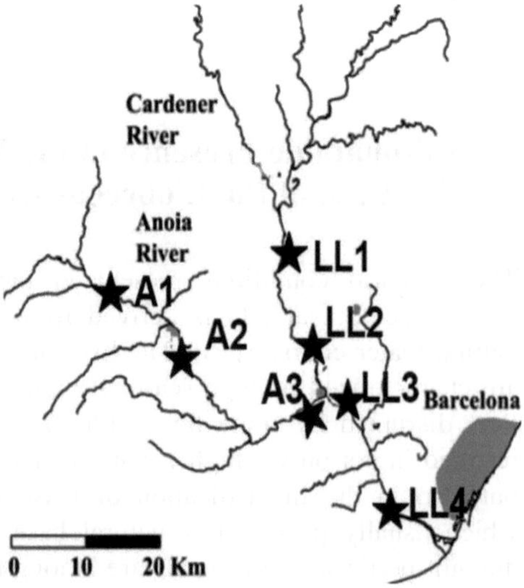

Fig. 1 Sampling sites studied in the Llobregat and Anoia River

concentrations are due to both frequent WWTP discharges and accumulation. SMX was present at a maximum concentration of 4,297 ng L^{-1}, followed by SMZ at 2,482 ng L^{-1}, and its acetylated metabolite that was present at a concentration of 695 ng L^{-1}. Furthermore, estimated values for SPY, sulfamethoxypyridazine (SMP), SDZ, and SPY were out of the analytical calibration range (>5,000 ng L^{-1}) in the sampling location. These values are over two orders of magnitude above the values obtained in continental rivers (Fig. 2, Table 2) and, as can be observed, correspond to SAs of both veterinary and human use. In a recent work, SAs have been detected in effluents of four different WWTPs along this basin, but their concentrations were never higher than 300 ng L^{-1} [74]. In the Anoia River, despite its lower flow and dilution factor exerted on the incoming pollutants, concentrations were markedly lower, results that can be explained in terms of the lower number of discharging WWTPs to this tributary in comparison with the Llobregat. Urban inputs play a major role in both the hydrology and the presence of pollutants in this basin, as demonstrated with SAs.

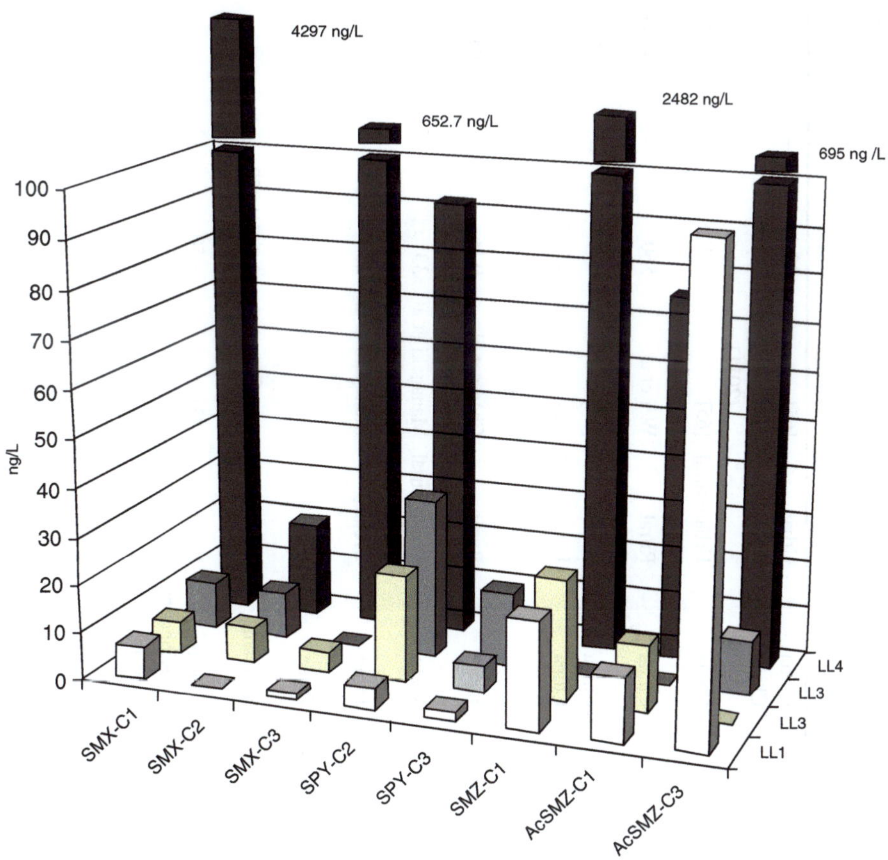

Fig. 2 Sulfonamide concentrations (ng L^{-1}) detected in the Llobregat and Anoia River

Table 2 Maximum sulfonamide concentrations detected in different river waters worldwide. Concentrations during high and low flow periods are given when possible (high flow/low flow)

Study site	Country	Area	Reference	SMX	AcSMX	SDZ	AcSDZ	SMZ	AcSMZ	SMM	SDM	SMT
Jiulong River	China	Urban	[65]	93.4	–	60.5	–	124.4	–	49	–	–
Jiulongjiang tributaries	China	Rural	Peng et al. (2011)	15.5/58.3	–	50.7/316[a]	–	55.6/775.5	–	–	nd/1.9	–
Jiulong estuary	China	Urban–rural	[65]	8.9/28.2	–	10.5/55.3[a]	–	19.3/281.4	–	31.1	0.8	–
Jiangsu area	China	Rural	Wei et al. (2011)	560	–	17,000	–		–	–	–	–
Haihe River	China	Urban	[98]		–	170	–	190	–	–	–	–
Liao River	China	Urban–rural	Jia et al. (2011)	173.2	268.5	30.5	3.3	26.4	11.5	35.1	1	–
Han River	Korea	Urban	Choi et al. (2007)	82	–		–		–	–		–
Koyama River	Japan	Rural	Chang et al. (2008)	0.56	–	0.05	–	0.14	–	–	0.17	0.07
Mekong River	Vietnam	Urban–rural	Managaki et al. (2007)	33	–	–	–	28	–	–	nd	–
Tamagawa River	Japan	Urban	Managaki et al. (2007)	23	–	–	–	nd	–	–	nd	–
Kim otsukigawa River	Japan	Rural	Managaki et al. (2007)	3	–	–	–	nd	–	–	nd	–
Red River delta	Vietnam	Rural	[96]	3,847/4,330	–	–	–	46.2/66.2	–	–	nd/nd	nd/nd
South-east England	United Kingdom	Urban	[73]	<50	239	–	–	–	–	–	–	–
Danshuei River/Gaoping River	Taiwan	Urban	[37]	369	–	–	–	–	–	–	–	–
Rio grande	New Mexico	Urban	Brown et al. (2006)	300	–	–	–	–	–	–	–	–

Study site	Country	Area	Reference	SMX	AcSMX	SDZ	AcSDZ	SMZ	AcSMZ	SMM	SDM	SMT
Elbe	Germany		[69]	70	–	–	–	–	–	–	–	–
	Czec Republic									–		
Cache La Poudre River	US	Urban	Yang et al. (2003)	320	–	–	–	20	–	–	40	–
Various	US	Urban	[65]	150	–	–	–	nd	–	–	10	nd
Seine River	France	Urban	[66]	26	–	–	–	–	–	–	–	–
Seine tributaries	France	Urban	[68]	19.8	–	–	–	<LOQ	–	–	–	–
Alzette River	Luxembourg	Rural	Pailler et al. (2009)	22	–	–	–	<LOQ	–	–	3	–
Mess River	Luxembourg		Meyer et al. (2011)	5	–	–	–	<LOQ	–	–	<LOQ	–
Alzette River	Luxembourg	Rural	Meyer et al. (2011)	118	–	–	–	19	nd	–	nd	–
Danube River	Croatia	Urban	Massey et al. (2010)	6	–	–	–	5	–	–	–	–
Illinois River	US	Rural	Campagnolo et al. (2002)	564	–	–	–	–	–	–	–	–
Ebro River	Spain	Urban	[31]	30.3	–	1.3	–	12.6	2.01	–	2.97	4.62
Ebro tributaries	Spain	Urban	[31]	35.6	–	6.4	–	65.2	20.2	–	23.1	nd
Llobregat River	Spain	Urban	[82]	4,297	–	[a]	–	2,481.8	695	–	136	5.6
Llobregat tributaries	Spain	Urban	[82]	167.9	–	nd	–	15.7	20.2	–	5.4	10.3
Jamaica bay estuary	US	Urban	Benotti et al. (2007)	80.7	–	–	–	–	–	–	–	–
Douro estuary	Portugal	Urban	[67]	53.3	–	–	–	–	–	–	–	–

(continued)

Table 2 (Continued)

Study site	STZ	SMR	SCP	SMP	SPY	AcSPY	SSX	SSD	SNT	SQX	SDX	SuSTZ	SBZ	SGD	STT	SFS
Jiulong River	–	–	–	–	–	–	–	–	–	–	–	–	–	–		–
Jiulongjiang tributaries	–	nd/nd	–	–	–	–	–	–	–	–	–	–	–	–	5.4/86.8	
Jiulong estuary	–	nd/nd	–	–	–	–	–	–	–	–	–	–	–	–	nd/14.9	–
Jiangsu area	–	–	–	–	–	–	–	–	–	640	630	–	–	–	–	–
Haihe River	–	–	210	–	–	–	–	–	–	–	–	–	–	–	–	–
Liao River	8.5	–	8.1	–	15.7	13.7	–	0.4	–	13.6	–	–	–	8	nd	–
Han River	–	–	–	–	–	–	–	–	–	–	–	–	–	–	–	–
Koyama River	6.6	nd	–	–	3	–	nd	–	–	8.9	–	–	–	–	nd	0.48
Mekong River	nd	nd	–	–	nd	–	–	–	–	–	–	–	–	–	–	–
Tamagawa River	nd	nd	–	–	132	–	–	–	–	–	–	–	–	–	–	–
Kim otsukigawa River	nd	nd	–	–	nd	–	–	–	–	–	–	–	–	–	–	–
Red River delta	nd/nd	nd/nd	–	–	nd/57.5	–	–	–	–	–	–	–	–	–	–	–
South-east England	–	–	–	–	–	–	–	–	–	–	–	–	–	–	–	–
Danshuei River/Gaoping River	–	–	–	–	–	–	–	–	–	–	–	–	–	–	–	–
Rio grande	–	–	–	–	–	–	–	–	–	–	–	–	–	–	–	–
Elbe	–	–	–	–	–	–	–	–	–	–	–	–	–	–	–	–
				–												
Cache La Poudre River	30	6	30	–	–	–	–	–	–	–	–	–	–	–	–	–
Various	nd	nd	nd	–	–	–	–	–	–	–	–	–	–	–	–	–
Seine River	–	–	–	–	–	–	–	–	–	–	–	–	–	–	–	–
Seine tributaries	–	–	–	–	–	–	–	–	–	–	–	–	–	–	–	–
Alzette River	<LOQ	–	–	–	–	–	–	–	–	–	–	–	–	–	–	–
Mess River	2	–	–	–	–	–	–	–	–	–	–	–	–	–	–	–
Alzette River	5	–	–	–	–	–	–	–	–	–	–	–	–	–	–	–
Danube River	nd	–	–	–	<LOQ	–	–	–	–	–	–	–	–	–	–	–
Illinois River	–	–	–	–	–	–	–	–	–	–	–	–	–	–	–	–

Study site	STZ	SMR	SCP	SMP	SPY	AcSPY	SSX	SSD	SNT	SQX	SDX	SuSTZ	SBZ	SGD	STT	SFS
Ebro River	9.6	30.4	–	2.09	19.2	–	0.73	10.9	64.7	22	36.3	16.7	9.85	–	–	–
Ebro tributaries	10.1	42.2		18.1	42.5	–	0.6	23.8	127	40.4	43.3	37	14.6	–	–	–
Llobregat River	960.6	–	–	164.9	91.8	–	24.7	–	–	–	–	–	–	–	–	–
Llobregat tributaries	3	–	–	nd	39.7	–	nd	–	–	–	–	–	–	–	–	–
Jamaica bay estuary	–	–	–	–	–	–	–	–	–	–	–	–	–	–	–	–
Douro estuary	–	–	–	–	–	–	–	–	–	–	–	–	–	–	–	–

SMX sulfamethoxazole, *AcSMX* N[4]-acetylsulfamethoxazole, *SDZ* sulfadiazine, *AcSDZ* N[4]-acetylsulfadiazine, *SMZ* sulfamethazine, *AcSMZ* N[4]-acetylsulfamethazine, *SMM* sulfamonomethoxine, *SMT* sulfamethizole, *STZ* sulfathiazole, *SMR* sulfamerazine, *SCP* sulfachloropyridazine, *SMP* sulfamethoxypyridazine, *SPY* sulfapyridine, *AcSPY* N[4]-acetylsulfapyridine, *SSX* sulfisoxazole, *SSD* sulfadimidine, *SNT* sulfanitran, *SQX* sulfaquinoxaline, *SDX* sulfadoxine, *SuSTZ* succinyl-sulfathiazole, *SBZ* sulfabenzamide, *SGD* sulfaguanidine, *STT* sulfameter, *SFS* sulfisomidine. Peng et al. (2011) J Environ Monitor 13:446–454; Wei et al. (2011) Chemosphere 82:1408–1414; Jiia et al. (2011) Environ Toxicol Chem 30:1252–1260; Choi et al. (2007) Chemosphere 66:977–984; Chang et al. (2008) Chi Sci Bull 53:514–520; Managaki et al. (2007) Environ Sci Technol 41:8004–8010; Brown et al. (2006) Sci Total Environ 366:772–783; Yang and Carlson (2003) Water Res 37(19):4654–4656; Pailler et al. (2009) Sci Total Environ 407:4736–4743; Meyer et al. (2011) Environ Monit Assess 180:127–146; Massey et al. (2010) Ecol Eng 36:930–938; Campagnolo et al. (2002) Sci Total Environ 299:89–95; Benotti et al. (2007) Environ Sci Technol 41:5795–5802. – no data, *nd* not detected, *<LOQ* under the method limit of quantification

[a]Values detected higher than the analytical calibration range

4 Ecotoxicological Effects of Sulfonamides in the Aquatic Environment

There is a substantial lack of ecotoxicological data regarding adverse effects of SAs and their metabolites, which is probably one of the main reasons for the absence of European regulation on maximum levels of this family of antibiotics in any environmental compartment. Nowadays, none of the PhPs detected in surface water are considered in any of the Drinking Water Directives worldwide [1]. Recently, different PhPs such as carbamazepine or diclofenac were considered to be included in the list of priority substances of the new European Directive 2008/105/EC on environmental quality standards, although they were finally withdrawn. Whereas SAs are probably not pharmacologically active in humans at the concentrations detected so far (usually at the ng L^{-1} level), they might be potential micropollutants to key living organisms in aquatic ecosystems (e.g., fish, aquatic invertebrates and unicellular algae). These different taxonomic groups, belonging to different trophic levels, may be exposed and negatively affected to different extents. For example, severe toxic effects in primary producers may imply loss of the whole food-chain structure, as they represent a significant portion of the total biomass of the ecosystem and are important as a source of carbon for the rest of the aquatic biosphere. Despite the lack of toxicity data available in the literature, it has been demonstrated that generally microalgae are more sensitive than crustaceans and fish to antibacterial agents (e.g., triclosan and ciprofloxacin). However, SAs have proved to hardly pose any toxicity against green algae [83, 84]; estimated inhibitory concentration (IC) values were much higher than those expected in surface waters and SAs have been considered unlikely to be toxic to algae at environmental concentrations. SMX, as one of the most consumed SAs in human medicine and most frequently detected in natural waters, has been the target of different toxicity evaluations. Median effective concentrations (EC_{50}) range from 80 mg L^{-1} against green algae [85] to values of 0.52 mg L^{-1} for algae and 0.21 mg L^{-1} for crustaceans [85], indicating that the risk posed by this substance should not be excluded in real environmental conditions. It has been demonstrated that aquatics plants [86], crustaceans and fish are also vulnerable to SAs; SMX also showed toxicity against rainbow trout (*Oncorhynchus mykiss*), but at concentrations so high that were not representative of the real situation in freshwaters [87]. Bioaccumulation of SMZ in sturgeon (*Acipenser schrenkii*) was also demonstrated, but considered of little environmental concern regarding presence in tissues consumed by humans or to biomagnification in fish consumed by fish predators [88]. On the other hand, toxicity and bioaccumulation in marine environment were observed in brine shrimp exposed to SDM, with the potential implications for the rest of the food chain in the marine community [89].

SAs are usually not detected as isolated drugs in the aquatic environment but together with other SAs, and synergistic effects could be expected when residues of different SAs are detected in the same study site [84, 90]. Belonging to the same family of compounds implies similar molecular structure and modes of action, so "concentration addition" is likely. Furthermore, it is necessary to take into account that the degradation products and metabolites of SAs may also be involved in the final toxic effects on the algae, making the interpretation of the toxic data more complex. Recently, EC_{50} values for *Vibrio fischerii* were calculated for SPY and its acetylated metabolite; concentrations of 27.4 mg L^{-1} and of 8.2 mg L^{-1} for SPY and the metabolite, respectively, after 15 min exposure were reported [46, 74]. According to the EU legislation (Directive 447 93/67/EEC) that categorizes the toxicity to aquatic organism depending on the EC_{50}, SPY would be classified as harmful, and its metabolite as toxic. To the author's knowledge, the only reference regarding harmful effects of acetylated SAs is that by Eguchi et al. [90], in which the metabolites of SDM, SMX, and SDZ showed much weaker growth inhibitory effects than the corresponding parent SA against microalgae, usually the more sensitive taxa. The simultaneous presence of the corresponding acetylated metabolites enhanced the inhibitory effect of the three SAs, and also the addition of the diaminopirimidine trimethoprim.

4.1 Ecotoxicity of Sulfonamide Intermediate Products

Whether or not SAs are biodegraded in the aquatic environment would settle the very first step for a complete environmental risk assessment (ERA). At the same time, the toxicity of the intermediate by-products of both biotic and abiotic degradation should be taken into account when evaluating the derived ecological risk. SAs undergo photocatalytic degradation [91, 92] and, if the photodegradation products generated are biodegradable, they can be removed during wastewater treatment using biological methods. If these products are persistent or not readily biodegradable, risks of ecotoxicity should be considered. Both inhibitory and stimulatory effects could be expected, as demonstrated by Baran et al. for sulfathiazole (STZ), SMX, SDZ, and sulfachloropyridazine (SCM) against green algae growth [93]. Photoenhanced toxicity under natural sunlight has already been demonstrated for three SAs (SMX, STZ, and SMZ) against crustacean *Daphnia magna*, suggesting that the photodegradation of the parent compound leads to the formation of more toxic by-products [94]. Also, the by-products of SMX after ozonation treatment were toxic against *D. magna* and *P. subcapicata* [95]. In both cases, the assayed concentrations of SAs that were acutely toxic to *D. magna* were much higher than levels detected in the environment and the ecological risks associated were considered to be limited.

4.2 Bacterial Resistance

So far, environmental research on antibiotics in general has focused mainly on the bacterial resistance acquired against antimicrobials in the different environmental compartments. Nowadays, the widespread presence of resistant bacterial strains has been demonstrated in several scientific works. In river ecosystems, the frequent presence of SMX has led to the detection of SMX-resistant bacteria belonging to *Aeromonas spp.*, typical waterborne bacteria [96]. The Acinetobacter genera were also affected by the presence of this SA [97], and a correlation was established between SMX environmental concentration and occurrence of SMX-resistant bacteria. SAs-resistant genes have been found not only in surface water but also in river sediments [98]. The concentration of these genes was up to 1,200 times higher in sediments, indicating that they can be considered as important antibiotic resistance genes (ARGs) reservoirs. SAs may have qualitative and quantitative effects upon the resident microbial community found in sediment, which can in turn affect the degradation of organic matter. WWTP effluents have been considered as ARG sources in different works too [4, 99, 100].

4.3 Environmental Risk Assessment for Sulfonamides in Surface Waters

As mentioned above, little information is available regarding the ecological effects of SAs and other PhPs, due mainly to the fact that such investigations are not legally required as part of the licensing procedures for human medicaments. The risk assessment guidelines set up by the European Medicines Agency (EMEA) for the marketing authorization of new medicinal products have been used in a few occasions to prioritize the risk from drugs that are already in use and to assess the potential impact of drugs yet to be released [31, 48, 101–106]. Although they are designed as part of the process for registering new drugs, they are used nowadays as the only restrictive measure established so far to evaluate environmental risk from drugs that are already being consumed and that are being excreted in aquatic or terrestrial environments. The ERA protocol is a two-phase tiered process that begins with an approximate calculation of the predicted environmental concentration (PEC) of the drug in water. These guidelines recommend that any drug exceeding 10 ng L^{-1} in surface water should progress to Phase II, where standard acute toxicity tests will be carried out in order to estimate predicted no-effect concentration (PNEC) or nonobserved effect concentration (NOEC) [107]. Finally, the ratio of the PEC to PNEC, known as the hazard quotient (HQ), indicates whether a potential environmental impact is implicit and further testing might be needed ($HQ > 1$). It is also recommended that when the total concentration of metabolites is a 10% greater than the concentration of the corresponding parent drug, the metabolites are also to be further investigated (phase II tier B) in order to determine their ecotoxicological effects. The EMEA Committee for Medicinal Products for Veterinary Use also established similar guidelines to assess the potential for veterinary medicines to affect nontarget species in the environment, including both aquatic and terrestrial species [108]. When PNEC values are not

Table 3 Estimation of hazard quotients (HQ) for the different sulfonamides present in surface waters, following the EMEA guidelines

SA	MEC	REF_{MEC}	PEC	Taxa	$PNEC_{acute}$	REF_{PNEC}	HQ
	–		0.95		0.146	[63]	*6.3*
	0.4	[104]			0.03	[105]	*13.4*
	–		0.31		0.027		*11.4*
			0.31		0.59[a]		*0.5*
			1.6	*Blue green algae*	0.027		*59.3*
	–		1.6		0.59[a]	[101]	*2.7*
	0.0356	[34]	–		0.027		*1.32*
	0.284	[75]	–		0.027		*10.52*
	4.3	[83]	–		0.027		*159.26*
SMX	0.0356	[34]	–		78.1		<0.001
	0.284	[75]	–	*V. fischerii*	78.1	Boxall et al. (2002)	0.004
	4.3	[83]	–		78.1		0.055
	0.0356	[34]	–				0.001
	0.284	[75]	–	*Daphnids*	25.2	Lutzhoft et al. (1999)	0.011
	4.3	[83]	–				0.171
	0.0356	[34]	–				<0.001
	0.284	[75]	–	*Fish*	562.5	Choi et al. (2007)	<0.001
	4.3	[83]	–				0.008
AcSMX	0.094	[75]		*Blue green algae*	101	[90]	<0.001
	0.042	[34]	–				0.002
SPY	0.177	[75]	–	*V. fischerii*	27.4	[74]	0.006
	0.092	[83]	–				0.003
ACSPY	0.522	[75]		*V. fischerii*	8.2	[74]	0.064
	0.065	[34]	–				<0.001
	0.0364	[75]	–	*V. fischerii*	344.7	Boxall et al. (2002)	<0.001
	2.48	[83]	–				0.007
	0.065	[34]	–				<0.001
SMZ	0.0364	[75]	–	*Daphnids*	147.5	Migliore et al. (1993)	<0.001
	2.48	[83]					0.017
	0.065	[34]	–				<0.001
	0.0364	[75]	–	*Fish*	110.7	[63]	<0.001
	2.48	[83]	–				0.022
	0.009	[34]	–				<0.001
	0.07	[75]	–	*Blue green algae*	16.32	Migliore et al. (1993)	0.004
	0.96	[83]	–				0.059
	0.009	[34]	–				<0.001
	0.07	[75]	–	*V. fischerii*	1,001	Boxall et al. (2002)	<0.001
	0.96	[83]	–				0.001
STZ	0.009	[34]	–				<0.001
	0.07	[75]	–	*Daphnids*	78.9	Migliore et al. (1993)	<0.001
	0.96	[83]	–				0.012
	0.009	[34]	–				<0.001
	0.07	[75]	–	*Fish*	101	[89]	<0.001
	0.96	[83]	–				0.010
SDZ	0.286	[75]	–	*Blue green algae*	1.225	Migliore et al. (1993)	0.233
AcSDZ	0.067	[75]	–	*Blue green algae*	101	[90]	<0.001

(continued)

Table 3 (continued)

SA	MEC	REF_{MEC}	PEC	Taxa	$PNEC_{acute}$	REF_{PNEC}	HQ
	0.023	[34]	–				0.010
	0.001	[75]	–	*Blue green algae*	2.3	[90]	<0.001
	0.136	[83]	–				0.059
	0.023	[34]	–				<0.001
	0.001	[75]	–	*V. fischerii*	501	Boxall et al. (2002)	<0.001
	0.136	[83]	–				<0.001
SDM	0.023	[34]	–				<0.001
	0.001	[75]	–	*Daphnids*	204.5	Boxall et al. (2002)	<0.001
	0.136	[83]	–				<0.001
	0.023	[34]	–				<0.001
	0.001	[75]	–	*Fish*	101	[63]	<0.001
	0.136	[83]	–				0.001

Boxall et al. (2002) Toxicol Lett 131:19–28; Lüzthoft et al. (1999) Arch Environ Contam Toxicol 36:1–6; Choi et al. (2007) Chemosphere 66:977–984; Migliore et al. (1996); Migliore et al. (1993) Int J Salt Lake Res 2:141–152.
MEC measured environmental concentrations (μg L^{-1}), *PNEC* predicted no-effect concentration, REF_{MEC} literature reference for the MEC value, REF_{PNEC} literature reference for the PNEC value

available, an alternative PNEC can be derived by dividing EC_{50} or median lethal concentration (LC_{50}) values (acute toxicity data) by an uncertainty factor of up to 1,000 [109], and so converting acute to chronic toxicity values, since data on chronic toxicity for SAs is lacking. Likewise, measured environmental concentrations (MECs) are used in the calculation instead of PECs. In order to set up a worst case scenario, maximum MECs and the lowest EC_{50} or LC_{50} values are used. In all cases, the MECs should be higher than the boundary value of 10 ng L^{-1} established by EMEA in Tier 1. Table 3 summarizes the HQ values reported to date in the literature. As can be observed, HQs $>$ 1 were detected only for SMX and only for blue green algae. The highest risk corresponded to the exposure to concentrations detected in the Llobregat River, highlighting once more the vulnerability of water courses located in the Mediterranean climate region.

Acknowledgments This work has been funded by the Spanish Ministry of Science and Innovation through the projects CEMAGUA (CGL2007-64551/HID) and SCARCE (Consolider Ingenio 2010 CSD2009-00065). MJ García acknowledges AGAUR (Generalitat de Catalunya, Spain) for economic support through an FI pre-doctoral grant.

References

1. Ternes T, Joss A (2006) Human pharmaceuticals, hormones and fragrances. The challenge of micropollutants in urban water management. IWA Publishing, London
2. Sarmah AK, Meyer MT, Boxall ABA (2006) A global perspective on the use, sales, exposure pathways, occurrence, fate and effects of veterinary antibiotics (vas) in the environment. Chemosphere 65(5):725–759
3. Kümmerer K (2004) Pharmaceuticals in the environment – sources, fate, effects and risks. Springer, Berlin, pp 27–44

4. Costanzo SD, Murby J, Bates J (2005) Ecosystem response to antibiotics entering the aquatic environment. Mar Pollut Bull 51(1–4):218–223
5. Amin MM, Zilles JL, Greiner J, Charbonneau S, Raskin L, Morgenroth E (2006) Influence of the antibiotic erythromycin on anaerobic treatment of a pharmaceutical wastewater. Environ Sci Technol 40(12):3971–3977
6. Jones OAH, Voulvoulis N, Lester JN (2002) Aquatic environmental assessment of the top 25 English prescription pharmaceuticals. Water Res 36:5013–5022
7. Löffler D, Römbke J, Meller M, Ternes TA (2005) Environmental fate of pharmaceuticals in water-sediment systems. Environ Sci Technol 39(14):5209–5218
8. Zuccato E, Calamari D, Natangelo M, Fanelli R (2000) Presence of therapeutic drugs in the environment. Lancet 355(9217):1789–1790
9. Chafer-Pericas C, Maquieira T, Puchades R, Company B, Miralles J, Moreno A (2010) Multiresidue determination of antibiotics in aquaculture fish samples by HPLC-MS/MS. Aquacult Res 41(9):e217–e225
10. Hamscher G, Priess B, Nau H (2006) A survey of the occurrence of various sulfonamides and tetracyclines in water and sediment samples originating from aquaculture systems in northern Germany in summer 2005. Untersuchung von teichwässern und -sedimenten in Niedersächsischen aquakulturen im sommer 2005 auf sulfonamide und tetracycline 57(4):97–101
11. Pérez-Trallero E, Iglesias L (2003) Tetracyclines, sulfonamides and metronidazole. Tetraciclinas, sulfamidas y metronidazol. Enferm Infec Micr Cl 21(9):520–529+533
12. UE (2003) Regulation 1831/2003/ec on additives for use in animal nutrition
13. Parfitt KE (1999) Martindale–the complete drug reference, 32nd edn. Pharmaceutical Press, London
14. Halling-Sorensen B, Nors Nielsen S, Lanzky PF, Ingerslev F, Holten Lutzhoft HC, Jorgensen SE (1998) Occurrence, fate and effects of pharmaceutical substances in the environment – a review. Chemosphere 36(2):357–393
15. Kwon JW (2011) Mobility of veterinary drugs in soil with application of manure compost. Bull Environ Contam Toxicol 87(1):40–44
16. Motoyama M, Nakagawa S, Tanoue R, Sato Y, Nomiyama K, Shinohara R (2011) Residues of pharmaceutical products in recycled organic manure produced from sewage sludge and solid waste from livestock and relationship to their fermentation level. Chemosphere 84(4):432–438
17. Pan X, Qiang Z, Ben W, Chen M (2011) Residual veterinary antibiotics in swine manure from concentrated animal feeding operations in shandong province, china. Chemosphere 84(5):695–700
18. Hamscher G, Pawelzick HT, Hoper H, Nau H (2005) Different behavior of tetracyclines and sulfonamides in sandy soils after repeated fertilization with liquid manure. Environ Toxicol Chem 24(4):861–868
19. Boxall ABA, Kolpin DW, Halling-Sorensen B, Tolls J (2003) Are veterinary medicines causing environmental risks? Environ Sci Technol 37(15):286A–294A
20. Schauss K, Focks A, Heuer H, Kotzerke A, Schmitt H, Thiele-Bruhn S, Smalla K, Wilke BM, Matthies M, Amelung W, Klasmeier J, Schloter M (2009) Analysis, fate and effects of the antibiotic sulfadiazine in soil ecosystems. Trends Anal Chem 28(5):612–618
21. Kotzerke A, Sharma S, Schauss K, Heuer H, Thiele-Bruhn S, Smalla K, Wilke BM, Schloter M (2008) Alterations in soil microbial activity and n-transformation processes due to sulfadiazine loads in pig-manure. Environ Pollut 153(2):315–322
22. Diaz-Cruz MS, Garcia-Galan MJ, Barcelo D (2008) Highly sensitive simultaneous determination of sulfonamide antibiotics and one metabolite in environmental waters by liquid chromatography-quadrupole linear ion trap-mass spectrometry. J Chromatogr A 1193 (1–2):50–59
23. Batt AL, Snow DD, Aga DS (2006) Occurrence of sulfonamide antimicrobials in private water wells in washington county, Idaho, USA. Chemosphere 64(11):1963–1971

24. Lindsey ME, Meyer M, Thurman EM (2001) Analysis of trace levels of sulfonamide and tetracycline antimicrobials, in groundwater and surface water using solid-phase extraction and liquid chromatography/mass spectrometry. Anal Chem 73(19):4640–4646
25. Sacher F, Lang FT, Brauch HJ, Blankenhorn I (2001) Pharmaceuticals in groundwaters: analytical methods and results of a monitoring program in Baden-wurttemberg, Germany. J Chromatogr A 938(1–2):199–210
26. Blackwell PA, Lutzhoft HCH, Ma HP, Halling-Sorensen B, Boxall ABA, Kay P (2004) Fast and robust simultaneous determination of three veterinary antibiotics in groundwater and surface water using a tandem solid-phase extraction with high-performance liquid chromatography-UV detection. J Chromatogr A 1045(1–2):111–117
27. Diaz-Cruz MS, Barcelo D (2006) Determination of antimicrobial residues and metabolites in the aquatic environment by liquid chromatography tandem mass spectrometry. Anal Bioanal Chem 386(4):973–985
28. Karthikeyan KG, Meyer MT (2006) Occurrence of antibiotics in wastewater treatment facilities in Wisconsin, USA. Sci Total Environ 361(1–3):196–207
29. Batt AL, Aga DS (2005) Simultaneous analysis of multiple classes of antibiotics by ion trap LC/MS/MS for assessing surface water and groundwater contamination. Anal Chem 77(9):2940–2947
30. Watanabe N, Bergamaschi BA, Loftin KA, Meyer MT, Harter T (2010) Use and environmental occurrence of antibiotics in freestall dairy farms with manured forage fields. Environ Sci Technol 44(17):6591–6600
31. García-Galán MJ, Díaz-Cruz MS, Barceló D (2011) Occurrence of sulfonamide residues along the ebro river basin. Removal in wastewater treatment plants and environmental impact assessment. Environ Int 37(2):462–473
32. Gobel A, McArdell CS, Joss A, Siegrist H, Giger W (2007) Fate of sulfonamides, macrolides, and trimethoprim in different wastewater treatment technologies. Sci Total Environ 372(2–3):361–371
33. Gros M, Petrović M, Barceló D (2007) Wastewater treatment plants as a pathway for aquatic contamination by pharmaceuticals in the Ebro river basin (Northeast Spain). Environ Toxicol Chem 26(8):1553–1562
34. Ye S, Yao Z, Na G, Wang J, Ma D (2007) Rapid simultaneous determination of 14 sulfonamides in wastewater by liquid chromatography tandem mass spectrometry. J Sep Sci 30(15):2360–2369
35. Topp E, Monteiro SC, Beck A, Coelho BB, Boxall ABA, Duenk PW, Kleywegt S, Lapen DR, Payne M, Sabourin L, Li H, Metcalfe CD (2008) Runoff of pharmaceuticals and personal care products following application of biosolids to an agricultural field. Sci Total Environ 396 (1):52–59
36. Miao XS, Bishay F, Chen M, Metcalfe CD (2004) Occurrence of antimicrobials in the final effluents of wastewater treatment plants in Canada. Environ Sci Technol 38(13):3533–3541
37. Lin AYC, Tsai YT (2009) Occurrence of pharmaceuticals in Taiwan's surface waters: impact of waste streams from hospitals and pharmaceutical production facilities. Sci Total Environ 407(12):3793–3802
38. Chang XS, Meyer MT, Liu XY, Zhao Q, Chen H, Chen JA, Qiu ZQ, Yang L, Cao J, Shu WQ (2010) Determination of antibiotics in sewage from hospitals, nursery and slaughter house, wastewater treatment plant and source water in Chongqing region of three gorge reservoir in China. Environ Pollut 158(5):1444–1450
39. Ahel M, Mikac N, Cosovic B, Prohic E, Soukup V (1998) The impact of contamination from a municipal solid waste landfill (Zagreb, Croatia) on underlying soil. Water Sci Technol 37(8):203–210
40. Schwarzbauer J, Heim S, Brinker S, Littke R (2002) Occurrence and alteration of organic contaminants in seepage and leakage water from a waste deposit landfill. Water Res 36(9):2275–2287
41. Bound JP, Voulvoulis N (2005) Household disposal of pharmaceuticals as a pathway for aquatic contamination in the united kingdom. Environ Health Perspect 113(12):1705–1711

42. Samuelsen OB, Lunestad BT (1996) Bath treatment, an alternative method for the administration of the quinolones flumequine and oxolinic acid to halibut Hippoglossus hippoglossus, and in vitro antibacterial activity of the drugs against some Vibrio sp. Dis Aquat Organ 27(1):13–18
43. Le-Minh N, Khan SJ, Drewes JE, Stuetz RM (2010) Fate of antibiotics during municipal water recycling treatment processes. Water Res 44(15):4295–4323
44. Garoma T, Umamaheshwar SK, Mumper A (2010) Removal of sulfadiazine, sulfamethizole, sulfamethoxazole, and sulfathiazole from aqueous solution by ozonation. Chemosphere 79(8):814–820
45. Lin AY-C, Lin C-F, Chiou J-M, Hong PKA (2009) O_3 and O_3/H_2O_2 treatment of sulfonamide and macrolide antibiotics in wastewater. J Hazard Mater 171(1–3):452–458
46. García Galán MJ, Frömel T, Müller J,Peschka M, Knepper T, Díaz Cruz S, Barceló D (2011) Biodegradation studies of N^4-acetylsulfapyridine and N^4-acetylsulfamethazine in environmental water applying mass spectrometry techniques. Anal Bioanal Chem 402(9):2885–2896
47. Gobel A, McArdell CS, Suter MJF, Giger W (2004) Trace determination of macrolide and sulfonamide antimicrobials, a human sulfonamide metabolite, and trimethoprim in wastewater using liquid chromatography coupled to electrospray tandem mass spectrometry. Anal Chem 76(16):4756–4764
48. Gros M, Petrovic M, Ginebreda A, Barceló D (2010) Removal of pharmaceuticals during wastewater treatment and environmental risk assessment using hazard indexes. Environ Int 36(1):15–26
49. Tambosi JL, de Sena RF, Favier M, Gebhardt W, Jose HJ, Schroder HF, Moreira R (2010) Removal of pharmaceutical compounds in membrane bioreactors (MBR) applying submerged membranes. Desalination 261(1–2):148–156
50. Tambosi JL, de Sena RF, Gebhardt W, Moreira R, Jose HJ, Schroder HF (2009) Physicochemical and advanced oxidation processes – a comparison of elimination results of antibiotic compounds following an MBR treatment. Ozone Sci Eng 31(6):428–435
51. García Galán MJ, Díaz-Cruz, M.S., Barceló, D. (2011) Removal of sulfonamide antibiotics upon conventional activated sludge and advance membrane bioreactors treatment. J Hazard Mat (accepted)
52. Radjenovic J, Petrovic M, Barceló D (2007) Analysis of pharmaceuticals in wastewater and removal using a membrane bioreactor. Anal Bioanal Chem 387(4):1365–1377
53. Dodd MC, Huang CH (2004) Transformation of the antibacterial agent sulfamethoxazole in reactions with chlorine: kinetics, mechanisms, and pathways. Environ Sci Technol 38(21):5607–5615
54. Huber MM, Canonica S, Park GY, Von Gunten U (2003) Oxidation of pharmaceuticals during ozonation and advanced oxidation processes. Environ Sci Technol 37(5):1016–1024
55. Nakada N, Shinohara H, Murata A, Kiri K, Managaki S, Sato N, Takada H (2007) Removal of selected pharmaceuticals and personal care products (PPCPs) and endocrine-disrupting chemicals (EDCs) during sand filtration and ozonation at a municipal sewage treatment plant. Water Res 41(19):4373–4382
56. Kosutic K, Dolar D, Asperger D, Kunst B (2007) Removal of antibiotics from a model wastewater by RO/NF membranes. Sep Purif Technol 53(3):244–249
57. Zwiener C (2007) Occurrence and analysis of pharmaceuticals and their transformation products in drinking water treatment. Anal Bioanal Chem 387(4):1159–1162
58. Watts CD, Crathorne B, Fielding M, Killops SD (1982) Nonvolatile organic compounds in treated waters. Environ Health Perspect 46:87–89
59. Feitosa-Felizzola J, Chiron S (2009) Occurrence and distribution of selected antibiotics in a small mediterranean stream (Arc river, Southern France). J Hydrol 364(1–2):50–57
60. Garcia-Galan MJ, Diaz-Cruz MS, Barcelo D (2008) Identification and determination of metabolites and degradation products of sulfonamide antibiotics. Trends Anal Chem 27(11):1008–1022
61. Thiele-Bruhn S, Aust MO (2004) Effects of pig slurry on the sorption of sulfonamide antibiotics in soil. Arch Environ Contam Toxicol 47(1):31–39

62. Kolpin DW, Furlong ET, Meyer MT, Thurman EM, Zaugg SD, Barber LB, Buxton HT (2002) Pharmaceuticals, hormones, and other organic wastewater contaminants in US Streams, 1999–2000: a national reconnaissance. Environ Sci Technol 36(6):1202–1211
63. Kim SC, Carlson K (2007) Temporal and spatial trends in the occurrence of human and veterinary antibiotics in aqueous and river sediment matrices. Environ Sci Technol 41(1):50–57
64. Kolpin DW, Skopec M, Meyer MT, Furlong ET, Zaugg SD (2004) Urban contribution of pharmaceuticals and other organic wastewater contaminants to streams during differing flow conditions. Sci Total Environ 328(1–3):119–130
65. Zheng S, Qiu X, Chen B, Yu X, Liu Z, Zhong G, Li H, Chen M, Sun G, Huang H, Yu W, Freestone D (2011) Antibiotics pollution in Jiulong river estuary: source, distribution and bacterial resistance. Chemosphere 84(11):1677–1685
66. Tamtam F, Mercier F, Le Bot B, Eurin J, Tuc Dinh Q, Clement M, Chevreuil M (2008) Occurrence and fate of antibiotics in the Seine river in various hydrological conditions. Sci Total Environ 393(1):84–95
67. Madureira TV, Barreiro JC, Rocha MJ, Rocha E, Cass QB, Tiritan ME (2010) Spatiotemporal distribution of pharmaceuticals in the Douro river estuary (Portugal). Sci Total Environ 408 (22):5513–5520
68. Tamtam F, Mercier F, Eurin J, Chevreuil M, Le Bot B (2009) Ultra performance liquid chromatography tandem mass spectrometry performance evaluation for analysis of antibiotics in natural waters. Anal Bioanal Chem 393(6–7):1709–1718
69. Wiegel S, Aulinger A, Brockmeyer R, Harms H, Löffler J, Reincke H, Schmidt R, Stachel B, von Tümpling W, Wanke A (2004) Pharmaceuticals in the river Elbe and its tributaries. Chemosphere 57(2):107–126
70. Tamtam F, Le Bot B, Dinh T, Mompelat S, Eurin J, Chevreuil M, Bonté P, Mouchel JM, Ayrault S (2011) A 50-year record of quinolone and sulfonamide antimicrobial agents in Seine river sediments. J Soils Sediments 11:852–859
71. Cai-Ming T, Qiu-Xin H, Yi-Yi Y, Xian-Zhi P (2009) Multiresidue determination of sulfonamides, macrolides, trimethoprim, and chloramphenicol in sewage sludge and sediment using ultrasonic extraction coupled with solid phase extraction and liquid chromatography-tandem mass spectrometry. Fenxi Huaxue/Chinese J Anal Chem 37(8):1119–1124
72. Yang JF, Ying GG, Zhao JL, Tao R, Su HC, Chen F (2010) Simultaneous determination of four classes of antibiotics in sediments of the Pearl river using RRLC-MS/MS. Sci Total Environ 408(16):3424–3432
73. Ashton D, Hilton M, Thomas KV (2004) Investigating the environmental transport of human pharmaceuticals to streams in the United Kingdom. Sci Total Environ 333(1–3):167–184
74. García Galán MJ, Díaz-Cruz MS, Barceló D (2011) Removal of selected sulfonamides and its metabolites during conventional activated sludge treatment. Evaluation of the potential environmental impact of wastewater effluents on the receiving ecosystems. Water Res (submitted)
75. Petrovic M, Postigo C, Lopez de Alda M, Ginebreda A, Gros M, Radjenovic J, Barcelo D (2010) Occurrence and fate of pharmaceuticals and illicit drugs under water scarcity. In: Sabater S, Barcelo D (eds) Water scarcity in the Mediterranean: perspectives under global change, vol 8, Handbook of environmental chemistry. Springer, Berlin, pp 197–228
76. Gasith A, Resh VH (1999) Streams in mediterranean climate regions: abiotic influences and biotic responses to predictable seasonal events. Annu Rev Eco Sys 30:51–81
77. Marti E, Aumatell J, Godé L, Poch M, Sabater F (2004) Nutrient retention efficiency in streams receiving inputs from wastewater treatment plants. J Environ Qual 33(1):285–293
78. Muñoz I, Lopez-Doval JC, Ricart M, Villagrasa M, Brix R, Geiszinger A, Ginebreda A, Guasch H, Lopez De Alda MJ, Romani AM, Sabater S, Barcelo D (2009) Bridging levels of pharmaceuticals in river water with biological community structure in the Llobregat river basin (northeast Spain). Environ Toxicol Chem 28(12):2706–2714

79. Köck-Schulmeyer M, Ginebreda A, Postigo C, López-Serna R, Pérez S, Brix R, Llorca M, Alda MLD, Petrovic M, Munné A, Tirapu L, Barceló D (2011) Wastewater reuse in mediterranean semi-arid areas: the impact of discharges of tertiary treated sewage on the load of polar micro pollutants in the Llobregat river (NE Spain). Chemosphere 82(5):670–678
80. Kuster M, López de Alda MJ, Hernando MD, Petrovic M, Martín-Alonso J, Barceló D (2008) Analysis and occurrence of pharmaceuticals, estrogens, progestogens and polar pesticides in sewage treatment plant effluents, river water and drinking water in the Llobregat river basin (Barcelona, Spain). J Hydrol 358(1–2):112–123
81. López-Roldán R, de Alda ML, Gros M, Petrovic M, Martín-Alonso J, Barceló D (2010) Advanced monitoring of pharmaceuticals and estrogens in the Llobregat river basin (Spain) by liquid chromatography-triple quadrupole-tandem mass spectrometry in combination with ultra performance liquid chromatography-time of flight-mass spectrometry. Chemosphere 80 (11):1337–1344
82. Garcia-Galan MJ, Villagrasa M, Diaz-Cruz MS, Barcelo D (2010) LC-QqLIT MS analysis of nine sulfonamides and one of their acetylated metabolites in the Llobregat river basin. Quantitative determination and qualitative evaluation by ida experiments. Anal Bioanal Chem 397(3):1325–1334
83. Pro J, Ortiz JA, Boleas S, Fernández C, Carbonell G, Tarazona JV (2003) Effect assessment of antimicrobial pharmaceuticals on the aquatic plant lemna minor. Bull Environ Contam Toxicol 70(2):290–295
84. Yang LH, Ying GG, Su HC, Stauber JL, Adams MS, Binet MT (2008) Growth-inhibiting effects of 12 antibacterial agents and their mixtures on the freshwater microalga Pseudokirchneriella subcapitata. Environ Toxicol Chem 27(5):1201–1208
85. Isidori M, Lavorgna M, Nardelli A, Pascarella L, Parrella A (2005) Toxic and genotoxic evaluation of six antibiotics on non-target organisms. Sci Total Environ 346(1–3):87–98
86. Brain RA, Ramirez AJ, Fulton BA, Chambliss CK, Brooks BW (2008) Herbicidal effects of sulfamethoxazole in lemna gibba: using p-aminobenzoic acid as a biomarker of effect. Environ Sci Technol 42(23):8965–8970
87. Laville N, Ait-Aissa S, Gomez E, Casellas C, Porcher JM (2004) Effects of human pharmaceuticals on cytotoxicity, EROD activity and ROS production in fish hepatocytes. Toxicology 196(1–2):41–55
88. Hou XL, Shen JZ, Zhang SX, Jiang HY, Coats JR (2003) Bioconcentration and elimination of sulfamethazine and its main metabolite in sturgeon (Acipenser schrenkii). J Agric Food Chem 51(26):7725–7729
89. Migliore L, Brambilla G, Grassitellis A, Dojmi di Delupis G (1993) Toxicity and bioaccumulation of sulphadimethoxine in artemia (crustacea, anostraca). Int J Salt Lake Res 2(2):141–152
90. Eguchi K, Nagase H, Ozawa M, Endoh YS, Goto K, Hirata K, Miyamoto K, Yoshimura H (2004) Evaluation of antimicrobial agents for veterinary use in the ecotoxicity test using microalgae. Chemosphere 57(11):1733–1738
91. Andreozzi R, Raffaele M, Nicklas P (2003) Pharmaceuticals in STP effluents and their solar photodegradation in aquatic environment. Chemosphere 50(10):1319–1330
92. Boreen AL, Arnold WA, McNeill K (2004) Photochemical fate of sulfa drugs in the aquatic environment: sulfa drugs containing five-membered heterocyclic groups. Environ Sci Technol 38(14):3933–3940
93. Baran W, Sochacka J, Wardas W (2006) Toxicity and biodegradability of sulfonamides and products of their photocatalytic degradation in aqueous solutions. Chemosphere 65 (8):1295–1299
94. Jung JYKY, Kim JK, Jung D-H, Choi K (2008) Environmental levels of ultraviolet light potentiate the toxicity of sulfonamide antibiotics in *Daphnia magna*. Ecotoxicology 17:37–45

95. MdM G-R, Mezcua M, Agüera A, Fernández-Alba AR, Gonzalo S, Rodríguez A, Rosal R (2011) Chemical and toxicological evolution of the antibiotic sulfamethoxazole under ozone treatment in water solution. J Hazard Mater 192(1):18–25
96. Hoa PTP, Managaki S, Nakada N, Takada H, Shimizu A, Anh DH, Viet PH, Suzuki S (2011) Antibiotic contamination and occurrence of antibiotic-resistant bacteria in aquatic environments of northern Vietnam. Sci Total Environ 409(15):2894–2901
97. Goñi-Urriza M, Pineau L, Capdepuy M, Roques C, Caumette P, Quentin C (2000) Antimicrobial resistance of mesophilic Aeromonas spp. Isolated from two european rivers. J Antimicrob Chemother 46(2):297–301
98. Luo Y, Mao D, Rysz M, Zhou Q, Zhang H, Xu L, Alvarez PJJ (2010) Trends in antibiotic resistance genes occurrence in the Haihe river, China. Environ Sci Technol 44 (19):7220–7225
99. Storteboom H, Arabi M, Davis JG, Crimi B, Pruden A (2010) Tracking antibiotic resistance genes in the south platte river basin using molecular signatures of urban, agricultural, and pristine sources. Environ Sci Technol 44(19):7397–7404
100. Storteboom H, Arabi M, Davis JG, Crimi B, Pruden A (2010) Identification of antibiotic-resistance-gene molecular signatures suitable as tracers of pristine river, urban, and agricultural sources. Environ Sci Technol 44(6):1947–1953
101. Ferrari B, Mons R, Vollat B, Fraysse B, Paxeaus N, Lo Giudice R, Pollio A, Garric J (2004) Environmental risk assessment of six human pharmaceuticals: are the current environmental risk assessment procedures sufficient for the protection of the aquatic environment? Environ Toxicol Chem 23(5):1344–1354
102. Ginebreda A, Muñoz I, de Alda ML, Brix R, López-Doval J, Barceló D (2010) Environmental risk assessment of pharmaceuticals in rivers: relationships between hazard indexes and aquatic macroinvertebrate diversity indexes in the llobregat river (NE Spain). Environ Int 36(2):153–162
103. Grung M, Källqvist T, Sakshaug S, Skurtveit S, Thomas KV (2008) Environmental assessment of Norwegian priority pharmaceuticals based on the EMEA guideline. Ecotoxicol Environ Saf 71(2):328–340
104. Huschek G, Hansen PD, Maurer HH, Krengel D, Kayser A (2004) Environmental risk assesssment of medicinal products for human use according to European Commission recommendations. Environ Toxicol 19(3):226–240
105. Park S, Choi K (2008) Hazard assessment of commonly used agricultural antibiotics on aquatic ecosystems. Ecotoxicology 17(6):526–538
106. Santos JL, Aparicio I, Alonso E (2007) Occurrence and risk assessment of pharmaceutically active compounds in wastewater treatment plants. A case study: Seville city (Spain). Environ Int 33(4):596–601
107. EMEA (CHMP) (2006) Guideline on the environmental risk assessment of medicinal products for human use
108. EMEA (CVMP) (2004) Guideline on environmental impact assessment for veterinary medicinal products phase II
109. Sanderson H, Johnson DJ, Wilson CJ, Brain RA, Solomon KR (2003) Probabilistic hazard assessment of environmentally occurring pharmaceuticals toxicity to fish, daphnids and algae by ECOSAR screening. Toxicol Lett 144(3):383–395

Perfluorinated Compounds' Analysis, Environmental Fate and Occurrence: The Llobregat River as Case Study

Marta Llorca, Francisca Pérez, Marinella Farré, Yolanda Picó, and Damià Barceló

Abstract Perfluorinated compounds are industrial chemicals widely used for more than 60 years. However, during the last decade, due to their high resistance to degradation, bioaccumulation attached to proteins, biomagnification to the food chain and their relation to toxicological effects of these compounds have gained scientific and regulatory attention.

In addition, the difficulty associated with their analysis in complex matrices such as biota, food and human fluids and tissues samples should be mentioned.

This chapter provides a comprehensive examination of the current knowledge on PFCs' analysis, environmental fate and occurrence in aquatic systems, using as a central example the Llobregat River.

Keywords Drinking water, Liquid chromatography, Llobregat River, Mass spectrometry, Perfluorinated compounds, Sediments, Surface water, Wastewater

M. Llorca, F. Pérez, and M. Farré (✉)
Department of Environmental Chemistry, Institute of Environmental Assessment and Water Studies (IDAEA), (CSIC) Jordi Girona 18-26, 08034 Barcelona, Spain
e-mail: mfuqam@cid.csic.es

Y. Picó
Nutrition and Food Chemistry Laboratory, University of Valencia, Valencia, Spain

D. Barceló
Department of Environmental Chemistry, Institute of Environmental Assessment and Water Studies (IDAEA), (CSIC) Jordi Girona 18-26, 08034 Barcelona, Spain

Catalan Institute of Water Research (ICRA), Girona, Spain

S. Sabater et al. (eds.), *The Llobregat: The Story of a Polluted Mediterranean River*, Hdb Env Chem (2012) 21: 193–238, DOI 10.1007/698_2012_147,

Contents

Abbreviations

AcH	Acetic acid
ACN	Acetonitrile
APCI	Atmospheric pressure chemical ionization
APPI	Atmospheric pressure photoionization
ASE	Accelerated solvent extractor
DMF	Dimethylformamide
EFSA	European Food Safety Authority
EPA	Environmental Protection Agency
EPI	Enhanced product ion
EQS	Environmental quality standards
ESI	Electrospray ionization source
Et-FOSA	Ethyl perfluorosulphonamide
EtOAc	Ethyl acetate
FID	Flame ionization detection
FoH	Formic acid
FOSA	Perfluorosulphonamide
FOSE	Perfluorooctane sulphonamide-ethanol
FTOH	Fluorotelomer alcohol
FTUCA	Fluorotelomer unsaturated carboxylate
GC	Gas chromatography
HCl	Hydrochloric acid
HLB	Hydrophilic lipophilic balance
ip-PFNA	Isopropyl perfluorononanoic acid
IT	Ion trap
LC/ARC	Liquid chromatography/accurate radioisotope counting
LC	Liquid chromatography
LC-MS/MS	Liquid chromatography coupled to tandem mass spectrometry
LC-MS	Liquid chromatography–mass spectrometry

LRET	Long-range environmental transport
MeOH	Methanol
MLOD	Method limit of detection
MLOQ	Limits of quantification
MS	Mass spectrometry
MS^2	Mass spectrometry/mass spectrometry
MS^3	Mass spectrometry/mass spectrometry/mass spectrometry
MTBE	Methyl *tert*-butyl ether
N_2	Nitrogen
NaAc	Sodium acetate
NaOH	Sodium hydroxide
N-EtFOSAA	2-(*N*-Ethyl perfluorooctane sulphonamido) acetic acid
NH_4Ac	Ammonium acetate
NH_4OH	Ammonium hydroxide
N-MeFOSAA	2-(*N*-Methyl perfluorooctane sulphonamido) acetic acid
OW	Office of Water
PAPs	Polyfluoroalkyl phosphates
PE	High density polyethylene
PEEK	Polyether ether ketone
PFASAs	Perfluorinated sulphonamides
PFASEs	Perfluorinated sulphonamide ethanols
PFASs	Perfluoro alkyl sulphonates
PFBA	Perfluorobutanoic acid
PFBS	Perfluorobutane sulphonate
PFC	Perfluorinated compounds
PFCAs	Perfluoro carboxylic acids
PFDA	Perfluorodecanoic acid
PFDoA	Perfluorododecanoic acid
PFDS	Perfluorodecane sulphonate
PFEtS	Perfluoroethyl sulphonate
PFHpA	Perfluoroheptanoic acid
PFHpS	Perfluoroheptane sulphonate
PFHxDA	Perfluorohexadecanoic acid
PFHxS	Perfluorohexane sulphonate
PFNA	Perfluorononanoic acid
PFOA	Perfluorooctanoic acid
PFODA	Perfluorooctadecanoic acid
PFOS	Perfluorooctane sulphonate
PFOSI	Perfluorooctane sulphinate
PFPeA	Perfluoropentanoic acid
PFPrA	Perfluoropropyl acid
PFPrS	Perfluoropropyl sulphonate
PFTeA	Perfluorotetradecanoic acid
PFTOHs	Perfluorotelomers alcohols

PFUnA	Perfluoroundecanoic acid
PHA	Provisional Health Advisories
PLE	Pressurized liquid extraction
POP	Persistent organic pollutant
POSF	Perfluorooctane sulphonyl fluoride
PP	Polypropylene
PTFE	Polytetrafluoroethylene
PVDF	Polyvinylidene fluoride
QqLit	Hybrid quadrupole linear ion trap
QqQ	Triple quadrupole mass spectrometer
QTOF	Hybrid quadrupole time of flight
RP	Reversed phase
RSD	Relative standard deviation
SCARCE project	Assessing and predicting effects on water quantity and quality in Iberian rivers caused by global change (2009–2014)
SPE	Solid phase extraction
SRM	Selected reaction monitoring
TBA	*tert*-Butyl alcohol
t-Bu-PFOS	*tert*-Butyl perfluorooctane sulphonate
TDI	Tolerable daily intake
TFA	Trifluoroacetic acid
THPFOS	Tetrahydro-perfluorooctane sulphonate
TOF	Time of flight
WAX	Weak anionic exchange
WWTPs	Wastewater treatment plants

1 Introduction: Uses, Production and Global Distribution

Perfluorinated compounds (PFCs) have been manufactured since 1950s. Because of their properties, these compounds are employed for many industrial applications including stain repellents, textile, paints, waxes, polishes, electronics, adhesives and food packaging. Two of the most important PFCs are perfluorooctane sulphonate (PFOS) salts, components of fire-fighting foam concentrates, and perfluorooctane acid (PFOA), primarily used as emulsifier in industrial applications, for example in the production of fluoropolymers such as polytetrafluoroethylene (PTFE). Furthermore, PFOS and PFOA as well as other perfluoro carboxylic acids (PFCAs) are stable degradation products and/or metabolites of neutral PFCs like fluorotelomers alcohols (PFTOHs), perfluorinated sulphonamides (PFASAs) and perfluorinated sulphonamide ethanols (PFASEs). Figure 1 shows the most commonly used PFCs structures.

Because of their use in different industrial applications, these compounds enter to the environment through removal facilities, which are not able to degrade PFCs, or, for example, by their application in agricultural soils or by irrigation with

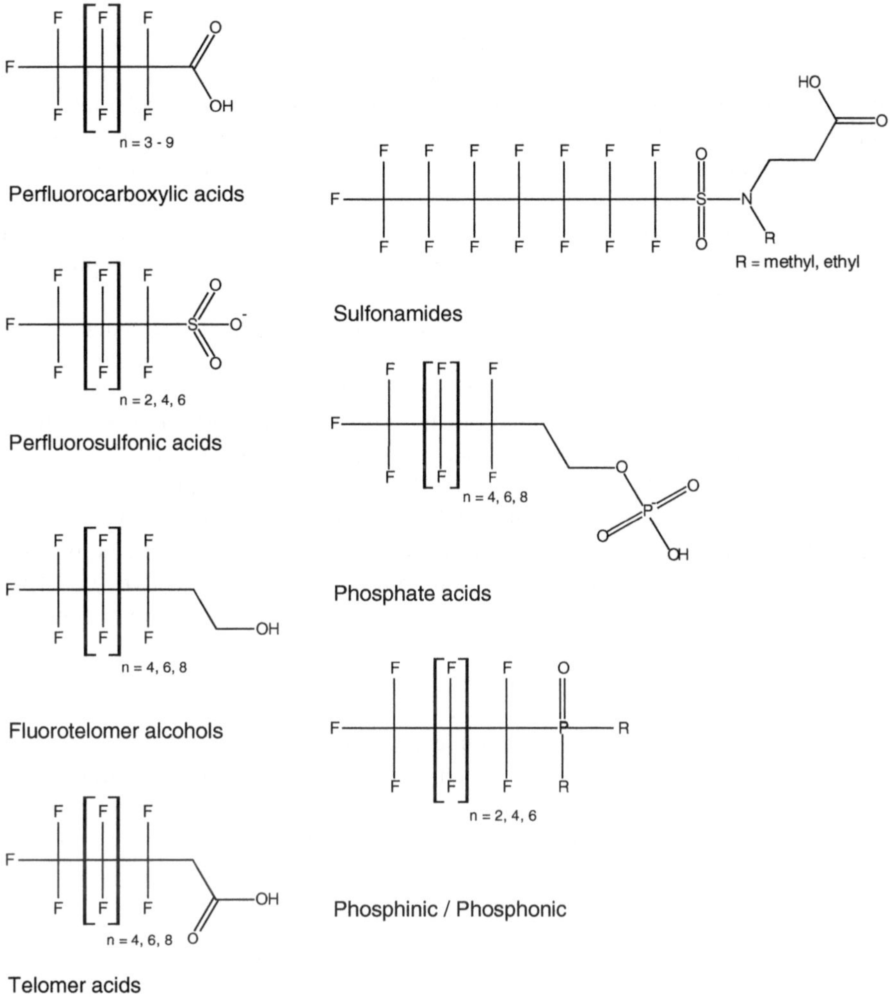

Fig. 1 The most common structures of PFCs

contaminated water. These analytes are widespread around the world in water and soils as well in organism due the high affinity to low weight proteins [1]. The bioaccumulation derives to biomagnification through the food chain and, finally, arrives to human through diet and drinking water [2, 3]. PFCs have been detected in environmental and biological samples. They are present in remote areas as the Arctic (atmosphere [4], Arctic Ocean [5], biological samples [6, 7] and few reviews have been published [8, 9]) or Antarctic (biological samples as penguins or seals [10, 11]). Regarding the presence in human matrices, these analytes have been reported in blood from donors from different countries [12, 13], liver [14], urine, human breast milk [14–17] and cord blood [18, 19] being the breastfeeding and pregnancy a possible transferability route. Fish is another biological matrix, which has been analysed due to its involvement into marine trophic chain. This is one of the main entrance routes of PFCs into human organisms with other daily products as milk, meat or vegetables [2, 3, 20–23]. In 2006, EPA established the tolerable daily intake (TDI) for PFOA and PFOS [24, 25], and in 2008 the EFSA established as well, TDI levels at 150 ng/kg bw per day in the case of PFOS and 1,500 ng/kg bw for PFOA.

PFCs are considered as emerging organic pollutants since they have not been regulated. However, in the last decade, there are some of them which are proposed to be under regulation [26].

In 2006, EPA and the eight major PFCs producer companies [Arkema, Asahi, BASF Corporation (successor to Ciba), Clariant, Daikin, 3M/Dyneon, DuPont and Solvay Solexis] launched the "PFOA Stewardship Program". The companies committed to phase out global emissions by 2015 [27]. Recently, PFOS has been included as a persistent organic pollutant (POP) under the Stockholm Convention for global regulation of production and use [28]. PFCs are also prime candidates for chemicals that will need authorization within the REACH regulation [26]. PFOS was added to the Annex III Substances subject to review for possible identification as priority substances or priority hazardous substances of the Directive 2008/105/EC of the European Parliament and Council of 16 December 2008 concerning the environmental quality standards (EQS) in the field water policy [29].

Wastewater has been also identified as a major source of PFCs in the environment since currently their elimination during wastewater treatments in wastewater treatment plants (WWTPs) is not completely achieved, and therefore, important amounts of these compounds can reach the natural environments through the treated effluents. Moreover, also sewage sludge produced in WWTPs is important source of PFCs' contamination, since there is a redistribution of PFOS and long carbon chain PFCs into the sludge [30–32]. Several works during last years have informed about concentrations in sludge in the range between low ng/g and μg/g. These high concentrations are of concern because sewage sludge can be partially used in agricultural lands generating an indirect source of PFCs via consumption of crops, air-borne transport, surface waters and ground waters draining from these sites [33–35]. Clarke et al. scored different groups of organic contaminants commonly found in sewage sludge with respect to their potential significance for agricultural utilization [36], and in this classification PFCs obtained 10 scores over 11, based on their persistence in soil (more than 6 months), their potential accumulation in human food chain, their potential bioaccumulation and their possible soil ecotoxicity [36]. On the other hand, it should be considered that PFCs in sludge amended soil can be mobilized by rainfall [37], reaching phreatic waters.

Drinking water has been identified as one of the major sources of human exposure [38, 39] to PFCs. For this reason, in 2009, the EPA's Office of Water (OW) have set a Provisional Health Advisories (PHA) maximum concentration values for PFOA and PFOS in drinking water. PHA values are 0.4 μg/L for PFOA and 0.2 μg/L for PFOS [40].

To protect human health and the environment against PFCs' contamination, there is a need to assess their presence in the environment and main sources of human exposure and assess possible damages involving their occurrence, bioaccumulation, as well as, their environmental fate and behaviour. In this context, this chapter provides a summary of the state-of-the-art in the analysis of PFCs in environmental samples, and the occurrence of PFCs will be presented and discussed in a typical Mediterranean river: the Llobregat River.

2 Analysis of PFCs

2.1 Sampling Process and Preservation

Storage and conservation of samples for PFCs' analysis present some critical steps because losses or contamination of the samples can easily occur.

Martin et al. [41] summarized the key challenges in trace analysis of PFCs. They include blank contamination issues, purity of reference standards and matrix effects in the ionization process of the mass spectrometer. Blank contamination is one of the most relevant problems associated with PFCs' analysis, and it is associated with fluoropolymer during sampling, storage and materials used in the laboratory. In order to avoid sources of contamination different measures have been suggested, as for example, pre-cleaning of the containers and materials prior sampling by rinsing with semi-polar solvents [38]. It is also important the materials involved during sampling, storage and the analytical process, being polypropylene (PP) containers [34, 42, 43], high density polyethylene (PE) bottles [44–46], or foil containers for solid samples [35] the recommended ones, because these materials cannot contaminate the samples. However, for solid samples losses by sorption to PP and PE containers can be considered negligible [46], whereas for aqueous samples in case of long-chain PFCs is higher [47, 48]. On the other hand, stored aqueous samples in glass have been widely discussed by different authors, whereas some works have reported sorption in glass when prepared samples were tested [49], this phenomena is not expected in real samples with complex matrices [50]. It is also important to consider possible losses due to volatilization of the PFTOHs and short C chain PFCs as PFBA. Therefore, it is recommended to avoid headspace in sampling bottles [51], and kept samples at low temperature after collection [44, 45, 52]. Szostek et al. studied the stability of PFTOHs in water under different storage conditions [53]. In this work it was concluded that aqueous samples can safely be stored in the freezer using glass vials and sealed with a septum lined with alumina foil. In addition, no biodegradation or biotransformation was observed under these conditions [53]. On the other hand, also it was studied the possible benefices associated with the use of biological inhibitors (such as formalin) to preserve the samples, but it was found that some suppress the MS responses during the analysis [54], and therefore should be avoided.

Drying procedures are usually applied for solid matrices using room temperature until a constant weight [42, 55], with soft temperature (40°C) along 3–4 days in porcelain bowls [46] or PP containers [34], and in an oven at 103°C overnight [52]. Other specific procedures consisted of direct freeze (−20°C) prior to any treatment in order to perform lyophilization [35, 43] or previous centrifugation to remove supernatant [56] and lyophilization [44]. Dried sludge or sediment is finely ground (<0.5 mm) [46] and homogenized with a mortar and pestle [35, 42–44, 55]. This homogenized sample is subsequently passed through a mesh sieve to remove pebbles or debris [43, 55]. Homogenized samples are kept frozen until analysis in PP containers [35, 42] or high-density PE bottles [44, 46]. Other authors interested in the analysis of PFCs in raw sewage sludge from Wets prepare the sample by centrifugation followed by filtration by GF/B glass filter and stored these filters at 4°C until extraction usually by solid–liquid extraction [45].

2.2 *Sample Pre-treatments*

Table 1 summarizes main sample pre-treatment that can be applied to the analysis of PFCs in different types of environmental matrices.

2.2.1 Water Samples

Extraction procedures for water analysis have been carried out using protocols based on solid phase extraction (SPE). Due to the different polarities of PFCs, different extraction SPE cartridges have been explored. Broadly, good recoveries were reported using Oasis WAX-SPE cartridges including short-chain (C_4–C_6) compound. These cartridges have been applied in many monitoring studies [39, 57, 58]. For longer-chain PFCs, less polar phases (C_{18} and Oasis HLB) may be applied [59–61]. Non-ionic PFCs may be extracted from the matrix by non-polar media (C_{18} SPE). Moderate polar media (Oasis HLB and Oasis WAX-SPE) have also been applied for extraction of non-ionic PFCs. However, one of the critical points in PFCs' analysis is background contamination in the analytical blanks [62–64]. One known source of procedural contamination is contact with laboratory materials made of, or containing, fluoropolymers [54, 62]. Water samples may be filtered [54, 65] to separate solids from the liquid phase. However, filtration can result in losses by adsorption of PFCs on the filters, or on the contrary levels can increase by contamination from the filters, as was found by Schultz et al. for fibre, nylon, cellulose acetate and polyethersulphone filters [54]. They applied centrifugation as an alternative for separating theliquid from the solids.

Controversial studies reported the cross-contamination of samples during PFCs' analysis using different SPE cartridges. Yamashita et al. [66] examined the source of blank contamination at various different steps, including sample collection, extraction and treatment of samples. PFOS and PFOA contamination in the SPE

Table 1 Sample pre-treatments and instrumental analysis of different published works

Analytes	Matrices	Pre-treatment	Instrumental analysis	Quality parameters	Reference
PFBS, PFPA, PFHxA, PFHpA, PFOA, PFNA, ip-PFNA, PFOS, PFDA, PFDS	Water	(1) Filtration (glass microfiber membrane 0.7 μm) (2) SPE by Oasis WAX (conditioning: 4 mL 0.1% NH_4OH MeOH, 4 mL MeOH, 4 mL water; loading: 200 mL; elution: 4 mL 0.1% NH_4OH MeOH (3) Evaporated under N_2 (4) Reconstituted in 0.5 mL MeOH	*LC–MS/MS(QqQ)* • Injection volume: 20 μL • Column: LiChroCARTLiChrospher 100 RP-18 (250 mm × 4 mm; 5 μm) • Mobile phase: water (20 mM NH_4Ac): methanol (20 mM NH_4Ac) • SRM mode	• MLOQ: 0.05 and 5 ng/L • Recoveries: 69–110%	[108]
PFBA, PFPeA, PFHxA, PFHpA, PFOA, PFBS, PFOS	Surface and deep water (lake) and precipitation water	(1) 300 mL sample + Surrogate IS (2) SPE by Oasis HLB (conditioning: 10 mL MeOH, 10 mL water; loading; elution: 10 mL MeOH) (3) Extract reduced to dryness under N_2 (4) Reconstitution in 1 mL MeOH and filtered (0.2 μm) Or (1) 50 L sample (2) XAD resins	*LC–MS/MS(QqLIT)* • Injection volume: 10 μL • Column: C8 Phenomenex guard column C8 Luna (3 μm, C8, 50 mm × 2 mm) • Mobile phase: water (10 mM NH_4Ac): methanol (10 mM NH_4Ac) • SRM mode *GC–MS (for PFCA anilides)* • Column: ZB-5 Zebron fused silica capillary column (30 m × 0.25 mm i.d.) • SIM mode	• MLOD: 0.5 ng/L • Recoveries (XAD-7): 25–160% (1–34 L tap water)	[117]
PFBA, PFPeA, PFHxA, PFHpA, PFOA, PFBS, PFOS	Surface river water	(1) 100 mL sample (pH 7–8 adjusted) + Surrogate IS (2) SPE by Strata-x (200 mg) (conditioning: 2 mL acetone: acetonitrile:FoH (50:50:1),	*LC–MS/MS(QqQ)* • Injection volume: 50 μL • Column: NUCLEODUR SPHINX-RP 2.0 mm × 150 mm, 3 μm	• MLOD: 2 ng/L • Recoveries: 11–117%	[38]

(continued)

Table 1 (continued)

Analytes	Matrices	Pre-treatment	Instrumental analysis	Quality parameters	Reference
		3 × 2 mL water (pH 8); loading; elution: 4 × 2 mL acetone:acetonitrile:FoH (50:50:1) (3) Extract reduced to dryness under N_2 (4) Reconstitution in 0.5 mL 10 mM NH_4Ac (water/MeOH, 75/25)	• Mobile phase: water/methanol, 75/25 (10 mM NH_4Ac): acetonitrile/methanol, 75/25 (10 mM NH_4Ac) • SRM mode		
PFOA, PFNA, PFOS, PFDA, PFUdA, FOSA	Surface water	(1) 1 L sample filtered (1 µm glass fibre) + Surrogate IS addition (2) SPE by Oasis HLB 6 cc (conditioning:5 mL MeOH; elution: 15 mL MeOH) (3) Dried under N_2 at 0.5 mL (4) IS addition	*LC–MS/MS(QqQ)* • Injection volume: 3 µL • Column: C_{18} column (5 µm particle size, 50 mm × 2.1 mm) • Mobile phase: water:methanol • SRM mode	• MLOD: 0.1–2.8 ng/L • Recoveries: 49–92%	[118]
PFOS, PFOA, PFNA, PFHxS, PFBS	Surface water	(1) Surrogate IS addition and incubate 12–18 h at 4°C (2) SPE by Oasis HLB 6 cc (conditioning:10 mL MeOH, 15 mL water; elution: 15 mL MeOH) (3) Dried under N_2 (4) Reconstituted in 0.2 mL 70:30 water:acetonitrile	*UPLC–MS/MS(QqQ)* • Injection volume: no data • Column: Acquity UPLC BEH C_{18} column (1.7 µm particle size, 50 mm × 2.1 mm) before injection LiChroCART HPLC RP-18e column (125 µm × 2 µm × 5 µm) • Mobile phase: water (2 mM NH_4Ac): acetonitrile • SRM mode	• MLOD: 0.6–3 pg/L (PFCs) • Recoveries: 60–122% (PFCs)	[59, 60]

Ionic pair extraction					
PFOA, PFNA, PFDA, PFUnA, PFOS, 4:2, 6:2, 8:2, 10:2 diPAP	Sudge from WWTPs	(1) 2 g sludge (2) 4 mL 0.25 M $NaCO_3$ + 1 mL 0.5 M TBA, pH 10 (3) Extraction with 5 mL aliquots of MTBE (4) MTBE aliquots combined (5) Dryness under N_2 (6) Reconstituted in 0.5 mL of MeOH and filtered by 0.2 μm nylon filter	*LC–(−ESI)-MS/MS(QqQ)* • Injection volume: no data • Column: Gemini C_{18} (50 mm × 4.6 mm, 3 μm) and Ascentis Express C_{18} (50 mm × 4.6 mm, 2.7 μm) • Mobile phase: water (10 mM NH_4Ac): MeOH (10 mM NH_4Ac) • SRM mode	• MLOQ: 0.0625 ng/g dw (PFCs) 0.375 ng/g dw (diPAPs) • Recoveries: 17–105% (PFCs) 38–53% (diPAPs)	[73]
PFPeA, PHxA, PFHpA, PFOA, PFNA, PFDoA, PFTeA, 6:2 FTUCA, 8:2 FTUCA, PFBS, PFHxS, PFOS, FOSA, N-MeFOSA, N-EtFOSA, N-MeFOSAA, N-EtFOSAA	Digested sewage sludge from WWTPs	(1) 0.5 g sludge (2) 0.5 mL TBA 0.5 M + 4 mL 0.25 M $NaCO_3$, pH 10) + 2× 5 mL MTBE (3) Evaporated under N_2 (4) Reconstitution in 5 mL MeOH (5) Clean-up by Envicarb cartridges	*LC–(−ESI)-MS/MS(QqQ)* • Injection volume: 20 μL • Column: 70 mm × 2 mm × 3 μm Nucleodur C_{18} gravity • Mobile phases: 2.5 mM NH_4Ac MeOH:water (95:5) and 2.5 mM NH_4Ac water:MeOH (95:5) • SRM mode with collision-induced fragmentation	• MLOQ: 0.6–30 ng/g • Recoveries: 70–169% • RSD: 2–20%	[74]
Solid–acid liquid extraction					
PFHxS, PFOS, PFDS, FOSAA, N-MeFOSAA, N-EtFOSAA, PFOA, PFNA, PFDA, PFUnA, PFDoA, PFTeA	Digested sludge from WWTP and surface sediments	(1) 0.1 g sludge (2) 7.5 mL 1% AcH (3) Vortexed, sonicated at 60°C (15 min) and centrifugated at 3,000 rpm, 10 min (4) Supernatant separation and repeat steps 2–4 twice; solid part + 1.7 mL MeOH:water (1% AcH) (9:1, v:v) and vortexed, sonicated, and centrifugated (5) Supernatants combination centrifuged 1 h	*HPLC–(−ESI)-MS/MS(QqQ)* • Injection volume: 20 μL • Column: 40 mm × 2.1 mm TargaSprite C_{18} (5 μm pore size) and C_{18} guard column • Mobile phase: water (2 mM NH_4Ac):MeOH • SRM mode	• MLOD: 0.7–2.2 ng/g dw (sludge), 0.041–0.246 ng/g dw (sediment) • Recoveries: 73–98% (sediment), 56–89% (aged sediment), 41–91% (digested sludge), 37–98% (primary sludge)	[75]

(continued)

Table 1 (continued)

Analytes	Matrices	Pre-treatment	Instrumental analysis	Quality parameters	Reference
		(6) SPE by 500 mg C_{18} (conditioning: 10 mL MeOH, 10 mL 1% AcH; rinsed: 10 mL water and 2 h vacuum; elution: 4 mL MeOH) (7) Concentrated under N_2 to 2 mL (8) 800 μL MeOH + 1,200 μL 0.01% NH_4OHaq + 6 mL MeOH:0.01% NH_4OHaq (70:30) addition (9) Stored at 4°C and before analysis: 500 μL extract + 50 μL IS			
PFBS, PFHxS, PFOS, FOSA, PFOA, PFNA, PFDA, PFUdA, PFDoA, PFTeA, PFHxDA, PFODA	Sludge from WWTPs	(1) 0.1 g sludge + surrogate (2) 7.5 mL (1% AcH) (3) Sonication 20 min at 60°C (4) Centrifugation at 3,500 rpm, 10 min (5) Supernatant separation and pellet with 1.7 mL MeOH (1% AcH, 90:10), sonication 20 min at 60°C and centrifugation (6) Repeat steps 2–5 twice and combine extracts (7) 7.5 mL 1% AcH addition (8) SPE by Oasis HLB (conditioning: MeOH, 1% AcH; washing: 20% MeOH; elution: 5 mL MeOH) (9) Concentration to 1 mL and filtration by 0.2 μm nylon filter	*HPLC–(−ESI)-MS/MS(QqQ)* • Injection volume: 10 μL • Column: Keystone Betasil C_{18} (50 mm × 2 mm; 5 μm) • Mobile phase: mater (2 mM NH_4Ac):MeOH • SRM mode	• Recoveries: 70–130% 37–65% (PFCAs > 11C) • LOQ < 25 ng/g dw	[42]

PFHxS, PFOS, FOSA, PFOA, PFNA, PFDA, PFUnA, PFDoA	Sludge from rural and urban areas WWTPs	(1) 0.1 g sludge + surrogate (2) 7.5 mL (1% AcH) (3) Sonication 20 min at 60°C (4) Centrifugation at 3,500 rpm, 10 min (5) Supernatant separation and pellet with 1.7 mL MeOH (1% AcH, 90:10), sonication 20 min at 60°C and centrifugation (6) Repeat steps 2–5 twice and combine extracts (7) 7.5 mL 1% AcH addition (8) SPE by Oasis HLB (conditioning: MeOH, 1% AcH; washing: 20% MeOH; elution: 5 mL MeOH) (9) Concentration to 1 mL and filtration by 0.2 μm nylon filter.	*HPLC–(−ESI)-MS/MS(QqQ)* • Injection volume: 10 μL • Column: Keystone Betasil C_{18} (50 mm × 2 mm; 5 μm) • Mobile phase: mater (2 mM NH_4Ac):MeOH • SRM mode	• Recoveries: 59–107% (sludge)	[76]
PFOA, PFOS	Sewage sludge samples from municipal WWTPs	(1) 0.1 g sludge (2) 7.5 mL 1% AcH (3) Vortexed, sonicated at 60°C, 15 min and centrifugation at 3,000 rpm, 10 min (4) Supernatant separation and repeat steps 2–4 twice (5) Solid separated part + 1.7 mL MeOH:water 1% AcH (9:1), vortexed, sonicated and centrifugated (6) Combined supernatants centrifuged 1 h; SPE by 500 mg C_{18} cartridges (conditioning: 10 mL MeOH, 10 mL 1% AcH; rinsed: 10 mL water and 2 h vacuum; elution: 4 mL MeOH)	*HPLC–(−ESI)-MS/MS(QTOF)* • Injection volume: 10 μL • Column: C_{18} reversed phase • Mobile phase: water (NH_4Ac 5 mM):MeOH • SRM mode	• MLOD: 0.5 ng/g (PFOS) 0.8 ng/g (PFOA) • Recoveries: 85–114% (PFOS) 71–98% (PFOA)	[52]

(continued)

Table 1 (continued)

Analytes	Matrices	Pre-treatment	Instrumental analysis	Quality parameters	Reference
		(7) Concentrated under N_2 to 2 mL (8) Addition of: 800 μL of MeOH + 1,200 μL of 0.01% NH_4OHaq + 6 mL of MeOH:0.01% NH_4OHaq (70:30) (9) Stored at 4°C (10) Prior analysis, 500 μL extract + 50 μL IS (20–50 ng/mL)			
PFOA, PFOS	Grab samples of primary, activated, secondary and anaerobically digested sludge from sewage treatment plants (STPs)	(1) 0.1 g freeze dried sludge in PP tube (2) 7.5 mL 1% AcH (3) Vortexed, sonicated (60°C) 15 min and centrifugation at 3,000 rpm 10 min (4) Supernatant collection (repeat steps 2–4 twice) (5) Solid part + 1.7 mL [MeOH: water 1% AcH (9:1)], vortexed, sonicated and centrifugated (6) Supernatants centrifuged 1 h (7) Combined extracts (35.1 mL) loaded onto SPE cartridges (Oasis HLB 500 mg) and rinsed with 10 mL water [conditioning: 5 mL MeOH, 10 mL water (1% AcH)] (8) Elution: 2× 2 mL MeOH	*HPLC–(ESI−)-MS/MS(QqQ)* • Injection volume: 10 μL • Column: Zobax Extend C_{18} (150 mm × 2.1 mm; 5 μm) and guard column XDB-C8 (2.1 mm i.d. × 2.5 mm; 5 μm) at 30°C • Mobile phase: water (2 mM NH_4Ac):MeOH • SRM mode	• MLOQ: 1 ng/g dw (PFOS) 5 ng/g dw (PFOA) • Recoveries: 84% (PFOS) 70% (PFOA)	[44]

		(9) Extracts diluted with 6 mL Cl_2CH_2 and loaded onto the silica cartridge [conditioning: 5 mL Cl_2CH_2/MeOH (60:40, v/v)] (10) Eluent concentrated to dryness under N_2 (11) Reconstitution in 1 mL MeOH:water (0.01% NH_4OH) (70:30, v/v) and filtration by 0.2 μm nylon syringe filter			
PFBS, PFPA, PFHxA, PFHpA, PFOA, PFNA, ip-PFNA, PFOS, PFDA, PFDS	Sediment	(1) 1 g homogenized sediment into PP tube (2) 10 mL 1% AcH, vortexed, 40°C sonication (15 min) (3) Centrifuged (4) supernatant mixed with 2× [2.5 mL MeOH:1% AcH (9:1), sonicated 15 min, centrifuged] (5) Supernatant + 10 mL 1% AcH (6) Combined extracts and volume adjusted to 200 mL with water (7) SPE by Oasis WAX (conditioning: 4 mL 0.1% NH_4OH MeOH, 4 mL MeOH, 4 mL water; loading: 200 mL; elution: 4 mL 0.1% NH_4OH MeOH) (8) Evaporated under N_2 (9) Reconstituted in 0.5 mL MeOH	*LC–MS/MS(QqQ)* • Injection volume: 20 μL • Column: LiChroCARTLiChrospher 100 RP-18 (250 × 4 mm; 5 μm) • Mobile phase: water (20 mM NH_4Ac): methanol (20 mM NH_4Ac) • SRM mode	• MLOQ 0.002 and 0.2 ng/g • Recoveries: 62–103%	[108]

(continued)

Table 1 (continued)

Analytes	Matrices	Pre-treatment	Instrumental analysis	Quality parameters	Reference
PFOS, PFOA, PFNA, PFHxS, PFBS	Sediments	(1) 1 g (<120 μm particle), in PP tube (2) Surrogate IS addition and incubate 18 h at 4°C (3) 9 mL methanol + 10 mL AcH 1%, mixed and ultrasonicated (4) Centrifuged and supernatant evaporated under N_2 (5) Reconstitution in 1 mL of acetonitrile and incubated in an ultrasonic bath (6) Purification by activated charcoal + 50 μL of AcH, mixed 1 min (7) Centrifuged and 0.15 mL filtered by 2 μm GHP Accrodisc (8) Addition of in 0.35 mL water	*UPLC–MS/MS(QqQ)* • Injection volume: no data • Column: acquity UPLC BEH C_{18} column (1.7 μm particle size, 50 mm × 2.1 mm) before injection LiChroCART HPLC RP-18e column (125 μm × 2 μm × 5 μm) • Mobile phase: water (2 mM NH_4Ac): acetonitrile • SRM mode	• Recoveries: 100% (PFOS) 108% (PFOA)	[59]
PFBS, PFHxS, PFHpS, PFOS, PFDS, PFHxA, PFHpA, PFOA, PFNA, PFDA, PFUnA, PFDoA	Sludge from municipal WWTPs, livestock WWTPs and industrial WWTPs	(1) 0.1 g sludge + IS + 7.5 mL 1% AcH (2) Sonication at 60°C, 20 min and centrifugation at 3,500 rpm, 5 min (3) Supernatant separation and solid with 1.7 mL MeOH: 1% AcH (9:1), sonication and centrifugation (4) Repeat step 3 for solid part twice and combine all the extracts (5) SPE by Oasis HLB (6) Extract evaporated under N_2 and reconstitution in 1 mL MeOH	*HPLC–(−ESI)-MS/MS(QqQ)* • Injection: 10 μL • Column: Betasil C_{18} (100 mm × 2.1 mm, 5 μm) and guard column 12.5 mm × 2.1 mm Narrow Bore C_{18} • Mobile phase: water (NH_4Ac 2 mM):MeOH • SRM mode	• MLOQ: 1–5 ng/g • Recoveries: 21%(PFDS) 69–119% the rest of PFCs	[43]

Solid–alkaline liquid extraction					
PFHxA, PFHpA, PFOA, PFNA, PFDA, PFUnA, PFDoA, PFTrA, PFTeA, PFHxS, PFOS	Sludge from WWTP	(1) 0.5 g dw + 0.5 mL NaOH 1 M (2) 30 min sonication/heat and overnight incubation (3) Neutralization (w/HCl) (4) 10 mL ACN:MeOH (1/1, v/v), shaked 1 h (5) Supernatant separation (repeat steps 4–5 one more time) (6) 2 mL of combined extracts (20 mL) + 98 mL water at pH4 (7) Sonicated 30 min (8) SPE by HLB cartridges (conditioning: 5 mL MeOH, 5 mL water; wash step: 5 mL 25% MeOH; elution: 5 mL MeOH) (9) Extract dried under N_2 (10) Reconstitution in 1 mL ACN: water (60:40) + IS addition	*UPLC–(−ESI)-MS/MS(QqQ)* • Injection volume: 20 μL • Column: Waters BEH C_{18} (100 mm × 2.1 mm; 2.1 μm) at 35°C and a Waters BEH C_{18} trapping cartridge • Mobile phase: water: ACN (pH4 with HAc) • SRM mode	• MLOQ: 1.8–6.8 ng/g dw • Recoveries: 82–104%	[77]
PFBS, PFHxS, PFHpS, PFOS, PFDS, PFBA, PFPeA, PFHxA, PFHpA, PFOA, PFNA, PFDA, PFUnA, PFDoA, PFTrA, PFTeA, FOSA, N-MeFOSA, N-EtFOSA	Sludge from different WWTPS	(1) 1 g sludge (5 g sediment) + 10 mL MeOH (1% NH_4OH) (2) Vortexed 30 s and sonicated 10 min at 60°C (3) Supernatant + water (1% AcH) (4) Preconcentration under N_2 to 5 mL (5) Purification by dispersed solvent (EnviCarb) and rinse with 2.5 mL MeOH (1% NH_4OH, 1% AcH) (6) Concentrate to 1 mL under N_2	*UPLC–(−ESI)-MS/MS(QqQ)* • Injection volume: 10 μL • Column: Waters BEH C_{18} (50 mm × 2.1 mm; 1.7 μm particle size) • Mobile phase: water/MeOH (2 mM NH_4Ac, 95/5):MeOH (2 mM NH_4Ac). UPLC system were replaced by PEEK • SRM mode	• MLOD: 0.14–1.43 ng/g (sludge) 0.03–0.29 ng/g (sediment) • Recoveries: 62–94% (sludge) • RSD: 2–7% (sludge)	[78]

(continued)

Table 1 (continued)

Analytes	Matrices	Pre-treatment	Instrumental analysis	Quality parameters	Reference
Solid–liquid extraction					
PFHxS, PFOS, PFDS, PFOA, 9,9,10,10,11,11,12,12,13,13,13-Undecylfluorotredecane-1-sulphonate, *N*-ethyl-*N*-(heptadecafluorooctane)-sulphonyl-glycinic acid, 5,5,6,6,7,7,8,8,9,9,10,10,11,11,12,12,12-Heptadecafluorodecane sulphonylamido polyethocylate and polyethocylate methyl ether, partly fluorinated alkyl-ethoxylated (2-perfluoroalkylethanol polyglycol ether), metabolites of partly fluorinated alkylethoxylates	Sludge from WWTP	(1) 2 g sludge (dw) (2) Soxhlet or hot vapour extraction, 6 h [solvent optimization: MeOH, EtOAc, MeOH–HCl] (3) (In parallel to Soxhlet) PLE by ASE [optimization extraction steps: (a) EtOAc–DMF (8:2), (b) MeOH–H_3PO_4 (95:5), (c) MeOH–H_3PO_4 (99:1) and (d) MeOH–H_3PO_4 (99:1)]	*LC–(±ESI or APCI)-MS/MS* • Injection volume: 10 μL • Column: Multospher 100 RP 5-5 (C8, 5 μm, spherical; 250 mm × 4.6 mm i.d.) or using a PF-C8 column (150 mm × 4.6 mm i.d.) filled with spherical perfluorinated RP-C8 material (5 μm) • Mobile phase: MeOH:water • SRM mode	*Soxhlet and hot vapour* • Recoveries: 48–52% *PLE* • MLOQ 10 μg/g (anionic) 20 μg/g (non ionic) • Recoveries: 105% (ionic) 119% (non ionic)	[83]
PFHxA, PFHpA, PFOA, PFNA, PFDA, PFUnA, PFDoA, PFTeA	Liquid sludge	(1) 5 g liquid sludge + surrogates (into PP tube) (2) 20 mL MeOH, wrist-action shaker, mixed 30 min (3) Settled 30 min or centrifugated at 3,000 rpm, 20 min (4) 1 mL supernatant + 25 mg Envi-Carb graphitized carbon adsorbent + 50 μL AcH, vortexed and centrifugated 10,000 rpm, 30 min (5) 500 μL supernatant + 500 μL water, mixed (6) Internal standard addition	*HPLC–(−ESI)-MS/MS(QqQ)* • Injection volume: 100–200 μL • Column: Zorbax Rx-C8 (15 cm × 2.1 mm i.d., 5 μm) analytical column and Luna C_{18}(2) (3 cm × 4.6 mm i.d., 3 μm) inserted in the HPLC between the pump and injector to delay any fluorochemicals originating from PTFE instrument components • Mobile phase: ACN (0.15% AcH) • SRM mode	• MLOQ: 1 ng/g • Recoveries: 70–120% • RSD: <20%	[79]

PFOA, PFNA, PFDA, PFUnA, PFOS, FOSA	Sludge	(1) 7 g sludge (dw) + surrogates (into PP tube) (2) 2× (10 mL MeOH, ultrasonic bath) (3) Extracts pooled + 1 L water (4) SPE by C_{18} (conditioning: MeOH, water; eluting: 10 mL MeOH) (5) Evaporated to dryness (6) Reconstitution in 1 mL MeOH:water (2 mM NH_4Ac) (1:1)	*HPLC–(−ESI)-MS/MS(QqQ)* • Injection volume: 20 µL • Column: C_{18} Betasil column (2.1 mm × 50 mm) • Mobile phase: H_2O (2 mM NH_4Ac):MeOH (2 mM NH_4Ac) • SRM mode	• MLOD: 0.4–1.7 ng/g (dw) • Recoveries: 83–105% (FOSA) 26% • RSD: 4–8%	[56]
N-EtFOSE, N-EtFOSAA, FOSAA, FOSA, N-EtFOSA, PFOSI, PFOA, PFOS	Aerobic batch assays with WWTPs materials	(1) Lyophilized solids + MeOH (2) Sonication 20 min, 60°C, and centrifugation at 3,600 rpm, 15 min (3) Extract poured into a new tube + 20 mL MeOH (4) Purification by dispersed solvent (ENVI-Carb) (5) Extracts diluted 1/10 with MeOH and again diluted with 3/5 water (0.01% NH_4OH) (6) 0.5 mL in glass autosampler vial (7) (In parallel) SPE from headspace sampling eluted with 10 mL of MeOH and diluted 3/5 with water (0.01% NH_4OH)	*HPLC–(−ESI)-MS/MS(QqQ)* • Injection volume: 30 µL • Column: Targa Sprite C_{18} (5-µm pore size equipped with a C_{18} guard column • Mobile phase: H_2O (2 mM NH_4Ac):MeOH • SRM mode	• MLOQ: 0.1 µg/L (vial) 46–49 pmol (microcosm) • Recoveries: 83–119%	[80]

(continued)

Table 1 (continued)

Analytes	Matrices	Pre-treatment	Instrumental analysis	Quality parameters	Reference
[1,2-^{14}C] 6:2 FTOH	Sassafras soil	(1) 5 g sample + 15 mL ACN (2–7 days) at 50°C (2) Centrifugation at 657 g, 20 min (3) Supernatant separation and soil with 15 mL ACN: 250 mM NaOH (9:1) at 50°C, shaked overnight (4) 80 μL 5 M HCl and centrifugation at 657 g, 20 min (5) Supernatants stored at −10°C (dark)/Bottle septum experiments with 50 mL ACN at 50°C, 2–7 days/C_{18} Cartridges experiments with 5 mL ACN elution for the first 7 days	• *Radioactivity* by Beckman LS5000 TD liquid scintillation counter • *Liquid* chromatography/ accurate radioisotope counting (LC/ARC) system for the soil CH_3CN extracts and C_{18} cartridge eluent		[82]
TFA, PFPrA, PFBA, PFPeA, PFHxA, PFHpA, PFOA, PFNA, PFDA, PFUnA, PFDoA, PFTeA, PFBS, PFHxS, PFOS	Waste sludge and sediments	(1) sonication solvent extraction was used for removing PFCAs from solid matrices (2) SPE was performed to concentrate PFCAs using WAX cartridges (3) The SPE eluent was cleaned up using dispersive carbon sorbent to remove the co-eluted interfering compounds	*HPLC–(−ESI)-MS/MS* • Injection volume: 10 μL • Column: Hypersil Gold C_{18} (150 mm × 2.1 mm; 3 μm pore size) at 30°C • Mobile phase: water (2 mM NH_4Ac):MeOH. HPLC tubing made up of PTFE replaced with PEEK tubing • SRM mode	• MLOQ: 0.10–0.50 ng/g • Recoveries: 66–111% (sediment) 73–112% (soils) 57–115% (sludge) • RSD 1–15% (sediment) 1–19% (soil) 2–18% (sludge)	[55]

PFBS, PFHxS, PFHpS, PFOS, PFDS, PFBA, PFPeA, PFHxA, PFHpA, PFOA, PFNA, PFDA, PFUnA, PFDoA, PFTrA, PFTeA	Anaerobically stabilized sewage sludge from municipal WWTPs	(1) 0.5 g sludge (dw) in a 15 mL PP tube + 0.5 mL water + IS (2) Sequentially extraction by 2.5, 1.5 and 1.0 mL of MeOH (shaking the slurry 10 min and sonication 20 min at 40°C) (3) Centrifugation at 3,500 rpm, 8 min (4) Combined extracts + Envicarb graphitized carbon adsorbent (300 mg) (5) Shaked 20 min, centrifugation at 3,500 rpm along 30 min (6) 4 mL supernatant + IS + 1 mL water (0.01% NH_4OH) (7) Stored at −4°C before analysis	*LC–(−ESI)-MS/MS(QqQ)* • Injection volume: 20 μL • Column: Nucleodur C_{18} gravity column (70 mm × 2 mm; 3 μm) • Mobile phase: 2.5 mM NH_4Ac MeOH:water (95:5) and 2.5 mM NH_4Ac water: MeOH (95:5) • SRM with collision-induced fragmentation	• MLOQ: 0.5–17 ng/g • Recoveries: 94–115% (PFSAs) 21–109% (PFCAs)	[81]
PFBA, PFPeA, PFHxA, PFHpA, PFOA, PFNA, PFDA, PFUnA, PFDoA, PFTrA, PFTeA, PFHxDA, PFODA, PFBS, PFHxS, PFOS, PFDS, FOSA	Sewage sludge from a domestic WWTP	(1) 0.5 g sludge (dw) + surrogates (2) PLE (MeOH solvent at 70°C, two cycles, 1 min static time, 100 bar) (3) Evaporation to 1 mL + 30 mL water (4) SPE by Oasis WAX (conditioning: 4 mL MeOH (NH_4OH), 4 mL water; loading: under gravity conditions; eluting: 4 mL MeOH(NH_4OH)) (5) Dried under N_2 (6) Reconstitution in initial mobile phase conditions + IS	*HPLC–(−ESI)-MS/MS(QqLIT)* • Injection volume: 10 μL • Column: LiChroCART® 125-2, Pusopher® STAR, RP-18e (5 μm) • Mobile phase: water (20 mM NH_4Ac):MeOH • SRM mode	• MLOQ: 50–2,772 pg/g • Recoveries: 65–111% • RSD: 4–30%	[35]

(continued)

Table 1 (continued)

Analytes	Matrices	Pre-treatment	Instrumental analysis	Quality parameters	Reference
PFPeA, PFHxA, PFHpA, PFOA, PFNA, PFDA, PFUnA, PFDoA, PFHxS, PFOS	Sludge from industrial WWTP	(1) Suspended solid and dried sludge (2) Extraction by PLE (MeOH solvent, three cycles, 15 min, 2,000 psi at 100°C) (3) 60–80 mL of extract diluted into 1 L of water (4) SPE extraction	*HPLC–(−ESI)-MS/MS(QqQ)* • Injection volume: 10 µL • Column: Agilent Eclipse XDB-C_{18} (2.1 mm × 100 mm; 5 µm) • Mobile phase: water (5 mM NH_4Ac):ACN • SRM mode		[84]
PFBS, PFHxS, PFOS, PFBA, PFPeA, PFHxA, PFHpA, PFOA, PFNA, PFDA, FOSA, N-MeFOSA, N-EtFOSA	Sewage sludge samples from WWTPs	(1) 1 g sludge + surrogates + 10 mL MeOH in a PP tube (2) Shaked 10 min, ultrasonic bath at 40°C for 30 min and centrifugation at 3,000 rpm (15 min) (3) Supernatant separated and solid part with 9 mL MeOH in an ultrasonic bath and centrifugation (4) Repeated step 3 one more time and combine all the extracts (5) Purification by disperse solvent (EnviCarb) (6) Eluted extract + 1 L water (7) SPE by Oasis WAX [conditioning: 12 mL MeOH (0.1% NH_4OH), 12 mL water; cleaning: 12 mL 25 mM NaAc (pH4); elution: 8 mL MeOH (0.1% NH_4OH)] (8) aliquot + IS	*HPLC–(−ESI)-MS/MS(QqQ)* • Injection volume: 20 µL • Column: Varian Polaris C_{18} analytical column (50 mm × 2.0 mm; 3 µm particle diameter) at 40°C • Mobile phase: water (2 mM NH_4Ac):MeOH • SRM mode	• MLOQ: 0.02–0.71 ng/g • Recoveries: 64–129% • RSD: 3–22%	[34]

Other extraction procedures					
PFHxA, PFOA, PFNA, PFOS, PFBS, PFUnA, FOSA, 6:2 FTS, 6:2 FTOH, 8:2 FTOH	Activated sludge from WWTP	(1) PFCAs were extracted from supernatant of the sludge experiments (2) SPE by C_{18} cartridges (conditioning: 10 mL MeOH, 10 mL water; elution: 10 mL MeOH) (3) Reduced under a gentle stream of N_2 and filtered before analysis	*HPLC–(−ESI)-MS* • Mobile phase: water (5 mM NH_4Ac):MeOH • FTOHs detected as acetate adducts	• Recoveries: ($^{13}C_2$-PFOA) > 60%	[119]
PFBS, PFHxS, PFOS, PFPeA, PFHxA, PFHpA, PFOA, PFNA, PFDA, PFUnA, PFDoA	Suspended soil and mixed liquor suspended solid from WWTPs	(1) Suspended soil and mixed liquor suspended soil filtered by GF/B filter (2) PLE (ASE200, Dionex, USA) of GF/B filter (three static cycles, MeOH solvent, 100°C, 2,000 psi, 10 min) (3) Extract dried to reduce <10 mL (60°C, water bath) (4) Reduced extract + 1 L water (5) SPE by Oasis HLB plus and PrepSepC-Agri (short) in tandem (6) Elution with 4 mL of MeOH under gravity conditions (7) Dryness under N_2 (8) Reconstitution in 1–2 mL of water:ACN (6:4) and vortexed	*HPLC–(ESI−)-MS/MS(QqQ)* • Injection volume: 10 μL • Column: Agilent Eclipse XDB-C_{18} (2.1 mm × 100 mm; 5 μm) at 40°C • Mobile phase: water (5 mM NH_4Ac):ACN • SRM mode	• MLOQ: 0.02–0.22 ng/g • Recoveries: 64–112% • CV <20%	[88]

cartridges, OASIS HLB and Sep-Pak (C_{18}), was evaluated. Both SPE cartridges were a cause of contamination by PFOS and PFOA. However, higher concentrations of PFOS and PFOA were reported for Sep-Pak cartridges. In the case of the Oasis HLB, PFOS, PFOA, PFHxS and PFBS were detected, but at lower concentrations than those found in the Sep-Pak cartridges. On the other hand, Taniyashu et al. [67] evaluated Oasis HLB and Oasis WAX columns for the extraction of PFCs. In this study, few target PFCs were detected in procedural blanks at a few pg/L in the final extract. However, PFOA, PFDA and PFUnA were still found at relatively high concentrations. In general, the performance of these columns was comparable. Recoveries were good (70–100%) for most compounds, but for short-chain PFCAs recoveries using Oasis WAX-SPE cartridges were higher. Losses due to evaporation during analysis and adsorption to the polypropylene sample container surface as discussed earlier were suggested causes for the lower recoveries.

2.2.2 Solid Samples

Sample pre-treatment in complex matrices as sludge or sediments is required to minimize matrix effects. It is important to be sure that transformation processes do not occur during this process. For example, the hydrolysis of fluorotelomer compounds to fluorotelomer alcohol during solvent extraction of soils was reported by Dasu et al. [68]. General approaches can be summarized in four different pre-treatments based on solid–liquid extraction or supernatant liquid extraction:

1. Ionic-pair extraction
2. Solid–acid liquid extraction
3. Solid–alkaline liquid extraction
4. Solid–liquid extraction

Table 1 summarizes the pre-treatments that can be found in some published works. The first four extraction procedures are based on the extraction of lyophilized or dried solid. The last one corresponds to non-dried solid extraction procedures.

Ionic-Pair Extraction

Ylinen et al. developed an ion-pair extraction procedure employing tetrabutyl ammonium (TBA) counter ions for the determination of PFOA in plasma and urine in combination with gas chromatography (GC) flame ionization detection (FID). Later on Hansen et al. [69] improved the sensitivity of the ion-pair extraction approach using methyl tertiary butyl ether (MTBE) and by inclusion of a filtration step to remove solids from the extract making it amenable for liquid chromatography coupled to tandem mass spectrometry (LC-MS/MS) determination. Ion-pair extraction procedure has been the basis of several procedures for biota [70, 71] and food samples [72]. However, this method has shown some limitations, such as

(1) co-extraction of lipids and other matrix constituents, (2) the absence of a clean-up step to overcome the effects of matrix compounds and (3) the wide variety of recoveries observed, typically ranging from <50% to >200%. Sludge samples usually contain high amounts of interferences and, once the sample is reconstituted, a filtration step previous to the analysis [73] or an additional clean-up step by Envicarb cartridges [74] is necessary. This pre-treatment improves the limits of quantification (MLOQ). Using this last approach, Eon et al. obtained recovery rates between 17% and 105% [73].

Solid–Acid Liquid Extraction

Current methods are based on the procedure described by Higgins et al. [75]. This method is based on the extraction of dried soils using acetic acid 1% at 60°C in an ultrasonic bath. Then, the mixture is centrifuged and the supernatant collected. The extraction of the solid residue is repeated twice; the extracts are combined and after centrifugation are cleaned up using SPE. Sometimes, a filtration step could be also required in order to eliminate the non-dissolved matter [44, 76]. This methodology has been used in the sludge characterization of different PFCs including PFCAs, PFSAs and fluorinated sulphonamides [43, 44, 52, 75, 76]. The recoveries were in most of the cases between 40% and 119%. In general, this treatment allows better quality parameters than ionic-pair extraction.

Solid–Alkaline Liquid Extraction

This procedure has been used for the analysis of acids, sulphates and sulphonamides fluorinated compounds by different authors. In general, alkaline methanol (with NaOH or NH_4OH) is used. After the alkaline treatment, a neutralization step with HCl (if NaOH is used) [77] or acetic acid at 1% (if the alkaline digestion has been carried out by NH_4OH) is needed [78]. Some more details can be found in Table 1. As a last clean-up step, SPE or dispersive-SPE (EnviCarb) is performed. The alkaline extraction allowed method limit of detection (MLOD) in the range of 0.14–1.43 ng/g dw in sludge [78] and MLOQ between 1.8 and 6.8 ng/g dw [77]. The recoveries ranged from 62% to 104% with RSDs between 2% and 7% (Table 1).

Solid–Liquid Extraction

This is the most commonly used methodology in the extraction of non-volatile PFCs from solid matrices due the facility and simplicity of the extraction and the required solvents. Some published works performed the extraction by methanol or acetonitrile with a shaker [79], in an ultrasonic bath with temperature between 40°C and 60°C along 20–30 min [34, 55, 56, 80, 81] or at 50°C along 2–7 days [82], depending on the analysed compounds.

In some other works, the extraction is performed using pressurized liquid extraction (PLE) by accelerated solvent extractor (ASE) [35, 83, 84]. The use of PLE instead of Soxhlet or hot vapour extraction allows decrease in the time of extraction due to the use of high pressure. The most frequent solvent extractor is methanol using a different number of cycles, temperatures (between 70°C and 100°C) and pressures (between 1,500 and 2,000 psi). After extraction, a clean-up process is in general required to decrease the matrix effects [35, 74], using SPE (C_{18}, anionic exchange (WAX)) or by using Envicarb graphitized carbon adsorbent. The methodology allows, in general, a MLOQ between 0.05 and 2.77 ng/g of PFCs, recoveries ranging from 57% to 120% in most of the cases and 2–30% of RSD (see Table 1).

2.3 Instrumental Analysis

During this step is important minimize possible sources of contamination due Teflon or PVDF tubing other materials as PEEK. Other precaution that can be considered is the addition of extra chromatographic column prior to the injector in order to delay PFCs peaks due to system contamination [79].

Liquid chromatography-mass spectrometry (LC-MS) or LC-MS/MS has been in general the techniques of choice for the analysis of PFCs. Therein detailed information about the main experimental conditions used for analysis such as LC-MS/MS precursor-product ion transitions.

LC separation of PFCs has been mainly carried out with C_{18} and C_8 columns. In spite of the wide use of RP-C_{18} columns for PFCs' analysis, the interference producing the enhancement of spectral signal has been reported. RP columns with shorter alkyl chain bonded phases (e.g., C_8, C_6, phenyl and phenylhexyl) also separated the branch isomers, but to a lesser extent. To minimize the separation of branched isomers, the authors increased the LC column temperature to 35°C or 40°C [85, 86]. Taniyasu et al. [87] explored the chromatographic properties and separation of short-chain PFAs on RP-C_{18} and ion-exchange columns. The results showed that using RP-C_{18}, the peaks of PFPrA and PFEtS were broad and not adequately resolved, whereas that of TFA was not retained in the analytical column eluting with the solvent front. This suggested that RP columns are not suitable for the analysis of short-chain PFAs, especially TFA. As a proper alternative, ion-exchange columns have superior retention properties for more hydrophilic substances enabling the analysis of short-chain PFCAs, TFA, PFPrA, PFBA, PFEtS, PFPrS and PFBS together with several long-chain PFCAs, in water samples.

Due to the complexity of environmental samples, it is possible that the co-occurrence of certain compounds can interfere the analyte determination. This problem has been partially solved using LC–MS/MS. However, certain interferences can affect the analyte ionization producing ion suppression or enhancement [35, 74]. The use of labelled PFCs during analytical process (surrogates or internal standards) helps to assess and normalize these instrumental effects.

LC-MS/MS performed using triple quadrupole mass spectrometer (QqQ) combined with multiple selected reaction monitoring (SRM) is one of the more widely applied analyzer [34, 43, 44, 73–78, 80, 81, 84, 88], as well as, to be one of the better suited for quantification of PFCs. Nowadays the performance of hybrid quadrupole linear ion trap (QqLit) [35] or hybrid quadrupole time of flight (QTOF) [52] has been also considered for trace quantification of PFCs. PFCs contain carboxylic, sulphonic, hydroxy or sulphonamide group. They have acidic properties and can therefore dissociate. Consequently, electrospray ionization in the negative mode (ESI(−)) has been the interface most widely used for the analysis of anionic perfluorinated surfactants, allowing limits of detection in the pg to ng/g range, although atmospheric pressure chemical ionization (APCI) and positive ESI have been employed for specific PFCs' analysis [83]. In addition, ESI has been optimized for the determination of neutral compounds such as the sulphonamides FOSA, Et-FOSA and *t*-Bu-PFOS. The use of atmospheric pressure photoionization (APPI) has been explored in few works [68–70]. Takino et al. [68] found as the main advantage of this technology the absence of matrix effects, but the limits of detection were considerably higher than those obtained by LC–ESI-MS/MS.

Pseudomolecular ions are formed such as $[M-K]^-$ for PFOS (*m*/*z* 499), $[M-H]^-$ for PFOA (*m*/*z* 413) and FOSA (*m*/*z* 498), which are generally selected as precursor ions for MS^2 experiments using ion trap and a triple quadrupole instruments. Berger et al. [89] have presented a comparison between IT, QqQ and TOF instruments. Tandem mass spectrometry showed excellent specificity, but the background is eliminated by the instrument, and thus it cannot be visualized. Applying TOF-MS gives an estimation of the amount of matrix left in the extract, which could impair the ionization performance and the high mass resolution of the TOF-MS instrument offers excellent specificity for PFCs' identification after a crude sample injection. Recently, the analytical suitability of three different LC–MS/MS systems: QqQ, conventional 3D-IT and QqLIT, to determine trace levels of PFCs in fish and shellfish was compared [90]. In this study, the accuracy was similar in the three systems, with recoveries always over 70%. Precision was better for the QqLIT and QqQ systems (7–15%) than for the IT system (10–17%). The QqLIT (working in SRM mode) and QqQ systems offered a linear dynamic range of at least three orders of magnitude, whereas that of the IT system was two orders of magnitude. The main advantage of QqLIT system is the high sensitivity, at least 20-fold higher than the QqQ system. Another advantage of QqLIT systems is the possibility to use enhanced product ion (EPI) mode and MS^3 modes in combination of SRM mode for confirmatory purposes of target analytes in complex matrices.

Other instrumental tools have been employed by Liu et al. [82] in the study of aerobic biodegradation of [^{14}C] 6:2 PFTOH in a flow-through soil incubation system. The instrumental analysis was carried out by radioactivity and liquid chromatography/accurate radioisotope counting (LC/ARC).

3 Environmental Fate of PFCs

Partitioning and reactivity properties are important to understand and model the environmental behaviour of PFCs. Just during the recent years, it has been initiated the study of the reactivity properties of these compounds, and in addition should be pointed out that some data continue being contradictory. A starting point to study the mechanistic properties of PFCs, as well as, to evaluate and assess properties of new emerging PFCs is the study of selected physicochemical properties.

The perfluoro alkyl sulphonates (PFASs) and PFCAs are strong acids that exist in equilibrium between the neutral form and the anionic form. In general, both the anionic and neutral forms, as it happens with PFOA, are soluble in water. Although the Henry's law constant values suggests partitioning to air for the neutral, protonated form, predicting the amount that partitions into air is complicated because there is uncertainty over the degree to which carboxylic and sulphonic acids partition from the water to atmosphere. The uncertainty arises with regard to the value of the acid dissociation constant (i.e. pK_a), or the fraction of the acid form present at environmentally relevant pH. PFCA and PFAS have been detected in air, water and soil samples collected throughout the world. The oceans have been suggested as the final sink and route of transport for perfluorinated carboxylic and sulphonic acids, where they have been detected on the surface and at depths over 1,000 m [91].

Some PFAS/PFCA have the potential for long-range environmental transport (LRET) by a combination of dissolved-phase ocean and gas-phase atmospheric transport; however, determining which is the predominant transport pathway is complicated by the uncertainty over water to atmosphere partitioning. Furthermore, there is evidence that transport and subsequent oxidation of volatile alcohol PFAS/PFAC precursors may contribute to the levels of PFAS/PFCA in the environment.

The evaluation of PFCs in remote areas such as the Antarctica peninsula is one of the very few forms of evaluation of LRET. The global fate of POPs is associated with different biogeochemical cycles and geophysical drivers. The occurrence of PFCs into remote areas such as the Antarctica could be partially explained by the theory of *cold condensation*, concerning the chemical movements or chemical transformations from sources under the impact of environmental forces, such as temperature, and interaction with other environmental compartments (soil, oceans, etc.) [92]. In addition, the physicochemical characteristics of PFCs should be considered, since these properties dictate their environmental behaviour [5].

Just few previous studies have reported the presence of PFCs in different biota samples from the Antarctica continent [11, 70, 93], whereas this information could be of importance to establish Global PFCs' distribution and also the basis of LRET of these compounds.

Different studies have shown that PFTOH can be degraded by microorganisms and by abiotic processes. 8:2 FTOH and FTOH of other chain lengths, and related chemicals in mixed microbial cultures, activated sludge and soil systems have been shown to be easily degraded to form PFOA and related perfluorinated acids. Some studies have also shown that $-CF_2-$ groups can be mineralized, forming shorter-chain perfluoro acids. If FTOH are absorbed from ingestion, inhalation, dermal or ocular exposure or formed in vivo by from other compounds they can be metabolized by mammals and other organisms to form perfluorinated acids and other fluorinated compounds. FTOH can be degraded by abiotic processes in water and air to produce PFCA and various intermediates. FTOH are fairly volatile. Based on atmospheric half-lives determined in chamber studies, FTOH can be transported globally. Deposition or degradation in areas far from the source can result in PFCA contamination in high latitudes and other remote locations and contribute to global background levels of PFCA and PFAS.

Therefore, two mechanisms should be considered to explain the LRET capabilities of PFCs. The first suggests atmospheric distribution of neutral, volatile compounds (*flyers*), such as fluorotelomer alcohols and perfluorinated sulphonamido alcohols. *Flyer* compounds are susceptible to suffer atmospheric long-range transport because of their partitioning properties (log K_{aw} values estimated between 0 and 1 and log K_{ow} around 5), which indicate that these classes of chemicals can be classified as *flyers* according to the Globo-POP model [5, 92]. This is also in agreement with the findings of Dreyer et al. (2009). Then, after their transport and cold condensation, these *flyer* compounds can biodegrade as it has already been indicated in previous studies [80] or suffer in situ oxidation to form ionic PFCs [94].

The second mechanism is related to the properties of ionic PFCs (negligible vapour pressure, water solubility and moderate sorption to solids), which predicts their accumulation in surface waters (*swimmers*) [95, 96]. Some studies have evaluated the influence of these mechanisms and have been revealed that the dominant phenomenon is the hydrospheric transport for PFOS, PFOA and PFNA [97]. For example, fluorotelomer alcohols have short atmospheric lifetimes in the order of 10–20 days [98]. The geographical isolation of Antarctica combined with both, short atmospheric lifetimes of fluorotelomer alcohols and the low yield of the oxidation pathway, significantly reduces the potential for effective atmospheric delivery to the Antarctic continent. Therefore, atmospheric input of flyer PFCs to the Antarctica is principally a function of rapid and direct delivery of contaminated wind masses.

Concerning to degradation processes, some recent data show that perfluorooctane sulphonyl fluoride (POSF) and its derivatives can be degraded under environmental conditions to form perfluoroalkyl sulphonates and carboxylic acids. Reaction of POSF ($CF_3(CF_2)_n$-SO_2F) with methyl or ethyl amines is used to produce *N*-ethyl or *N*-methyl perfluorooctane sulphonamidoethanols. Similar reactions are used to make shorter- and longer-chain analogues to POSF and POSF derivatives. FOSE compounds, such as *N*-methyl and *N*-ethyl FOSEs can be degraded though a series of intermediates to form both PFCAs and perfluoroalkyl sulphonates. Other chemical intermediates produce other FOSA derivatives, including phosphate esters, fatty acids esters, silanes, carboxylates and polymers with acrylate, urethane and other linkages.

Longer- and shorter-chain perfluoro sulphonyl derivatives have also been produced intentionally and as unintended reaction products. Based on existing data from the open literature and CBI data, it is expected that most, if not all, of these POSF and other chain length sulphonyl fluorides and their derivatives will be degraded to carboxylic acids and/or sulphonate over time. Most of these compounds will have environmental and metabolism half-lives of weeks to months. Some will be degraded faster and some will degrade more slowly, but all will eventually be degraded.

Very little data are available on the behaviour of other perfluorochemicals in the environment and in vivo but the existing data suggest that they will also be degraded to form PFAC. For example, recent studies have shown that ingested mono- and di-polyfluoroalkyl phosphates (PAPs) can be degraded in rats to form PFOA and other PFAC in the body. They can also be degraded by microbial processes in soil and wastewater to form perfluorinated acids [99].

A limited number of studies on the degradation of fluorotelomer-based polymers have been submitted, but some studies have shown that fluorotelomer-based polymers are subject to hydrolysis, photolysis and biodegradation to some extent. Studies have shown half-lives of a few days to hundreds of years. In addition, preliminary research on degradation of fluorotelomers has shown that some urethanes and acrylates biodegrade; however, half-lives and kinetics of the fluorotelomers are not yet well defined.

These studies have shown that the perfluorinated portion of some polymers is released as the polymer is degraded by microbial or abiotic processes to form telomer alcohols or other intermediates and that they eventually form PFCA. Polymers based on POSF and other chain length chemistries show similar degradation rates and release intermediates that further degrade to form perfluorinated acids and sulphonates. Studies have shown that some polymers can undergo indirect photolysis in soil and in aquatic systems and be degraded with half-lives of days to several years.

4 Occurrence of PFCs: The Llobregat River as Case Study

During the last decade, an important work has been carried out in order to assess the occurrence of PFCs in the aquatic environment. In this section, the occurrence of PFCs in river basins will be revised taken as central example the Llobregat River, as example of a Mediterranean river suffering a high industrial pressure.

Different examples of works assessing the occurrence of PFCs in river water are summarized in Table 2.

Due to their persistency and wide use in the past PFOS and PFOA are the best-studied compounds. Currently, the use and production of these two compounds are almost stopped; however, due to their high resistance to degradation and because they are the end products of other PFCs in use, PFOS and PFOA continue being present in high concentrations in surface waters and sediments. Due to their physicochemical characteristics PFOA is mainly found in water, whereas PFOS is retained in higher proportion into the sediments.

Table 2 PFCs – published works

Analytes	Origin	Matrices	Results		Reference
Llobregat River					
PFBS, PFHxS, PFOS, PFOA, PFNA	Rivers: Muga, Fluvià, Ter, Besós, Llobregat, Ebro Different WWTPs (2009) Catalonia (Spain)	Surface river water and effluent water from WWTP	*Surface river*: PFBS = 0.07–0.88 ng/L PFHxS = 0.03–0.64 ng/L PFOS = 1.09–9.56 ng/L PFOA = 0.79–9.63 ng/L PFNA = 0.06–1.62 ng/L	*Effluent WWTP*: PFBS = 0.07–2.03 ng/L PFHxS = 0.03–25.3 ng/L PFOS = 0.03–72.1 ng/L PFOA = 3.47–61.9 ng/L PFNA = 0.06–14.1 ng/L	[60]
PFBA, PFPeA, PFHxA, PFHpA, PFOA, PFNA, PFDA, PFUnA, PFDoA, PFTrA, PFTeA, PFHxDA, PFODA, PFBS, PFHxS, PFOS, PFDS, FOSA	WWTPs from Catalonia, Spain (2010)	Sewage sludge ($n = 5$)	PFBA ≤ MLOD – 22.6 ng/g dw PFPeA ≤ MLOQ – 17.2 ng/g dw PFHxA ≤ MLOD – 4.8 ng/g dw PFHpA ≤ MLOQ – 4.5 ng/g dw PFOA = 7.0–30.3 ng/g dw PFNA = 1.0–2.4 ng/g dw PFDA = 6.1–23.5 ng/g dw PFUnA ≤ MLOQ – 12.2 ng/g dw PFDoA = 2.7–11.3 ng/g dw	PFTrA ≤ MLOQ PFTeA ≤ MLOQ – 5.0 ng/g dw PFHxDA ≤ MLOQ – 4.9 ng/g dw PFODA ≤ MLOD – 0.9 ng/g dw PFBS ≤ MLOD – 7.6 ng/g dw PFHxS ≤ MLOQ PFOS = 53–121.1 ng/g dw PFDS ≤ MLOD – 7.5 ng/g dw FOSA ≤ MLOD – 10.7 ng/g dw	[35]
Other Spanish areas					
PFOS, PFHxS, PFBS, PFOA, PFNA	Northern Spain: Asturias, Cantabria and Basque Country (Spain)	Sea emissaries ($n = 3$) Ports ($n = 5$) Sewage treatment plant effluent ($n = 3$) Industrial WW effluent ($n = 1$)	PFOS = 0.01–5.11 ng/L PFHxS ≤ MLOD – 0.31 ng/L PFBS ≤ MLOD – 5.08 ng/L	PFOA = 0.03–3.53 ng/L PFNA = 0.01–1.40 ng/L	[59]
PFBS, PFHxA, PFHpA, PFHxS, THPFOS, PFOA, PFNA, PFOS, FOSA, PFDA, PFUnA, PFDS, PFDoA, PFTrA	Ebro River (Garcia and Mora) Francolí River Cortiella River (Spain)	Surface river water	PFBS ≤ MLOQ PFHxA ≤ MLOQ PFHpA = MLOQ – 3.38 ng/L PFHxS = MLOQ – 0.78 ng/L THPFOS ≤ MLOQ PFOA = MLOQ – 24.9 ng/L PFNA = MLOQ – 0.64 ng/L	PFOS = MLOQ – 5.88 ng/L FOSA = MLOQ – 0.20 ng/L PFDA = MLOQ – 0.82 ng/L PFUnA ≤ MLOQ PFDS ≤ MLOQ PFDoA ≤ MLOQ PFTrA ≤ MLOQ	[39]

(continued)

Table 2 (continued)

Analytes	Origin	Matrices	Results		Reference
PFBS, PFPeA, PFHxA, PFHpA, PFOA, PFNA, ip-PFNA, PFOS, PFDA, PFDS	L'Albufera de Valencia (Spain)	Surface water	PFBS ≤ MLOD − 5.50 ng/L PFPeA ≤ MLOD − 5.40 ng/L PFHxA ≤ MLOQ − 6.90 ng/L PFHpA ≤ MLOQ − 18.4 ng/L PFOA = 0.99–120.2 ng/L	PFNA = 0.02–18.5 ng/L ip-PFNA ≤ MLOD − 5.44 ng/L PFOS = 0.94–58.1 ng/L PFDA ≤ MLOD − 10 ng/L PFDS ≤ MLOD − 1.29 ng/L	[108]
PFBS, PFHxA, PFHpA, PFHxS, THPFOS, PFOA, PFNA, PFOS, FOSA, PFDA, PFUnA, PFDS, PFDoA, PFTrA	Reus, Tarragona, Tortosa and Valls (Spain)	Drinking tap water	PFBS ≤ MLOQ PFHxA ≤ MLOQ PFHpA = MLOQ − 0.40 ng/L PFHxS ≤ MLOQ THPFOS ≤ MLOQ PFOA = MLOQ − 0.67 ng/L PFNA = MLOQ − 0.20 ng/L	PFOS ≤ MLOQ FOSA ≤ MLOQ PFDA = MLOQ − 0.63 ng/L PFUnA ≤ MLOQ PFDS ≤ MLOQ PFDoA ≤ MLOQ PFTrA ≤ MLOQ	[39]
PFOS, PFHxS, PFBS, PFOA, PFNA	Northern Spain (Asturias, Cantabria and Basque Country)	Sediment emissaries ($n = 3$) Port sediment ($n = 4$) River sediment ($n = 3$)	PFOS ≤ MLOD − 0.13 ng/g PFHxS ≤ MLOD PFBS = 0.01 ng/g PFOA ≤ MLOD − 0.06 ng/g PFNA ≤ MLOD − 0.08 ng/g		[59]
PFBS, PFPeA, PFHxA, PFHpA, PFOA, PFNA, ip-PFNA, PFOS, PFDA, PFDS	L'Albufera de Valencia, Spain	Sediment	PFBS ≤ MLOD − 0.02 ng/g PFPeA ≤ MLOD − 0.02 ng/g PFHxA ≤ MLOQ − 0.10 ng/g PFHpA ≤ MLOQ − 0.95 ng/g PFOA = 0.03–10.9 ng/g	PFNA ≤ MLOD − 1.24 ng/g ip-PFNA ≤ MLOD − 1.52 ng/g PFOS = 0.10–4.80 ng/g PFDA ≤ MLOD − 1.25 ng/g PFDS ≤ MLOD − 2.00 ng/g	[108]
PFBS, PFHxS, PFOS, PFBA, PFPeA, PFHxA, PFHpA, PFOA, PFNA, PFDA, FOSA, N-MeFOSA, N-EtFOSA	Twenty WWTPs from Spain (2006)	Sewage sludge	PFPeA ≤ 0.05–4.69 ng/g dw PFHxA ≤ 0.03–2.60 ng/g dw PFHpA ≤ 0.01–2.04 ng/g dw PFOA ≤ 0.03–7.94 ng/g dw	PFNA ≤ 0.01–10.23 ng/g dw PFDA ≤ 0.04–24.29 ng/g dw PFHxS ≤ 0.01–18.20 ng/g dw PFOS ≤ 0.01–286.81 ng/g dw	[34]
Other countries					
PFOA, PFNA, PFOS, PFDA, PFUnA, FOSA	Conasauga River, Altamaha River and streams and ponds of Dalton (Georgia, USA)	Surface river	PFOA = 2.6–1,280 ng/L PFNA = 0.6–456 ng/L PFOS = 0.2–368 ng/L	PFDA = 0.1–160 ng/L PFUdA = 0.1–117 ng/L FOSA = 10.7–420 ng/L	[118]

PFOS	Rivers: Tone, Arakawa, Tama (Tokyo, Japan)	Surface river water	PFOS = 0.5–58 ng/L	[61]
PFOS, PFOA, PFHpA, PFNA, PFDA, PFDoA, PFHxS	Rivers from Northern of China	Surface river water	PFOS ≤ MLOD – 31 ng/L PFOA = 0.43–82 ng/L PFHpA ≤ MLOD – 35 ng/L PFNA ≤ MLOD – 4.9 ng/L PFDA ≤ MLOD – 5.7 ng/L PFDoA ≤ MLOD – 0.29 ng/L PFHxS ≤ MLOD – 5.8 ng/L	[120]
PFBA, PFPeA, PFHxA, PFHpA, PFOA, PFBS, PFOS	Rhine river, Mohene river and tributaries, Ruhr area	Surface river, lake water and drinking water	*Rhine*: PFBA = 2–3 ng/L PFPeA = 2–42 ng/L PFHxA = 2–77 ng/L PFHpA = 2–11 ng/L PFOA = 2–48 ng/L PFBS = 2–46 ng/L PFOS = 2–26 ng/L *Ruhr area*: PFBA = 2–143 ng/L PFPeA = 2–1,638 ng/L PFHxA = 2–1,248 ng/L PFHpA = 25–148 ng/L PFOA = 9–3,640 ng/L PFBS = 3–71 ng/L PFOS = 5–193 ng/L *Mohene river and tributaries*: PFBA = 9–200 ng/L PFPeA = 25–2,670 ng/L PFHxA = 73–3,040 ng/L PFHpA = 46–989 ng/L PFOA = 11–33,900 ng/L PFBS = 17–1,450 ng/L PFOS = 2–5,900 ng/L *Drinking water*: PFBA = 2–11 ng/L PFPeA = 3–77 ng/L PFHxA = 3–56 ng/L PFHpA = 2–23 ng/L PFOA = 2–519 ng/L PFBS = 3–20 ng/L PFOS = 3–22 ng/L	[38]
PFBS, PFHxS, PFHpS, PFOS, PFDS, PFBA, PFPeA, PFHxA, PFHpA, PFOA, PFNA, PFDA, PFUnA, PFDoA, PFTrA, PFTeA, FOSA, N-MeFOSA, N-EtFOSA	WWTPs from Hong Kong • Plants (A) and (B): secondary treatment by activated sludge method • Plant (C): chemically enhanced primary treatment (2008)	Sludge	PFBA = 3.1–111.4 ng/g PFPeA = 0.5–10.1 ng/g PFHxA = 0.3–27.8 ng/g PFHpA = 0.4–4 ng/g PFOA = 1.3–15.7 ng/g PFNA = 0.5–23 ng/g PFDA = 0.3–15.2 ng/g PFUnA = 0.4–7.8 ng/g PFDoA = 0.6–8.6 ng/g PFTrA = 0.2–19 ng/g PFTeA = 0.2–46 ng/g FOSA = nd N-MeFOSA = nd N-EtFOSA = nd PFBS = 0.6–6.4 ng/g PFHxS = nd PFHpS = 106.6 ng/g (one sample) PFOS = 3.1–7304.9 ng/g PFDS = nd	[78]

(continued)

Table 2 (continued)

Analytes	Origin	Matrices	Results	Reference
PFPeA, PHxA, PFHpA, PFOA, PFNA, PFDoA, PFTeA, 6:2 FTUCA, 8:2 FTUCA, PFBS, PFHxS, PFOS, FOSA, N-MeFOSA, N-EtFOSA, N-MeFOSAA, N-EtFOSAA	WWTPs from Zürich, Switzerland (2008)	Digested sewage sludge	∑PFCAs = 16.9–21.6 ng/g dw (PFOA = 5.0–9.1 ng/g dw) (PFOA > PFDoA > PFHxA > PFNA > PFHpA) PFOS = 117–670 ng/g dw 6:2 FTUCA = 2.1–3.4 ng/g dw 8:2 FTUCA = 5.4–14.8 ng/g dw FOSA = 2–5 ng/g dw	[74]
PFPeA, PFHxA, PFHpA, PFOA, PFNA, PFDA, PFUnA, PFDoA, PFHxS, PFOS	WWTPs from industrial zones in Thailand	Samples from: • Activated sludge • Sludge	*Activated sludge concentrations* PFPeA = 7.6–29.4 ng/L PFHxA = 0.5–79.4 ng/L PFHpA = 1.7–43.3 ng/L PFOA = 13.7–142.0 ng/L PFNA = 12.1–308.4 ng/L PFDA = 4.8–81.3 ng/L PFUnA = 11.9–338.2 ng/L PFDoA = 7.6–48.4 ng/L PFHxS = 26.8–27.5 ng/L PFOS = 348.7–672.9 ng/L *Sludge concentrations* PFPeA = 2.9–3.3 ng/g PFHxA = 0.3–99.9 ng/g PFHpA = 1.6–52.6 ng/g PFOA = 11.3–136.0 ng/g PFNA = 5.1–512.8 ng/g PFDA = 3.8–327.7 ng/g PFUnA = 45.2–78.2 ng/g PFDoA = ND-310.6 ng/g PFHxS = 36.6–157.7 ng/g PFOS = 396.9–552.6 ng/g	[84]

4.1 Waste Water Treatment Plants

Wastewater treatment plants (WWTPs) are major sources of PFCs to the natural environment, through treated effluents and also when contaminated sludge is used in agricultural lands. The first work assessing the occurrence of PFCs in sewage sludge was performed by Higgins et al. [75] who studied the occurrence PFCs in sediments and sludge from WWTPs in San Francisco (1998–2004) [75]. Concentrations from 1.2 to 2,610 ng/g dw, were reported, being PFOS the compound at higher concentrations. Following this work, several works have been devoted to assess the content of PFCs in sewage sludge. PFOS is exceptionally stable chemical compound that is highly resistant to degradation and due to its higher partition coefficient in comparison with other PFCs, especially PFCAs, is present in high concentrations in sewage sludge worldwide. Furthermore, PFOS is the end-point of the degradation of fluorochemicals used in a variety of industrial and commercial applications. Compounds that may be transformed to PFOS are 2-(*N*-ethyl perfluorooctane sulphonamido) acetic acid (N-EtFOSAA) and 2-(*N*-methyl perfluorooctane sulphonamido) acetic acid (N-MeFOSAA), among others. These compounds have been also identified in general in WWTP sewage sludge at levels often exceeding PFOS. This could indicate that part of the PFOS is directly generated in the degradation process of related products.

Llorca et al. [35] investigated the presence of PFCs in sewage sludge from five WWTPs along the Llobregat River. The results showed that PFCs longer than 10 C chains were at lower ng/g concentration levels, or below. In general, the concentrations of perfluorocarboxylic acids were ranging from 0.4 to 30.3 ng/g. PFOA, PFNA, PFDA and PFDoA were present in all the samples at concentrations higher than 1.0 μg/kg. These high concentrations were in agreement with other works. For example, Zhang et al. [100], Guo et al. [101], Li et al. [55] or Ma et al. [102]. On the other hand, in most of the samples the long-chain acidic compounds were not detected and just PFOA was found to be in high amounts, but this concentration can be associated with the biodegradation of other long-chain congeners currently in use [103, 104]. This predominance of shorter C chains is supported by Ma et al. [78]. The authors found a dominance of even-chain length PFCAs in all of the WWTP sludge samples investigated. It is suggested that a strong aerobic degradation of fluorotelomer alcohols in WWTPs ends in shorter fluorinated compounds. In addition, developing substitute materials to replace long-chain PFCs, or new processes to eliminate their presence as impurities in other products, has been a significant technical challenge. There has been considerable progress in the development and introduction of substitutes and alternatives. Many substitutes are shorter-chain compounds that still provide the needed functionality, but lack the bioaccumulation potential of the long-chain PFCs. In this sense, considerable amounts of these products can also reach sludge of WWTP, partially contributing to these concentrations. In spite of the lack of data reporting the profile of PFCs present in sewage sludge during the past, available data seem to show a strong decrease in the presence of long-chain PFCs, and at the same time an increase in short C-chain compounds. In the studies of Llorca et al. in the sewage sludge of WWTP discharging into the Llobregat River, FOSA was another of the more frequently found compounds, with concentrations ranging from 0.3 to 10.7 μg/kg. There was not found perfluorosulphonates at higher levels than MLOQ with an exception of PFOS, which was detected at concentrations ranging from 53 to 121 ng/g, being the compound that was present at higher levels, as it was expected. Picó et al. within the Framework of the SCARCE project also analysed sewage sludge from the WWTP of Igualada in the Anoia tributary of Llobregat and also found PFOS at concentration as high as 1,790 ng/g. Zhou et al. [30] reported the sorption of PFCs on the heterogeneous protein composition of activated sludge and the different sorption kinetics according to their carbon chain length and different functional groups [105], which could explain the high concentrations of PFOS found by Llorca et al. [35] and other authors. There is a general agreement among results in sewage sludge were the concentrations of PFOS is in general three to ten times higher than the concentrations of PFOA. This difference could be associated with the different sorption kinetics in function the different functional groups, in agreement with Zhou et al. [105].

The lack of total elimination of PFCs in wastewater treatments has been proved, and many works have reported high concentrations of PFCs in treated effluents, being therefore one of the main inputs to receiving waters.

PFCs have been studied in effluent water from Llobregat WWTP located in El Prat de Llobregat by Sanchez-Avila et al. [60]. The sampling point was located near to river surface water sampling point location which was also investigated by the same author. The WWTP effluent values were below 0.77 ng/L for PFBS, below 0.03 ng/L for PFHxS, 14.1 ng/L for PFOS, 61.9 ng/L for PFOA and below 0.06 ng/L for PFNA. Comparing the levels found in surface river water and in effluent water, the concentrations in this last one were higher. These results suggested that PFCs are discharged into the river through WWTPs effluents arriving to potable water treatment plants and, finally, to humans through tap water. However, the WWTPs processes redistribute some of the PFCs from influent water to sludge. The study realized by Zhou et al. [30] showed the favourable sorption of PFCs on the heterogeneous protein composition of activated sludge which could be explain the highest levels found by Llorca et al. [35] in WWTPs from Llobregat River. Zhou et al., in another published work, studied the different sorption kinetics in function of carbon chain length and different functional groups [105]. The calculated distribution coefficients indicate that PFOS had a higher sorption tendency to activated sludge than PFOA. Becker et al. [106] studies supported this last result. The authors showed that, in WWTP, the calculated mass flow of PFOA was fully discharged into the river while about half of PFOS was retained in the sewage sludge. The study carried out by Pico et al. in the WWTP of Anoia also showed that PFOS was accumulated in the sludge whereas PFCAs including PFHpA, PFOA, PFNA and PFDA were mostly in water.

4.2 Surface Waters

In general, concentration gradients can vary up to several orders of magnitude between different areas along the same river or lake, reflecting in general proximity to known industrial sources and WWTPs concentrated near populated regions. For example, different Japanese works found PFOS and PFOA in surface river samples with concentrations in the range from 0.30 to 157 ng/L for PFOS, or between 1.6 and 104 ng/L also for PFOS and 3.8–311 ng/L for PFOA [65]. The study carried out by Skutlared et al. [38] in the Ruhr River in Europe should be mentioned. In this study the occurrence of 12 PFCs was assessed including PFBA, PFPeA, PFHxA, PFHpA, PFOA, PFBS and PFOS. The results showed extremely high concentrations of some compounds. PFOA was present at concentrations till 33,900 ng/L in the Moehne River, and the authors found the main source of contamination in an agricultural area near Brilon-Scharfenberg. In addition, it was proved that this source of contamination leads to the consecutive pollution of Lake Moehn, the Ruhr River and corresponding drinking waters. In another example in China Wang et al. [107] studied the environment around a manufacturing facility. The authors also observed a decreasing trend of the PFOS, PFOA and PFHxS concentrations in soils, water and chicken eggs with the increased distance to the production factory, indicating the production site to be the primary source of PFCs in the region.

Regarding the occurrence of PFCs in the Llobregat River, very few works have assessed the content of PFCs. In 2009 Sanchez-Avila et al. [60] investigated the levels of PFBS, PFHxS, PFOS, PFOA, PFNA, in surface waters from Catalonia, including a sample from an industrialized area of the Llobregat River. In this sampling site, PFOS and PFOA were the compounds found at higher concentrations, but should be mentioned that all the compounds investigated were also found at quantifiable concentrations: 0.88 ng/L for PFBS, 0.64 ng/L for PFHxS, 9.13 ng/L for PFOS, 9.63 ng/L for PFOA and 1.62 ng/L for PFNA. The concentration levels reported in this study were comparable with the data reported for other river waters from industrialized areas. However, the profile of compounds was quite different than the one found in other European countries, such as Germany, where the occurrence of PFOA in surface water is generally found in higher concentrations than PFOS.

Recently, under the frame of the SCARCE project Picó et al. have investigated the presence of 21 PFCs in different sampling sites along the Llobregat River during 2 sampling campaigns. In this case, PFOA was one of the more frequently found compounds. However, PFOS was found at higher concentrations up to 2.7 μg/L. The presence of these compounds showed an important spatial distribution. In agreement with data reported in other European rivers. However, the Llobregat River is affected in a great manner by climate episodes, such as flows, which can re suspends contaminants in general contained in the sediments. The higher concentrations were found near the mouth as it was expected, because it corresponds to a heavily populated and industrialized area. The compound found in higher concentration was PFHpA with concentrations around 30 ng/L in more polluted samples. In general, more frequent compounds and also those in higher concentration were short-chain compounds, indicating a tendency to replace more persistent long-chain PFCs by new short-chain ones.

4.3 Sediments

There are few available data reporting the levels of PFCs in sediment samples from Llobregat River basin within the SCARCE project. Different PFCAs including PFBA, PFPeA, PFHxA, PFHpA, PFOA, PFNA, PFDA, PFUnA and PFDoA were detected in sediment samples but at concentrations up to 7.5 ng/g. Although the number of PFSAs is restricted to PFBS and PFOS, they reach concentrations up to 11 ng/g. However, other related works indicated that these compounds are distributed between the two compartments, water and sediment. An example was published by Picó et al. [108]. The authors investigated different points from l'Albufera de Valencia (Spain) assessing the presence of PFCs between MLOD and 10.9 ng/g where the highest values corresponded to PFOA and PFOS. In the same study, the levels of PFCs in surface waters were assessed showing the distribution of these compounds between water and sediments. Another study performed by Gómez et al. [59] focused the investigation on the analysis of different sediments from Cantabrian Sea samples (North of Spain). The results of PFCs were in most of the cases below MLOD in sediment river samples.

4.4 Drinking Water

In order to elucidate the possible source of PFCs in drinking water, some authors have compared the levels found in the catchment sites in surface river or lakes and in tap water, and it was showed that certain relations can be established. This reveals the ineffective removal of PFCs by the purification processes performed at water purification plants [38, 109]. For example, Skutlarek et al. [38] showed extremely high concentrations for PFOA in tap water in the zone of Ruhr area, which is in agreement with the concentrations found in environmental surface waters of the same areas. In another example PFOS and PFOA were measured at concentration levels around 9 and 3 ng/L, respectively, in The Lake Maggiore (Switzerland), and the results of the analysis of drinking water produced from the lake gave almost identical results revealing the poor performance of sand filtration and chlorination which is applied by the local waterworks [109].

Similar conclusion can be extracted from the work carried out by Takagi et al. [110], who studied the presence of PFCs in different waters including raw water and drinking water from Japan. In raw water the results for PFOS and PFOA were between 0.26–22 ng/L and 5.2–92 ng/L, respectively, and in tap water similar results were also obtained in the ranges between 0.16–22 ng/L and 2.3–84 ng/L for PFOS and PFOA, respectively. In addition, other PFCs that in general are not assessed and that are generated during the water treatment processes should be considered [48, 103, 111, 112].

Regarding the study of the Llobregat River, the occurrence of PFCs in drinking water produced from Llobregat River has been assessed by Llorca et al. under the frame of the SCARCE project. Among the 21 compounds considered in this study, the presence of PFBA, PFPeA, PFOA, PFNA, PFBS, PFHxS and PFOS has been found in final drinking water at concentrations between 0.07 and 35 ng/L. The more polluted drinking waters were those corresponding to catchment locations in more industrialized and polluted areas, such as Barcelona city, as it was expected.

These results were in agreement with a previous work carried out by Ericson et al. [39, 57] who studied the presence of these contaminants in drinking water from different areas of Catalonia. In the study performed by Ericson et al., the concentrations of PFCs were in the range between 0.02 and 69 ng/L in tap waters.

5 Future Trends

In recent years, the research has been focused on the study of degradation mechanisms of fluorochemicals as PFOA and PFOS. Cheng et al. found that sonolysis (sonochemical) is able to decompose PFOS and PFOA present in groundwater beneath a landfill following a pseudo first-order rate constant [113].

Because of the poor degradability of these ones in the treatment facilities, these compounds are discharged directly into the rivers. Once these recalcitrant compounds reach the environment, they can arrive to the drinking water through the drinking water prepared from surface water [114] or enter into food chain through the irrigation of agricultural lands with contaminated waters [115] or by the bioaccumulation, and consequent biomagnification, through the food chain [3]. The study of these compounds in river waters as well in flora and fauna is of high importance since they are not regulated and should be under control in order to elucidate possible focuses of PFCs into, for example, the Llobregat River [116].

Acknowledgement This work was funded by the project "Assessing and predicting effects on water quantity and quality in Iberian Rivers caused by global change" SCARCE (CSD-2009-00065).

References

1. Prevedouros K (2006) Sources, fate and transport of perfluorocarboxylates. Environ Sci Technol 40(1):32–44
2. Kantiani L (2010) Emerging food contaminants: a review. Anal Bioanal Chem 398(6): 2413–2427

3. Pico Y Perfluorinated compounds in food: a global perspective. Crit Rev Food Sci Nutr 51:605–625
4. Shoeib M, Harner T, Vlahos P (2006) Perfluorinated chemicals in the arctic atmosphere. Environ Sci Technol 40(24):7577–7583
5. Wania F (2007) A global mass balance analysis of the source of perfluorocarboxylic acids in the Arctic Ocean. Environ Sci Technol 41(13):4529–4535
6. Sonne C (2010) Health effects from long-range transported contaminants in Arctic top predators: an integrated review based on studies of polar bears and relevant model species. Environ Int 36(5):461–491
7. Martin JW (2004) Identification of long-chain perfluorinated acids in biota from the Canadian Arctic. Environ Sci Technol 38(2):373–380
8. Stemmler I, Lammel G (2010) Pathways of PFOA to the Arctic: variabilities and contributions of oceanic currents and atmospheric transport and chemistry sources. Atmos Chem Phys 10(20):9965–9980
9. Butt CM (2010) Levels and trends of poly- and perfluorinated compounds in the arctic environment. Sci Total Environ 408(15):2936–2965
10. Schiavone A (2009) Perfluorinated contaminants in fur seal pups and penguin eggs from South Shetland, Antarctica. Sci Total Environ 407(12):3899–3904
11. Tao L (2006) Perfluorooctanesulfonate and related fluorochemicals in albatrosses, elephant seals, penguins, and polar skuas from the southern ocean. Environ Sci Technol 40(24):7642–7648
12. Kubwabo C, Vais N, Benoit FM (2004) A pilot study on the determination of perfluorooctanesulfonate and other perfluorinated compounds in blood of Canadians. J Environ Monit 6(6): 540–545
13. Ericson I (2007) Perfluorinated chemicals in blood of residents in Catalonia (Spain) in relation to age and gender: a pilot study. Environ Int 33(5):616–623
14. Kärrman A (2009) Biomonitoring perfluorinated compounds in Catalonia, Spain: concentrations and trends in human liver and milk samples. Environ Sci Pollut Res 1–9
15. Tao L (2008) Perfluorinated compounds in human breast milk from several Asian countries, and in infant formula and dairy milk from the United States. Environ Sci Technol 42 (22):8597–8602
16. Llorca M (2010) Infant exposure of perfluorinated compounds: levels in breast milk and commercial baby food. Environ Int 36(6):584–592
17. So MK (2006) Health risks in infants associated with exposure to perfluorinated compounds in human breast milk from Zhoushan, China. Environ Sci Technol 40(9):2924–2929
18. Inoue K (2004) Perfluorooctane sulfonate (PFOS) and related perfluorinated compounds in human maternal and cord blood samples: assessment of PFOS exposure in a susceptible population during pregnancy. Environ Health Perspect 112(11):1204–1207
19. Monroy R (2008) Serum levels of perfluoroalkyl compounds in human maternal and umbilical cord blood samples. Environ Res 108(1):56–62
20. Carabias-Martínez R (2005) Pressurized liquid extraction in the analysis of food and biological samples. J Chromatogr A 1089(1/2):1–17
21. EFSA (2008) Perfluorooctane sulfonate (PFOS), perfluorooctanoic acid (PFOA) and their salts Scientific Opinion of the Panel on Contaminants in the Food chain. EFSA Guidelines 653:2–131
22. Ericson I (2008) Human exposure to perfluorinated chemicals through the diet: intake of perfluorinated compounds in foods from the Catalan (Spain) market. J Agric Food Chem 56(5): 1787–1794
23. Lacina O (2011) Simple, high throughput ultra-high performance liquid chromatography/tandem mass spectrometry trace analysis of perfluorinated alkylated substances in food of animal origin: milk and fish. J Chromatogr A 1218(28):4312–4321

24. Committee on Toxicity of Chemicals in Food Environment (2006) COT statement on the tolerable daily intake for perfluorooctane sulfonate. Available on line at http://www.food.gov.uk/multimedia/pdfs/cotstatementpfos200609.pdf
25. Committee on Toxicity of Chemicals in Food, C.P.a.t.E. (2006) COT statement on the tolerable daily intake for perfluorooctanoic acid. Available on line at http://www.food.gov.uk/multimedia/pdfs/cotstatementpfoa200610.pdf
26. Commission decision of 12 August 2002 implementing Council Directive 96/23/EC concerning the performance of analytical methods and the interpretation of results. Off J Eur Union L221/8
27. USEPA (2006) 2010/15 Stewardship Program. Environmental Protection Agency. Available on line at http://www.epa.gov/oppt/pfoa/pubs/stewardship/index.html
28. UNEP (2010) New POPs SC-4/17: listing of perfluorooctane sulfonic acid, its salts and perfluorooctane sulfonyl fluoride. In: United Nations Environment Programme: Stockholm Convention on Persistent Organic Pollutants (POPs). Geneva, Switzerland
29. European Union Directive 2008/105/EC on the environmental quality standards in the field of water policy, amending and repealing Council Directives: 82/176/EEC, 83/513/EEC, 84/156/EEC, 86/280/EEC and amending Directive 2000/60/EC of the European Parliament and Council-348/84, Brussels, 2008
30. Zhou Q (2010) Sorption of perfluorooctane sulfonate and perfluorooctanoate on activated sludge. Chemosphere 81(4):453–458
31. Rayne S, Forest K (2009) Perfluoroalkyl sulfonic and carboxylic acids: a critical review of physicochemical properties, levels and patterns in waters and wastewaters, and treatment methods. J Environ Sci Health A Tox Hazard Subst Environ Eng 44(12):1145–1199
32. Ruan T (2010) Presence and partitioning behavior of polyfluorinated iodine alkanes in environmental matrices around a fluorochemical manufacturing plant: another possible source for perfluorinated carboxylic acids? Environ Sci Technol 44(15):5755–5761
33. Yoo H (2011) Quantitative determination of perfluorochemicals and fluorotelomer alcohols in plants from biosolid-amended fields using LC/MS/MS and GC/MS. Environ Sci Technol 45:7985–7990
34. Navarro I, Sanz P, Martínez MÁ (2011) Analysis of perfluorinated alkyl substances in Spanish sewage sludge by liquid chromatography-tandem mass spectrometry. Anal Bioanal Chem 400(5):1277–1286
35. Llorca M (2011) Analysis of perfluorinated compounds in sewage sludge by pressurized solvent extraction followed by liquid chromatography-mass spectrometry. J Chromatogr A 1218(30):4840–4846
36. Clarke BO, Smith SR (2011) Review of 'emerging' organic contaminants in biosolids and assessment of international research priorities for the agricultural use of biosolids. Environ Int 37(1):226–247
37. Gottschall N (2010) Polybrominated diphenyl ethers, perfluorinated alkylated substances, and metals in tile drainage and groundwater following applications of municipal biosolids to agricultural fields. Sci Total Environ 408(4):873–883
38. Skutlarek D, Exner M, Färber H (2006) Perfluorinated surfactants in surface and drinking waters. Environ Sci Pollut Res 13(5):299–307
39. Ericson I (2008) Levels of perfluorochemicals in water samples from Catalonia, Spain: is drinking water a significant contribution to human exposure? Environ Sci Pollut Res 15(7):614–619
40. USEPA (2009) Provisional Health Advisories (PHA) for PFOA and PFOS. Environmental Protection Agency. Available on line at http://www.epa.gov/oppt/pfoa/pubs/pfoainfo.html
41. Martin JW (2004) Analytical challenges hamper perfluoroalkyl research. Environ Sci Technol 38(13):248A–255A
42. Sinclair E, Mayack DT, Roblee K, Yamashita N, Kannan K (2006) Arch Environ Contam Toxicol 50:398–410

43. Guo R, Sim W-J, Lee E-S, Lee J-H, Oh J-E (2010) Evaluation of the fate of perfluoroalkyl compounds in wastewater treatment plants. Water Res 44:3476–3486
44. Yu J, Hu J, Tanaka S, Fujii S (2009) Perfluorooctane sulfonate (PFOS) and perfluorooctanoic acid (PFOA) in sewage treatment plants. Water Res 43:2399–2408
45. Hu J, Yu J (2010) An LC-MS-MS method for the determination of perfluorinated surfactants in environmental matrices. Chromatographia 72:411–416
46. Sun H (2011) Long-chain perfluorinated chemicals in digested sewage sludges in Switzerland. Environ Pollut 159(2):654–662
47. Loveless SE (2006) Comparative responses of rats and mice exposed to linear/branched, linear, or branched ammonium perfluorooctanoate (APFO). Toxicology 220(2–3):203–217
48. Frömel T, Knepper TP (2010) Fluorotelomer ethoxylates: sources of highly fluorinated environmental contaminants. Part I: biotransformation. Chemosphere 80(11):1387–1392
49. Holm A (2004) Determination of perfluorooctane sulfonate and perfluorooctanoic acid in human plasma by large volume injection capillary column switching liquid chromatography coupled to electrospray ionization mass spectrometry. J Sep Sci 27(13):1071–1079
50. Kärrman A (2006) Perfluorinated chemicals in relation to other persistent organic pollutants in human blood. Chemosphere 64(9):1582–1591
51. Liu J, Lee LS (2005) Solubility and sorption by soils of 8:2 fluorotelomer alcohol in water and cosolvent systems. Environ Sci Technol 39(19):7535–7540
52. Guo R, Zhou Q, Cai Y, Jiang G (2008) Determination of perfluorooctanesulfonate and perfluorooctanoic acid in sewage sludge samples using liquid chromatography/quadrupole time-of-flight mass spectrometry. Talanta 75:1394–1399
53. Szostek B, Prickett KB, Buck RC (2006) Determination of fluorotelomer alcohols by liquid chromatography/tandem mass spectrometry in water. Rapid Commun Mass Spectrom 20(19): 2837–2844
54. Schultz MM, Barofsky DF, Field JA (2006) quantitative determination of fluorinated alkyl substances by large-volume-injection liquid chromatography tandem mass spectrometry characterization of municipal wastewaters. Environ Sci Technol 40(1):289–295
55. Li F (2010) Quantitative characterization of short- and long-chain perfluorinated acids in solid matrices in Shanghai, China. Sci Total Environ 408(3):617–623
56. Bossi R (2008) Perfluoroalkyl compounds in Danish wastewater treatment plants and aquatic environments. Environ Int 34(4):443–450
57. Ericson I, Domingo J, Nadal M, Bigas E, Llebaria X, van Bavel B, Lindström G (2009) Levels of perfluorinated chemicals in municipal drinking water from Catalonia, Spain: public health implications. Arch Environ Contam Toxicol 57(4):631–638
58. D'Eon JC (2009) Perfluorinated phosphonic acids in Canadian surface waters and wastewater treatment plant effluent: discovery of a new class of perfluorinated acids. Environ Toxicol Chem 28(10):2101–2107
59. Gómez C (2011) Occurrence of perfluorinated compounds in water, sediment and mussels from the Cantabrian Sea (North Spain). Mar Pollut Bull 62(5):948–955
60. Sánchez-Avila J, Meyer J, Lacorte S (2010) Spatial distribution and sources of perfluorochemicals in the NW Mediterranean coastal waters (Catalonia, Spain). Environ Pollut 158(9): 2833–2840
61. Takazawa Y (2009) Occurrence and distribution of perfluorooctane sulfonate and perfluorooctanoic acid in the rivers of Tokyo. Water Air Soil Pollut 202(1–4):57–67
62. Villagrasa M, López De Alda M, Barcelo D (2006) Environmental analysis of fluorinated alkyl substances by liquid chromatography–(tandem) mass spectrometry: a review. Anal Bioanal Chem 386(4):953–972
63. Suja F, Pramanik BK, Zain SM (2009) Contamination, bioaccumulation and toxic effects of perfluorinated chemicals (PFCs) in the water environment: a review paper. Water Sci Technol 60:1533–1554
64. Taniyasu S (2003) A survey of perfluorooctane sulfonate and related perfluorinated organic compounds in water, fish, birds, and humans from Japan. Environ Sci Technol 37(12):2634–2639

65. Saito N (2003) Perfluorooctane sulfonate concentrations in surface water in Japan. Arch Environ Contam Toxicol 45(2):149–158
66. Yamashita N (2004) Analysis of perfluorinated acids at parts-per-quadrillion levels in seawater using liquid chromatography-tandem mass spectrometry. Environ Sci Technol 38(21):5522–5528
67. Taniyasu S (2005) Analysis of fluorotelomer alcohols, fluorotelomer acids, and short- and long-chain perfluorinated acids in water and biota. J Chromatogr A 1093(1–2):89–97
68. Dasu K (2010) Hydrolysis of fluorotelomer compounds leading to fluorotelomer alcohol production during solvent extractions of soils. Chemosphere 81(7):911–917
69. Hansen KJ (2001) Compound-specific, quantitative characterization of organic fluorochemicals in biological matrices. Environ Sci Technol 35(4):766–770
70. Giesy JP, Kannan K (2001) Global distribution of perfluorooctane sulfonate in wildlife. Environ Sci Technol 35(7):1339–1342
71. Yeung LWY (2009) A survey of perfluorinated compounds in surface water and biota including dolphins from the Ganges River and in other waterbodies in India. Chemosphere 76(1):55–62
72. Guruge KS (2008) Species-specific concentrations of perfluoroalkyl contaminants in farm and pet animals in Japan. Chemosphere 73(Suppl 1):S210–S215
73. Crozier P, Furdui V, Reiner E, Libelo EL, Mabury S (2009) Observation of a commercial fluorinated material, the polyfluoroalkyl phosphoric acid diesters, in human sera, wastewater treatment plant sludge, and paper fibers. Environ Sci Technol 43:4589–4594
74. Zhang T, Sun H, Gerecke AC, Kannan K, Müller CE, Alder AC (2010) Comparison of two extraction methods for the analysis of per- and polyfluorinated chemicals in digested sewage sludge. J Chromatogr A 1217:5026–5034
75. Higgins CP (2005) Quantitative determination of perfluorochemicals in sediments and domestic sludge. Environ Sci Technol 39(11):3946–3956
76. Loganathan BG (2007) Perfluoroalkyl sulfonates and perfluorocarboxylates in two wastewater treatment facilities in Kentucky and Georgia. Water Res 41(20):4611–4620
77. Yoo H, Washington JW, Jenkins TM (2009) Analysis of perfluorinated chemicals in sludge: method development and initial results. J Chromatogr A 1216:7831–7839
78. Ma R, Shih K (2010) Perfluorochemicals in wastewater treatment plants and sediments in Hong Kong. Environ Pollut 158:1354–1362
79. Powley CR (2005) Matrix effect-free analytical methods for determination of perfluorinated carboxylic acids in environmental matrixes. Anal Chem 77(19):6353–6358
80. Rhoads KR (2008) Aerobic biotransformation and fate of N-ethyl perfluorooctane sulfonamidoethanol (N-EtFOSE) in activated sludge. Environ Sci Technol 42(8):2873–2878
81. Sun H, Gerecke AC, Giger W, Alder AC (2011) Long-chain perfluorinated chemicals in digested sewage sludges in Switzerland. Environ Pollut 159(2):654–662
82. Liu J, Wang N, Buck RC, Wolstenholme BW, Folsom PW, Sulecki LM, Bellin CA (2010) Aerobic biodegradation of [14C] 6:2 fluorotelomer alcohol in a flow-through soil incubation system. Chemosphere 80:716–723
83. Schröder HF (2003) Determination of fluorinated surfactants and their metabolites in sewage sludge samples by liquid chromatography with mass spectrometry and tandem mass spectrometry after pressurised liquid extraction and separation on fluorine-modified reversed-phase sorbents. J Chromatogr A 1020(1):131–151
84. Kunacheva C (2011) Mass flows of perfluorinated compounds (PFCs) in central wastewater treatment plants of industrial zones in Thailand. Chemosphere 83(6):737–744
85. Kuklenyik Z (2004) Automated solid-phase extraction and measurement of perfluorinated organic acids and amides in human serum and milk. Environ Sci Technol 38(13):3698–3704
86. Calafat AM (2006) Perfluorochemicals in pooled serum samples from United States residents in 2001 and 2002. Environ Sci Technol 40(7):2128–2134
87. Taniyasu S (2008) Analysis of trifluoroacetic acid and other short-chain perfluorinated acids (C2–C4) in precipitation by liquid chromatography–tandem mass spectrometry: comparison to patterns of long-chain perfluorinated acids (C5–C18). Anal Chim Acta 619(2):221–230

88. Shivakoti BR (2010) Occurrences and behavior of perfluorinated compounds (PFCs) in several wastewater treatment plants (WWTPs) in Japan and Thailand. J Environ Monit 12(6):1255–1264
89. Berger U, Haukås M (2005) Validation of a screening method based on liquid chromatography coupled to high-resolution mass spectrometry for analysis of perfluoroalkylated substances in biota. J Chromatogr A 1081(2):210–217
90. Llorca M (2010) Study of the performance of three LC-MS/MS platforms for analysis of perfluorinated compounds. Anal Bioanal Chem 398(3):1145–1159
91. Yamashita N (2005) A global survey of perfluorinated acids in oceans. Mar Pollut Bull 51 (8–12):658–668
92. Lohmann R (2007) Global fate of POPs: current and future research directions. Environ Pollut 150(1):150–165
93. Bengtson Nash S (2010) Perfluorinated compounds in the Antarctic region: ocean circulation provides prolonged protection from distant sources. Environ Pollut 158(9):2985–2991
94. Wallington TJ (2006) Formation of C7F15COOH (PFOA) and other perfluorocarboxylic acids during the atmospheric oxidation of 8:2 fluorotelomer alcohol. Environ Sci Technol 40(3):924–930
95. Yamashita N (2008) Perfluorinated acids as novel chemical tracers of global circulation of ocean waters. Chemosphere 70(7):1247–1255
96. Ahrens L (2010) Sources of polyfluoroalkyl compounds in the North Sea, Baltic Sea and Norwegian Sea: evidence from their spatial distribution in surface water. Mar Pollut Bull 60(2):255–260
97. Armitage J (2006) Modeling global-scale fate and transport of perfluorooctanoate emitted from direct sources. Environ Sci Technol 40(22):6969–6975
98. Ellis DA (2003) Atmospheric lifetime of fluorotelomer alcohols. Environ Sci Technol 37(17):3816–3820
99. D'Eon JC, Mabury SA (2007) Production of perfluorinated carboxylic acids (PFCAs) from the biotransformation of polyfluoroalkyl phosphate surfactants (PAPS): exploring routes of human contamination. Environ Sci Technol 41(13):4799–4805
100. Zhang T (2010) Comparison of two extraction methods for the analysis of per- and polyfluorinated chemicals in digested sewage sludge. J Chromatogr A 1217(31):5026–5034
101. Guo R (2010) Evaluation of the fate of perfluoroalkyl compounds in wastewater treatment plants. Water Res 44(11):3476–3486
102. Ma R, Shih K (2010) Perfluorochemicals in wastewater treatment plants and sediments in Hong Kong. Environ Pollut 158(5):1354–1362
103. Lee H, Deon J, Mabury SA (2010) Biodegradation of polyfluoroalkyl phosphates as a source of perfluorinated acids to the environment. Environ Sci Technol 44(9):3305–3310
104. Frömel T, Knepper TP (2010) Biodegradation of fluorinated alkyl substances. Rev Environ Contam Toxicol 208:161–177
105. Zhou P (2009) Fluorine bonding – how does it work in protein-ligand interactions? J Chem Inf Model 49(10):2344–2355
106. Becker AM (2010) Perfluorooctanoic acid and perfluorooctane sulfonate released from a waste water treatment plant in Bavaria, Germany. Environ Sci Pollut Res 17(9):1502–1507
107. Wang Y (2010) Distribution of perfluorooctane sulfonate and other perfluorochemicals in the ambient environment around a manufacturing facility in china. Environ Sci Technol 44(21):8062–8067
108. Pico Y (2011) Occurrence of perfluorinated compounds in water and sediment of L'Albufera Natural Park (València, Spain). Environ Sci Pollut Res 1–12.
109. Loos R, Wollgast J, Huber T, Hanke G (2007) Polar herbicides, pharmaceutical products, perfluorooctanesulfonate (PFOS), perfluorooctanoate (PFOA), and nonylphenol and its carboxylates and ethoxylates in surface and tap waters around Lake Maggiore in Northern Italy. Anal Bioanal Chem 387(4):1469–1478

110. Takagi S, Adachi F, Miyano K, Koizumi Y, Tanaka H, Mimura M, Watanabe I, Tanabe S, Kannan K (2008) Perfluorooctanesulfonate and perfluorooctanoate in raw and treated tap water from Osaka, Japan. Chemosphere 72(10):1409–1412
111. Dinglasan MJA (2004) Fluorotelomer alcohol biodegradation yields poly- and perfluorinated acids. Environ Sci Technol 38(10):2857–2864
112. Wang N (2005) Fluorotelomer alcohol biodegradation – direct evidence that perfluorinated carbon chains breakdown. Environ Sci Technol 39(19):7516–7528
113. Cheng J (2008) Sonochemical degradation of perfluorooctane sulfonate (PFOS) and perfluorooctanoate (PFOA) in landfill groundwater: environmental matrix effects. Environ Sci Technol 42(21):8057–8063
114. Houtman CJ (2011) Emerging contaminants in surface waters and their relevance for the production of drinking water in Europe. J Integr Environ Sci 7(4):271–295
115. Ferré-Huguet N (2008) Assessment of metals from consuming vegetables, fruits and rice grown on soils irrigated with waters of the Ebro River in Catalonia, Spain. Biol Trace Elem Res 123(1):66–79
116. Petrovic M (2011) Combined scenarios of chemical and ecological quality under water scarcity in Mediterranean rivers. Trends Anal Chem 30(8):1269–1278
117. Scott BF (2006) Analysis for perfluorocarboxylic acids/anions in surface waters and precipitation using GC-MS and analysis of PFOA from large-volume samples. Environ Sci Technol 40(20):6405–6410
118. Konwick BJ (2008) Concentrations and patterns of perfluoroalkyl acids in Georgia, USA surface waters near and distant to a major use source. Environ Toxicol Chem 27(10): 2011–2018
119. Saez M, De Voogt P, Parsons JR (2008) Persistence of perfluoroalkylated substances in closed bottle tests with municipal sewage sludge. Environ Sci Pollut Res 15(6):472–477
120. Wang T (2011) Perfluorinated compounds in surface waters from Northern China: comparison to level of industrialization. Environ Int

Illicit Drugs and Metabolites in the Llobregat River Basin

Cristina Postigo, Nicola Mastroianni, Miren López de Alda, and Damià Barceló

Abstract Recently, the study of illicit drugs and metabolites in the aquatic environment has become a matter of scientific interest. An increasing number of studies have been carried out worldwide in this area of research in the last years. The Llobregat River basin has been one of the investigated areas in Spain. Its water quality has been shown to be affected by the presence of this type of emerging contaminants, often to a larger extent than other rivers, due to its marked Mediterranean character and urban and industrial pressures. This chapter reviews the occurrence of illicit drugs and their metabolites in both wastewaters and surface waters along the Llobregat River basin, and the analytical methodologies developed for their determination. Measured levels of these substances in the Llobregat River basin are compared with the levels found in other Spanish and European areas. Since treated wastewaters constitute the main source of illicit drugs and metabolites to the natural receiving waters, and surface waters are used for water supply purposes, the reported removal of these substances in wastewater treatment plants and drinking water treatment plants along the basin is also reviewed. Finally, the use of influent wastewater levels to estimate illicit drug use in riverine populations is also discussed.

Keywords Illicit drugs, Llobregat River, Psychoactive substances, Surface water, Wastewater

---

C. Postigo, N. Mastroianni, and M.L. de Alda (✉)
Department of Environmental Chemistry, Institute of Environmental Assessment and Water Research, (IDAEA-CSIC), C/ Jordi Girona, 18-26, 08034 Barcelona, Spain
e-mail: mlaqam@cid.csic.es

D. Barceló
Department of Environmental Chemistry, Institute of Environmental Assessment and Water Research, (IDAEA-CSIC), C/ Jordi Girona, 18-26, 08034 Barcelona, Spain

Catalan Institute for Water Research (ICRA), Parc Científic i Tecnològic de la Universitat de Girona, Edifici H2O, 17003 Girona, Spain

S. Sabater et al. (eds.), *The Llobregat: The Story of a Polluted Mediterranean River*, Hdb Env Chem (2012) 21:239–262, DOI 10.1007/698_2012_146,

Contents

Abbreviations

6ACM	6-Acetyl-morphine
ACN	Acetonitrile
AM	1-Phenylpropan-2-amine or amphetamine
BE	Benzoylecgonine
CAS	Conventional activated sludge
CE	Cocaethylene
COC	Cocaine
CODE	Codeine
DUI	Drug use indicator
DWTP	Drinking water treatment plant
EDDP	2-Ethylene-1,5-dimethyl-3,3-diphenylpyrrolidine
EPH	Ephedrine
ESI	Electrospray
FENTA	Fentanyl
HER	Heroin
IS	Internal standard
KETA	Ketamine
LC	Liquid chromatography
LOQ	Limit of quantification
LSD	Lysergic acid diethylamide
MA	*N*-Methyl-1-phenylpropan-2-amine or methamphetamine
MDA	3,4-Methylenedioxyamphetamine
MDEA	3,4-Methylenedioxyethamphetamine
MDMA	3,4-Methylenedioxymethamphetamine or ecstasy
MeOH	Methanol
METH	Methadone
MOR	Morphine
MS	Mass spectrometry
MS/MS	Tandem mass spectrometry

nor-CODE	Nor-codeine
nor-LSD	Nor-LSD and nor-iso-LSD
nor-MOR	Nor-morphine
OH-THC	11-Hydroxy-Δ^9-tetrahydrocannabinol
O-OH-LSD	2-Oxo-3-hydroxy LSD
PCP	Phencyclidine
QqLIT	Hybrid quadrupole-linear ion trap
QqQ	Triple quadrupole
RO	Reverse osmosis
RSD	Relative standard deviations
SPE	Solid phase extraction
SRM	Selective reaction monitoring
SW	Surface water
THC	Δ^9-Tetrahydrocannabinol
THC-COOH	11-Nor-9-carboxy- Δ^9-tetrahydrocannabinol
UF	Ultrafiltration
UPLC	Ultra-performance liquid chromatography
WW	Wastewater
WWTP	Wastewater treatment plant

1 Introduction

Illicit drugs and their metabolites have been recently recognized as emerging environmental contaminants of concern. Since 2004, year in which Jones-Lepp and coworkers [1] reported for the first time the presence of this type of compounds in waters, an increasing number of studies that confirm the presence of these substances in water matrices [2–4] and also in other environmental matrices, such as atmospheric particles [5–11], and sludge and river sediments [12, 13] have been published. The analysis of illicit drugs and their metabolites in environmental matrices requires the use of highly sensitive and selective reliable techniques capable of detecting the low levels at which these compounds are present in the environment. To this end, the determination of these compounds in concentrated extracts has been mainly carried out by means of liquid chromatography coupled to tandem mass spectrometry (LC-MS/MS), since the use of gas chromatographic techniques requires the derivatization of the analytes in order to increase their volatility.

The main source of these substances to the environment is their consumption and production. Unlike pharmaceuticals, direct deposition of illicit drugs is less likely. After consumption, different amounts of the consumed drug and its metabolization products are excreted via urine and feces. Under the best case scenario, these substances undergo physical–chemical and/or biological transformation in wastewater treatment plants (WWTPs). However, there is already enough evidence about their incomplete removal during wastewater treatment [14], and the proportion of biotransformation and mineralization that they experience is still unknown. Yet, wastewater treatment is an important process to reduce the levels of these substances before their release into the aquatic environment. This, together with the dilution that treated wastewaters experience when discharged into natural water masses, helps to attenuate the potential negative effects that these substances may pose to aquatic ecosystems and also to reduce their presence in drinking water sources. In this respect, Mediterranean river basins are more vulnerable to chemical pollution than other European catchments, because they experience drought periods. Water scarcity is directly related to an increase of the surface water levels of polar micropollutants present in discharged treated wastewaters (e.g., pharmaceuticals, illicit drugs, and metabolites), which are the main component of the river discharge in catchments with high industrial and urban pressures, and population densities, as it is the case of the Llobregat River basin. The study of illicit drugs and their metabolites in the Llobregat River basin is also justified because surface waters of the main river constitute the main apportionment source of water supply for the city of Barcelona and surrounding urban areas, thus, the presence of these substances represents a potential threat to public health. To date, various studies have been performed in this line [15–21], in some cases with a double objective: (1) to evaluate the occurrence of these compounds in wastewaters and surface waters of the Llobregat River basin and (2) to use environmental levels to estimate illicit drug use in the investigated areas as proposed by Daughton in 2001 [22], and implemented for the first time by Zuccato and coworkers [23] in 2005.

In this context, the objectives of the present chapter are to review the analytical techniques developed to analyze illicit drugs and metabolites in the Llobregat River basin and the levels reported in wastewaters and surface waters in this area, to compare these levels with those observed in other Spanish and European river basins, to assess the efficiency of removal of these compounds during wastewater treatment and drinking water production, and to examine the drug use estimations derived for various populations from the corresponding raw wastewater concentrations of drugs.

2 Analysis of Illicit Drugs and Metabolites in Waters

All analytical methodologies developed to investigate the presence of illicit drugs and metabolites in waters from the Llobregat River basin are based on solid phase extraction (SPE) of the target analytes present in the samples and further detection with LC-MS/MS [17, 19, 21]. They all cover illicit drugs and metabolites belonging to different chemical classes and their main features are summarized in Table 1.

Prior to analysis, water samples were filtered to remove suspended solids, and further spiked with mixtures of the deuterated analogues at known concentrations, which allows accurate quantification of the target analytes. In order to increase

Table 1 Review of analytical methods used for determination of illicit drugs and metabolites in waters from the Llobregat River basin

Ref.	Target analytes[a]	Sample pretreatment[b]	Liquid chromatography[c]		Mass spectrometry[d]	Accuracy method recovery	Sensitivity[e] method LOQ (ng/L)	Precision RSD (%)
			Chromatographic column	Mobile phase	Ionization- MS detector (acquisition mode)			
[17]	MOR, 6ACM, nor-MOR, HER, METH, EDDP, CODE, nor-CODE, FENTA, THC, THC-COOH	200 mL (WW, SW) Filtration (1.6 μm) IS addition (50 ng/L) SPE (Oasis HLB 200 mg)	Acquity BEH C_{18} (100 × 2.1 mm, 1.7 μm)	400 μL/min (2 runs × 9 min) $MeOH/H_2O$ (50 mM NH_4HCO_2) (pH 3.8)	(+) ESI-QqQ (2–3 SRM/ analyte)	44–99 (SW) 42–96 (WW)	0.1–12.5 (SW) 0.3–25.0 (WW)	2–7 (SW) 2–8 (WW)
[19]	COC, BE, AM, MA, MDA, MDEA, MDMA, LSD, KETA, PCP, FENTA	100 mL (WW, SW) Filtration (1.6 μm) IS addition (12.5 ng/L) SPE (Oasis HLB 200 mg)	Acquity BEH C_{18} (100 × 2.1 mm, 1.7 μm)	500 μL/min (1 run × 6.5 min) ACN/H_2O (30 mM NH_4HCO_2) (pH 3.5)	(+) ESI-QqQ (2 SRM/analyte)	75–99 (SW) 70–101 (WW)	0.1–0.9 (SW) 0.2–2.1 (WW)	4–11 (SW) 6–11 (WW)
[21]	COC, BE, CE, AM, MA, MDMA, EPH, LSD, nor-LSD, O-OH-LSD, MOR, 6ACM, HER, THC, THC-COOH, OH-THC	2 × 5 mL (WW) Filtration (1 μm + 0.45 μm) IS addition (20 ng/L) *On-line* SPE (PLRPs and Oasis HLB – 10 × 2 mm)	Purospher star RP-18 (125 × 2.0 mm, 5 μm)	300 μL/min (2 runs × 35 min) ACN/H_2O	(+/−) ESI-QqLIT (2 SRM/analyte)	5–59 (WW)	0.7–6.0 (WW)	2–14 (WW)

[a]*6ACM* 6-acetyl morphine, *AM* 1-phenylpropan-2-amine (amphetamine), *BE* benzoylecgonine, *CE* cocaethylene, *COC* cocaine, *CODE* codeine, *EDDP* 2-ethylene-1,5-dimethyl-3,3-diphenylpyrrolidine, *EPH* ephedrine, *FENTA* fentanyl, *HER* heroin, *KETA* ketamine, *LSD* lysergic acid diethylamide, *MA* N-methyl-1-phenylpropan-2-amine (methamphetamine), *MDA* 3,4-methylenedioxyamphetamine, *MDEA* 3,4-methylenedioxyethamphetamine, *MDMA* 3,4-methylenedioxymethamphetamine, *METH* methadone, *MOR* morphine, *nor-CODE* nor-codeine, *nor-LSD* nor-LSD and nor-iso-LSD, *nor-MOR* nor-morphine, *O-OH-LSD* 2-oxo-3-hydroxy LSD, *OH-THC* 11-hydroxy- Δ^9-tetrahydrocannabinol, *PCP* phencyclidine, *THC* Δ^9-tetrahydrocannabinol, *THC-COOH* 11-nor-carboxy-Δ^9-tetrahydrocannabinol

[b]*WW* wastewater, *SW* surface water, *IS* internal standard, *SPE* solid phase extraction

[c]*ACN* acetonitrile, *MeOH* methanol

[d]*ESI* electrospray, *QqQ* triple quadrupole, *QqLIT* quadrupole-linear ion trap, *SRM* selective reaction monitoring transition

[e]*LOQ* limit of quantification

[f]*RSD* relative standard deviation

method sensitivity and reduce matrix interferences, samples were preconcentrated onto polymeric SPE sorbents, such as Oasis HLB [17, 19, 21] and PLRPs [21] cartridges. Both *off-line* [17, 19, 21] and *on-line* SPE processes have been described [21]. The use of *on-line* SPE resulted in a fully automated methodology capable of achieving very low detection limits (in the pg or low ng/L level) with very low sample volumes (2 × 5 mL), and with minimal sample handling, high throughput, and time and labor saving as other clear advantages over *off-line* approaches. Conversely, the main drawbacks as compared to the *off-line* procedures are that no extract remains for further analysis and that the matrix may potentially interfere to a higher extent in the analysis, since the selection of washing and eluting solvents is somewhat less flexible than in *off-line* protocols.

Chromatographic separation was satisfactorily achieved with both LC in its classical version [21] and ultra-performance liquid chromatography (UPLC) [17, 19], by means of C_{18} columns (see Table 1). However, the use of UPLC columns (1.7 μm of particle size) allowed shortening the analytical chromatographic time and hence the consumption of solvents. Analytes were eluted from the column with a binary mobile phase consisting of acetonitrile [21] or methanol [17, 19] and water, by applying an organic solvent gradient at a constant flow rate. The mobile phase was in some cases buffered to an acidic pH (3.5–3.8) with ammonium formate/formic acid (see Table 1), in order to reduce chromatographic peak tailing and to improve ionization of the analytes determined in the positive mode [17, 19].

The investigated illicit drugs and metabolites were exclusively ionized by means of electrospray (ESI). With the exception of cannabinoids, which provide a better MS response under the negative ionization mode [24], all target analytes preferably produce positive ions in ESI (see Table 1). This atmospheric pressure ionization source, though highly versatile, is strongly affected by matrix interferences, which may negatively affect analyte recoveries in water samples, being more noticeable in highly polluted waters or with high organic matter content, e.g., influent wastewaters. The addition of isotopic-labeled analogues at the beginning of the analytical process allows correcting for matrix effects and also for potential analyte losses that may take place during the analysis due to evaporation, degradation, etc.

Tandem mass spectrometric determination was performed with triple quadrupole (QqQ) [17, 19] and hybrid quadrupole-linear ion trap (QqLIT) [21] instruments in the selective reaction monitoring (SRM) mode, by acquiring at least two transitions per target compound (see Table 1), which provides the best sensitivity and selectivity and allows obtaining four identification points [25]. On the contrary, just one SRM was registered to determine the isotope-stable internal standards used in the quantification process, since these compounds do not occur naturally in the environment.

3 Occurrence of Illicit Drugs and Metabolites in Wastewaters

The above presented methodologies were applied to investigate the levels of selected illicit drugs and metabolites in influent and effluent wastewaters of various WWTPs located along the Llobregat River basin [15, 18, 21]. Wastewater treatment in all studied WWTPs is based at least on a preliminary clarification step, which is followed by a biological treatment, which in most cases consists of conventional activated sludge processes (CAS), and a secondary clarification step. The studied WWTPs present different sizes, giving service to populations between 400 and 1,300,000 inhabitants. Their size and location, as well as the origin of the wastewaters treated, are shown in Fig. 1 [15, 18, 21]. Most of them were sampled only once, whereas five of them (see WWTPs 8, 10–13 in Fig. 1)

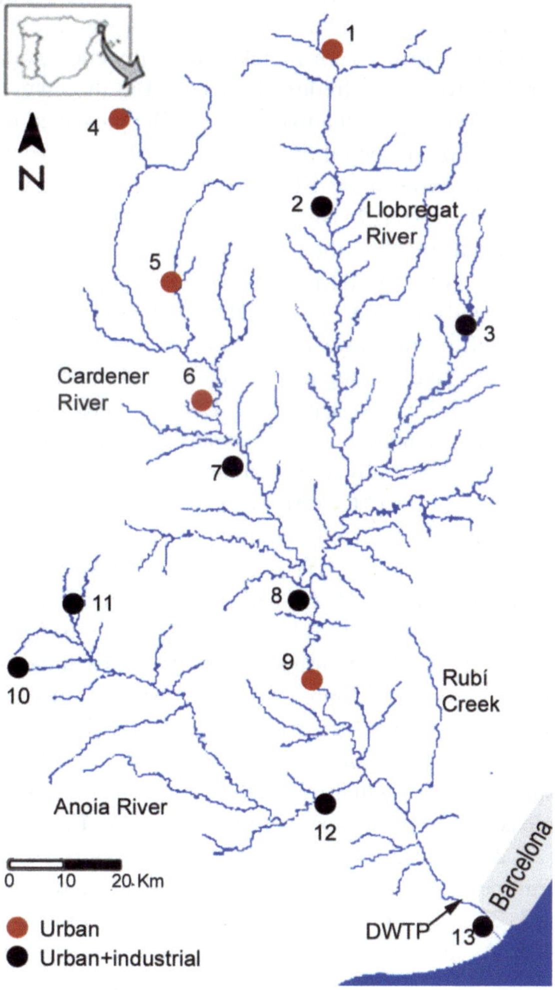

Fig. 1 Location of the WWTPs investigated in the Llobregat River basin (WWTP size: 1: 2,300 inhabitants, 2: 20,900 inh., 3: 5,700 inh., 4: 400 inh., 5: 16,500 inh., 6: 9,200 inh. 7: 20,200 inh., 8: 53,000 inh., 9: 3,200 inh., 10: 64,000 inh., 11: 151,500 inh., 12: 317,400 inh., 13: 1,300,000 inh.) [15, 18, 21]

were sampled in several occasions, in order to study removal efficiencies of illicit drugs and metabolites and/or the daily variation of the levels observed throughout the week.

3.1 Levels in Influent Wastewaters

Influent wastewaters, as expected, presented the highest levels of illicit drugs and metabolites in the investigated water matrices. Based on the reported data, the most ubiquitous and abundant compound in this matrix is the main cocaine metabolite, benzoylecgonine (BE), which is usually present at levels in the μg/L range (maximum concentration detected: 6 μg/L in WWTP 13 [21]). Cocaine (COC) and its metabolic product cocaethylene (CE), the amphetamine like compounds ephedrine (EPH), ecstasy (MDMA) and amphetamine (AM), the opioids morphine (MOR) and codeine (CODE), and the synthetic opioid used to treat heroin addiction, methadone (METH), and its main metabolite 2-ethylene-1,5-dimethyl-3,3-diphenylpyrrolidine (EDDP) were also frequently detected in the investigated samples but at comparatively lower levels than BE. Only COC presented occasionally levels above 1 μg/L (maximum concentration: 1,236 ng/L in WWTP 11 [18]). In terms of abundance, COC is followed by EPH, MOR, and CODE, with maximum concentrations of 725 ng/L, 356 ng/L, and 314 ng/L, respectively, in WWTP 13 [15, 21], (which are mainly attributed to their therapeutic use), and MDMA, whose maximum concentration raised up to 302 ng/L in WWTP 12 [18]. The other most frequently detected compounds, i.e., CE, AM, EDDP, and METH were usually detected at levels below 100 ng/L in influent wastewaters.

Among the investigated substances, 3,4-methylenedioxyethamphetamine (MDEA), heroin (HER), and fentanyl (FENTA) were not detected in any sample, and the remaining analytes, i.e., Δ^9-tetrahydrocannabinol (THC) and its metabolites –11-hydroxy- Δ^9-tetrahydrocannabinol (OH-THC) and 11-nor-9-carboxy- Δ^9-tetrahydrocannabinol (THC-COOH)– , LSD and its metabolites – nor-LSD, nor-iso-LSD (nor-LSD), and 2-oxo-3-hydroxy LSD (O-OH-LSD) – , the heroin metabolite 6-acetyl-morphine (6ACM), ketamine (KETA) and the amphetamine like-compounds methamphetamine (MA), and 3,4-methylenedioxyamphetamine (MDA), were rarely observed, and presented levels usually below 10 ng/L.

Levels of illicit drugs and metabolites measured in influent wastewaters of WWTPs of the Llobregat River basin are in-line with those reported for these substances in the peer-reviewed literature. Figure 2 summarizes the concentrations found in influent and effluent wastewaters of the Llobregat River basin of those substances for which more data are available, and compares them with the levels measured in other Spanish areas, e.g., the Eastern Coast [21, 26], North Western [27, 28], the Ebro River basin [29], Catalonia [18] and South Eastern [30], and other European countries, like Switzerland [31], Ireland [32], Belgium [33–36], Italy [23, 37, 38], Germany [39], Croatia [40], France [41], and United Kingdom [42, 43]. As it can be observed in Fig. 2, levels of these compounds in influent

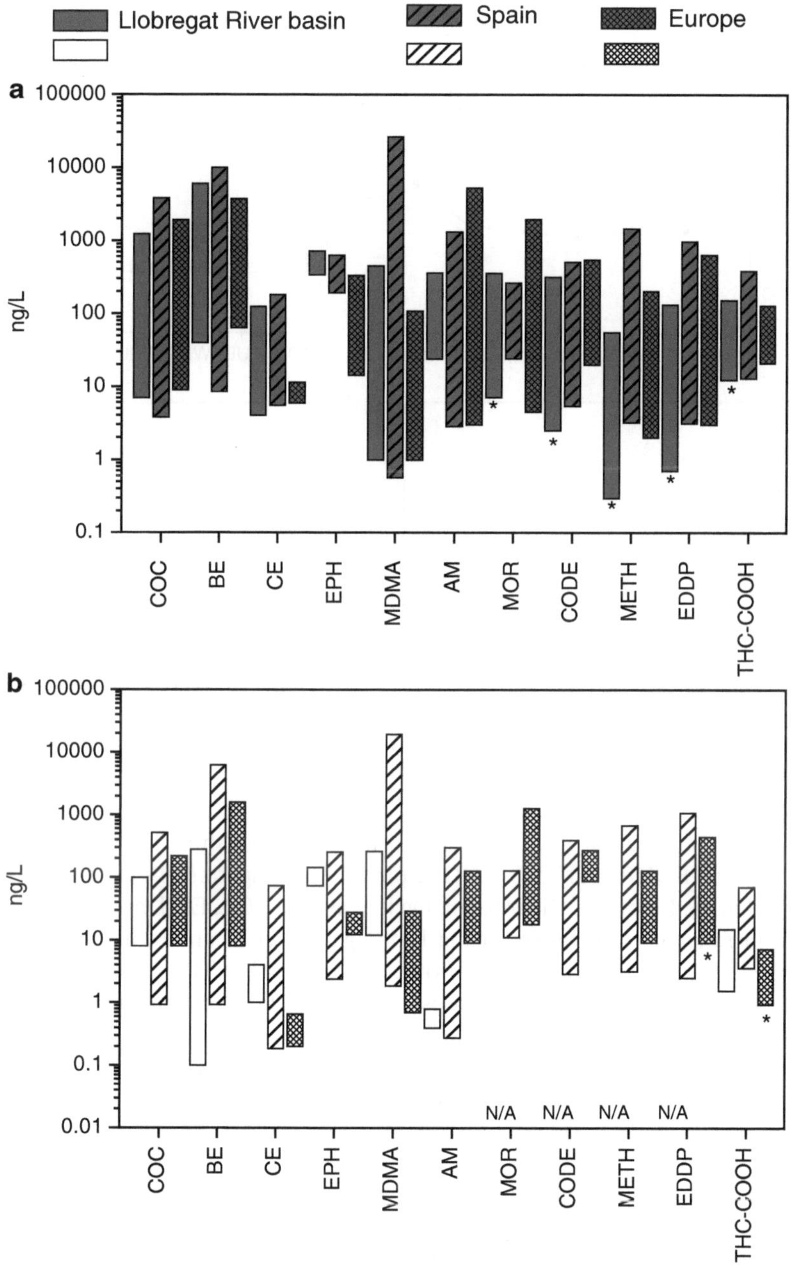

Fig. 2 Concentration ranges (expressed in ng/L in logarithmic scale) of the most investigated illicit drugs and metabolites in (a) influent and (b) effluent wastewaters from the Llobregat River basin and other Spanish and European areas (*N/A*: data not available, *asterisk*: just one value reported, minimum value is the method LOQ)

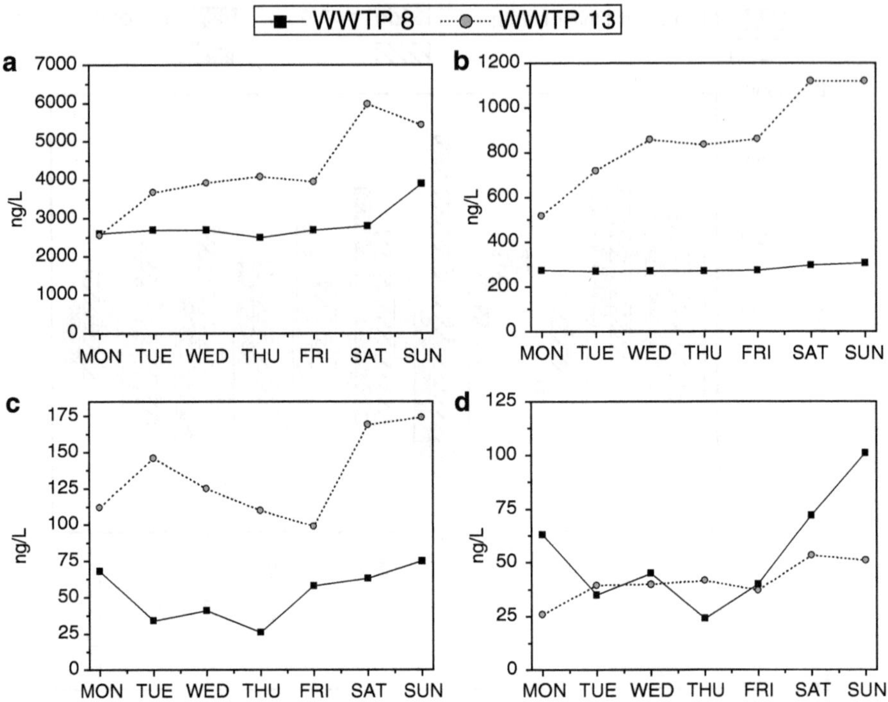

Fig. 3 Daily variation of (a) BE, (b) COC, (c) MDMA, and (d) AM levels in influent wastewaters of WWTP 8 and WWTP 13 (see location in Fig. 1)

wastewaters are similar to those reported in other Spanish areas, but for MDMA, AM, METH, EDDP, and THC-COOH, for which comparatively lower levels were reported in the Llobregat River basin. On the contrary, CE, EPH, and MDMA levels found in influent wastewaters of the Llobregat River basin are higher than those measured in influent wastewaters of other European WWTPs.

Huerta-Fontela et al. [18] and Postigo et al. [21] investigated the day-to-day variation of illicit drug levels throughout the week in influent and effluent wastewaters from WWTPs 8 and 13, respectively (see Fig. 1). Figure 3 shows the week profile of the concentrations of those compounds that were positively identified on a daily basis in both WWTPs, that is, COC and its main metabolite BE and the amphetamine-like compounds AM and MDMA. As it is shown, these substances presented higher levels at the weekend (Saturday and Sunday) than during the week (Monday to Friday), which may indicate their recreational use. Larger differences were observed for COC and BE levels in the largest WWTP (labeled as 13 in Fig. 1), being the average levels measured during the weekend 1.5 and 1.6 times larger, respectively, than the average levels observed during the working days. Other cocaine metabolite, CE, investigated only in WWTP 13, showed the same increasing trend, with average concentrations during the working days and the weekend of 59 ng/L and 125 ng/L, respectively. Regarding MDMA and AM, larger differences in their concentrations were observed in WWTP 8, with average levels 1.5 and 2.1 times higher during the weekend than during the working days for MDMA and AM, respectively. MA levels were above the method LOD in waters from WWTP 8 only during the weekend; however, levels observed in WWTP 13 throughout the week do not show relevant variations. In the same way, levels of EPH and the opioid MOR, compounds for which the week profile was only studied in WWTP 13, remained constant during the week, which may be indicative of their therapeutic use.

3.2 *Levels in Effluent Wastewaters*

The most abundant and ubiquitous illicit drugs and metabolites in effluent wastewaters of the Llobregat River basin were COC and its main metabolite BE and the amphetamine-like compounds MDMA and EPH, which were also among the most abundant and ubiquitous substances in influent wastewaters (see Fig. 2). However, comparatively lower levels, usually below 100 ng/L, were found, due to their partial removal during wastewater treatment processes [18, 21]. The range of concentrations measured for these compounds in this matrix in the Llobregat River basin and in other wastewater treatment facilities located in other areas of Spain and in other European countries is summarized in Fig. 2. As it can be observed in the figure, levels of illicit drugs and metabolites in effluent wastewaters of the Llobregat River basin are in general comparatively lower than those measured in other Spanish and European areas; only EPH and MDMA show distinctly lower levels in Europe.

HER, MOR, FENTA, and THC were not detected in any effluent wastewater sample [15, 21], and data to assess the presence of CODE, METH, and EDDP in this matrix have not been specifically reported for the Llobregat River basin [15, 17]. The remaining investigated compounds were occasionally detected at a few ng/L or even in the pg/L range.

Contrary to what it was observed in influent wastewaters, the levels of illicit drugs and metabolites in effluent wastewaters remained constant throughout the week [18, 21]. Additionally, the highest concentrations measured at the WWTP outlet did not correspond with the highest levels observed at the WWTP inlet, and therefore, effluent levels seem to be determined by diverse working parameters, other than the contaminant load entering the WWTP.

3.3 *Removal During Wastewater Treatment Processes*

Huerta-Fontela et al. [18] and Postigo et al. [21] calculated the removal efficiencies of illicit drugs and metabolites from waters in several WWTPs of the Llobregat River basin (plants 8, 10, 11, 12, and 13 in Fig. 1), based on the levels of these

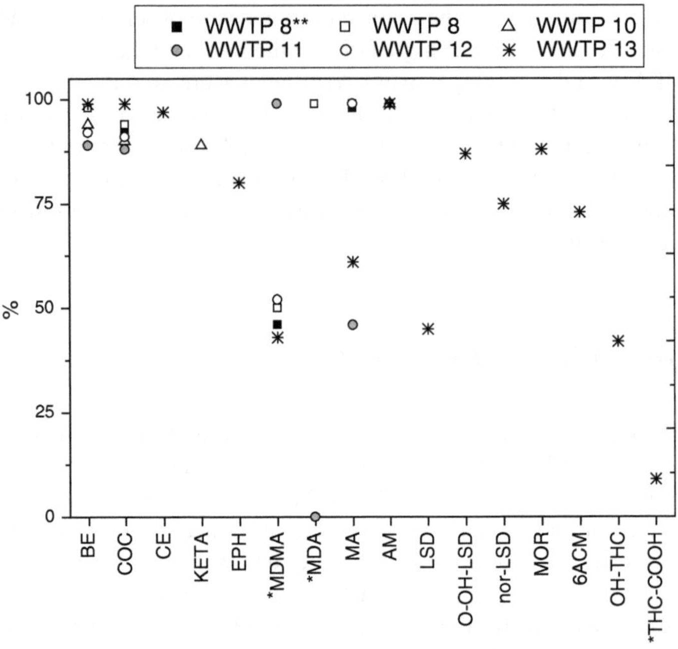

Fig. 4 Reported removals of illicit drugs and metabolites in WWTPs located along the Llobregat River basin. *WWTP 8***: average value of 7 consecutive days; *WWTP 8, 10–12*: average value of 4 days, *WWTP 13*: average value of 11 days. *Compounds occasionally observed at higher levels in effluent compared to influent

substances measured in influent and effluent composite samples, which were collected taking into account the hydraulic retention time of each WWTP. All investigated WWTPs use conventional activated sludge as secondary biological treatment. Figure 4 summarizes the reported data. As it can be observed, COC and its metabolites BE and CE, KETA, the amphetamine-like compounds EPH and AM, O-OH-LSD, and MOR showed the best elimination rates (above 80% on average) in all investigated WWTPs [18, 21]. Additionally, the day-to-day variation of the removal efficiencies of these compounds in WWTP 13 ($n = 11$) presented relative standard deviations (RSD) values below 3% for BE, COC, CE, AM, and MOR, and of 6% and 9% for EPH and O-OH-LSD, respectively.

Removal efficiencies observed for the remaining amphetamine-like compounds, i.e., MDMA, MDA, and MA, were observed to vary among the different investigated WWTPs [18] and within the same WWTP. For instance, MA and MDMA elimination rates presented RSD of 23% and 42%, respectively, in WWTP5. MDA and MDMA were determined occasionally at higher concentrations in effluent than in influent waters. In the case of MDA, this was attributed to N-demethylation of MDMA during wastewater treatment processes [18], whereas for MDMA, the process that leads to an increase of the effluent levels, which was also observed in WWTPs from the Ebro River basin [29], has not been elucidated yet.

Regarding the removal efficiency of the investigated opioids, other than MOR, in the Llobregat River basin, only elimination rates for 6ACM could be assessed, being 73% on average in WWTP 13. Boleda et al. [15] reported poor removal of METH and its metabolite EDDP, nor-morphine (nor-MOR), CODE, and nor-codeine (nor-CODE) in different WWTPs of Catalonia, two of them located in the Llobregat River basin, and occasionally higher concentrations in the effluents than in the influents. Persistence of EDDP and METH has been also reported in WWTPs of other European countries [32, 37]. In the case of nor-MOR and nor-CODE, N-demethylation of MOR and CODE, respectively, has been proposed as the mechanism that contributes to increase their levels during wastewater treatment [15].

Among the investigated cannabinoids, THC-COOH is removed comparatively worse than THC and OH-THC. This compound has also been measured occasionally at higher concentration in effluents than in influent wastewaters, which could be attributed to the cleavage of conjugated forms during treatment; however, it has not been investigated yet [15, 17, 21].

Overall, WWTPs based on CAS treatments, like the ones investigated in the Llobregat River basin, provide better elimination rates for illicit drugs and metabolites than wastewater treatment facilities that operate with only primary treatment [44] or with a secondary biological treatment based on biological filters [29, 41, 43, 45]. However, removal efficiency of illicit drugs and metabolites from wastewaters may be improved by incorporating advanced tertiary treatments to the existing wastewater treatment facilities. In this respect, reverse osmosis membranes have been observed to efficiently remove amphetamine-like compounds from municipal secondary treatment effluent wastewaters [46], and photocatalytic processes, such as photo-Fenton and heterogeneous photocatalysis with TiO_2, were able to mineralize cocaine, methadone, and their corresponding major metabolites in simulated municipal effluent wastewaters [47, 48].

3.4 *Estimation of Illicit Drug Use*

Back-calculation of illicit drug use at the community level from the levels of illicit drugs and metabolites measured in influent wastewaters is straightforward. To estimate illicit drug use by means of what is so-called the sewage epidemiology approach, concentrations of a drug use indicator (DUI) have to be normalized across the water volume entering the plant, the people served by the WWTP, and corrected by a factor that takes into account the molar mass ratio between the drug itself and the DUI, and the average excretion rate of the DUI.

Levels found throughout the week in wastewaters entering WWTP 13 were used to estimate illicit drug use in the area served by this facility, which covers about 1,300,000 people [21]. DUIs selected were the drug itself in the case of amphetamine-like compounds (AM, MA, and MDMA), and a drug metabolite in the case of cocaine (BE), heroin (6ACM), and cannabis (OH-THC). Results

Table 2 Estimations of illicit drug use, expressed in mg/day/1,000 people, in the Llobregat River basin and in other Spanish and European areas obtained with the sewage epidemiology approach

Studied area	Ref.	COC	MDMA	AM	MA	THC	HER
WWTP 13 – Llobregat River basin (Spain)	[21]	1,303–3,060	83–147	19–37	5–14	71–123	184–362
Catalonia (NE Spain)	[15, 18]	1,400	200	n/a	n/a	3,466[a]	138[a]
Ebro River basin (Spain)	[29]	1793[a]	60[b]	460[a]	2[b]	680[a]	24[a]
River Po Basin (Italy)	[23]	700	n.i.	n.i.	n.i.	n.i.	n.i.
Milano (Italy)	[44]	909	6	9	10	3,040	70
Lugano (Switzerland)	[44]	622	11	n.d.	n.d.	6,536	100
London (UK)	[44]	690	5	79	6	7,600	210
South Wales (UK)	[45]	900	n.i.	2,500	n.i.	n.i.	n.i.
Belgium	[33–35]	40–1,289	n.i	n.i	n.i	n.i	n.i
Zagreb (Croatia)	[40]	166	3.6	9.7	n.i.	3,690	262
Paris (France)	[41]	110–979[b]	1.6–15.4[b]	n.d.	n.i.	n.i.	n.i.

n.i. not investigated, *n.d.* compound not detected in influent wastewaters, *n/a* not reported data
[a]mg/day/1,000 adult people (15–64 years)
[b]mg/day/1,000 young people (15–34 or 15–44 years)

obtained, expressed in mg/day/1,000 people, are summarized in Table 2 and compared with illicit drug use figures obtained by applying the sewage epidemiology method in other investigated areas [15, 18, 23, 29, 33–35, 40, 41, 49, 50]. Note that figures shown are not fully comparable, since variations may exist in the DUI used in the back-calculations, e.g., MOR vs. 6ACM for heroin; THC-COOH vs. OH-THC for cannabis, or in the average excretion rate of the DUI considered. Moreover, different segments of the population have been occasionally considered in the peer-reviewed studies.

According to the figures obtained in the WWTP located in the lower part of the Llobregat River basin, the most consumed drug in this area is cocaine (2,200 mg/day/1,000 people on average), followed by heroin (244 mg/day/1,000 people on average), ecstasy (113 mg/day/1,000 people on average), cannabis (88 mg/day/1,000 people on average), amphetamine (30 mg/day/1,000 people on average), and methamphetamine (9 mg/day/1,000 people on average). These figures slightly differ from illicit drug use official estimates, which point out to cannabis as the most consumed drug followed by cocaine, ecstasy, amphetamine and methamphetamine, and heroin [51]. There are three explanations to these differences: (1) the existence of a local pattern of illicit drug use different to the national one, (2) the presence of biases in the sewage epidemiology approach applied, and (3) the presence of biases in the official methods.

The sewage epidemiology approach represents a very useful tool for illicit drug use estimation; however, it is still in full development and is subject to some limitations, which have been critically reviewed by Van Nuijs et al. [52]. For instance, in the case of cannabis, underestimation may occur because only the levels of the cannabis consumption indicator in the aqueous phase are considered, and these substances, due to their physical–chemical properties (log $K_{ow} > 5$) are likely to adsorb onto the wastewater suspended solids. On the other hand, THC is highly metabolized, and therefore, its main excretion products are excreted at a very low rate (<2%), which may lead to a high uncertainty of the figures obtained. On the other hand, the estimation of the consumption of amphetamine-like compounds requires the analysis of their enantiomeric forms to discern between their licit and illicit use, which has only been addressed in one of the peer-reviewed methodologies developed to determine these compounds in water [53]. Regarding estimation of heroin use, if MOR is used as DUI, it is necessary to correct for the amount of MOR excreted into the sewage system due to its therapeutic use. If 6ACM is used as DUI, its low stability and low excretion rate, may lead to underestimation of heroin real use.

Overall, uncertainty of the estimation figures obtained can be reduced by further research on metabolization patterns of illicit drugs, accurate measurements of the water flow entering the wastewater facilities and the development of methodologies that allow calculating the real number of people served by the WWTP.

4 Occurrence of Illicit Drugs and Metabolites in Surface Waters

Incomplete elimination of illicit drugs and metabolites during wastewater treatment leads to their continuous release into receiving surface waters. The seasonal and spatial variation of the levels of these substances in the Llobregat River basin has been evaluated in the main river, which also constitutes one of the main water sources for drinking water production, its two main tributaries, the Anoia River and the Cardener River (see Fig. 1), and one additional stream, the Rubí Creek. [15, 20]. Sampling points were selected in order to evaluate industrial and urban pressures on the chemical quality of the natural waters of the Llobregat River basin; this is, to assess the effects of WWTPs discharges and river flow diversion after heavily populated and industrialized areas on illicit drug levels in surface waters. The most investigated point in the Llobregat River is located at the entrance of a drinking water treatment plant (DWTP) that is located in the lower part of the basin and supplies drinking water to part of Barcelona and its metropolitan area [15, 16, 20].

4.1 *Levels in the Llobregat River and Its Tributaries*

The most abundant and ubiquitous compound in surface waters of the Llobregat River basin was one of the two major COC metabolites, BE, which is present in almost all analyzed samples and measured at a maximum level of 1,350 ng/L at the intake of the DWTP [20]. COC, MDMA, and AM were also frequently detected in surface waters, and their maximum concentrations (120, 190, and 90 ng/L, respectively) were also measured at the intake of the DWTP [20].

According to the results obtained from various monitoring studies carried out at the intake of this DWTP, levels of COC and amphetamine-like compounds increase in the summer period, which could be attributed to a lower river discharge. However, the highest concentrations were observed in winter, being 4-fold, 6-fold, and 13-fold higher for BE, COC, and MDMA, respectively, than in fall, which could be related, as suggested by Huerta-Fontela et al. [20], to a low degradation rate under the environmental conditions prevailing in winter. Regarding daily variation of the levels of these compounds throughout the week, the highest levels were observed during the weekend, which may be indicative of their recreational use, as aforementioned for influent wastewaters [20].

Among the most ubiquitous and abundant opioids in surface waters of the Llobregat River basin were CODE, EDDP, and METH, whose levels raise up to 251, 61, and 18 ng/L, respectively. MOR presented a lower frequency of detection and was quantified at a maximum concentration of 31 ng/L. The levels of CODE and EDDP in surface waters were also observed to increase in winter compared to spring and fall, whereas no relevant variations were observed for the levels of METH and MOR [15].

The THC metabolite, THC-COOH, was determined also in most of the analyzed samples and presented a maximum concentration of 80 ng/L [15]. Regarding its seasonal variation, levels of this compound were only measurable in winter, which may be attributed to a lower river flow [15].

LSD, phencyclidine (PCP), nor-MOR, and HER were always below the method limit of detection (12.5 ng/L for nor-MOR and 1.5 ng/L for the other compounds). The remaining investigated compounds were sporadically (KETA, 6ACM, FENTA, nor-CODE, and THC) or with a low frequency (MA, MDA) detected, presenting in all cases levels below 14 ng/L [15, 16, 20].

Figure 5 compares the levels of the most frequently detected compounds in surface waters of the Llobregat River basin with those determined in other river catchments from Spain and Europe. Overall, measured levels of illicit drugs and metabolites in surface waters of the Llobregat River basin are in agreement with those determined in other river water catchments from Spain [27, 29, 30, 54] and Europe [23, 31, 32, 35, 39, 42, 43, 55]. As it can be observed in Fig. 5, only the amphetamine-like compounds MDMA, MA, and AM and the cannabinoid THC-COOH presented distinctly higher levels in surface waters from the Llobregat River basin than in other rivers investigated in Europe.

The chemical pressure regarding the presence of illicit drugs and metabolites was observed to increase downstream the main river, which is likely related to the increase in population density and WWTP discharges. This behavior was not observed for the detected cannabinoid, THC-COOH, which presented the highest levels in the sampling points located upstream in the Llobregat and Cardener rivers. A relevant increase in the concentrations of cocaine-related and amphetamine-like compounds was detected in the Llobregat River waters collected after the mouths of the Cardener and Anoia tributaries (63% and 31% of increase in the total load of illicit drugs and metabolites detected, respectively), which may be directly related to the discharges of these rivers into the main stream. In this respect, it is important to remark that 90% of the Anoia River discharge is diverted due to its high pollution levels. Relevant loads of these substances were also found in the Rubí Creek (89 g/day

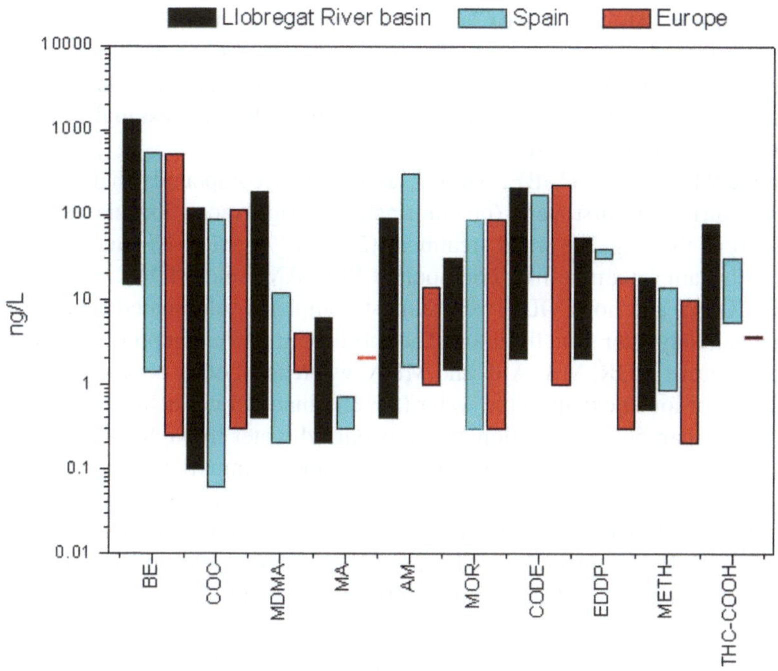

Fig. 5 Concentration range (expressed in ng/L) of the most frequently detected illicit drugs and metabolites in surface waters from the Llobregat River Basin and other Spanish and European water catchments

of cocaine-related and amphetamine-like compounds and 687 ng/L of opioids and cannabinoids), which flows through a densely populated and highly industrialized area. However, these concentrations will not end up into the main river, because the flow of the Rubí creek is fully diverted in order to prevent the dumping of high contaminant loads into the Llobregat River.

4.2 *Removal During Drinking Water Treatment*

The Llobregat River basin constitutes an important source for drinking water production in the Barcelona province. Thus, the occurrence of illicit drugs and metabolites may potentially have public health implications if they are still present in the finished water. The elimination of illicit drugs and metabolites in a DWTP located in the lowest part of the main river, which supplies drinking water to about 1 million people living in Barcelona and its metropolitan area, was evaluated by Boleda et al. [15, 16] and Huerta-Fontela et al. [20].

Water treatment in this DWTP consists of a conventional pre-oxidation step with chlorine until the break-point is achieved, followed by coagulation with alumcoagulants, e.g., $Al_2(SO_4)_3$, Al_xCl_3 or Al_2O_3, and flocculation. The clarified water is then sand filtered and afterward diluted with groundwater in variable proportions to improve its quality. Water is further treated with ozone, granular activated carbon filtration, and chlorine, the latter in order to ensure a chlorine residual concentration through the distribution system.

Unlike MDMA, COC, BE, METH, and EDDP, compounds that showed poor removal during the first steps (prechlorination, coagulation, flocculation, and sand filtration) of the drinking water treatment (23%, 13%, 9%, 54%, and 28%, respectively), the amphetamine-like compounds MA, AM, and MDA, and the opioids MOR, CODE, and nor-CODE were almost completely eliminated (above 90% on average) already after sand filtration. The subsequent ozonation process was found to entirely remove MOR, MA, AM, and MDA, whereas all other compounds remained to a higher or lower extent in the water (the elimination rates achieved by ozonation were lower than 56%). Filtration of the ozonated water through granular activated carbon further removed CODE and nor-CODE and an important proportion of other drugs (>99% COC, 72% BE, and 88% MDMA). However, only between 52 and 59% of METH and EDDP was removed during this treatment step. These substances, together with BE, were in fact still detected at the very end of the process in the finished drinking water. The average global removals observed for METH, EDDP, and BE were 91%, 88%, and 89%, respectively [15]. Despite these high elimination rates, final concentrations of these compounds in treated water reached up to 130 ng/L in the case of BE and were below 3 ng/L in the case of METH and EDDP.

The target cannabinoids and FENTA were also eliminated completely through the drinking water treatment process; however, this assessment is based on only one observation, since these compounds were present in only one raw water sample.

Groundwaters used to dilute river waters during drinking water production contained only traces of METH and EDDP, with maximum concentrations of 0.5 ng/L and 2.3 ng/L, respectively. Therefore, the contribution of this water to the illicit drug loads observed in finished waters could be considered negligible.

Boleda et al. [16] compared elimination efficiencies of illicit drugs and metabolites obtained with a conventional drinking water treatment process, such as the one aforementioned, and an advanced treatment. The latter consisted of ultrafiltration (UF) followed by ultraviolet disinfection, reverse osmosis (RO), and remineralization with calcite. According to this study, the compounds determined in raw waters, i.e., COC, BE, nor-BE, CODE, nor-CODE, METH, EDDP, MDMA, and KETA, presented satisfactory elimination rates (above 89%) with the conventional treatment. However, removal of these compounds was slightly improved with the treatment based on UF/RO (above 97%). In this respect, major elimination rates were observed after RO. In fact, UF and UV disinfection were not very effective in compound removal.

The main problem related to conventional drinking water treatment is the formation of disinfection by-products when the natural organic matter present in the water and potential reactive contaminants react with disinfection agents, such as chlorine, chloramine, and chlorine dioxide. The disinfection by-products generated when illicit drugs and metabolites react with chlorine or ozone still needs investigation. In this respect, the amphetamine-like compounds present amine moieties that are capable of reacting with chlorine, generating the potential human carcinogens *N*-nitrosodimethylamine-related compounds [56].

5 Conclusions

The application of several analytical methodologies to determine illicit drugs and metabolites in aqueous matrices to wastewaters and surface waters of the Llobregat River basin revealed the occurrence of this group of emerging contaminants in this area. The highest levels of these compounds were measured in influent wastewaters, whereas their concentrations were between one and two orders of magnitude lower in effluent wastewaters and surface waters. Potential ecotoxicity effects of environmental levels have not been investigated yet. The most frequently detected compounds in all investigated aqueous matrices were cocaine and its metabolite BE, the amphetamine-like compound MDMA, the opioids CODE and METH, and the metabolite of the latter, EDDP.

Levels found in influent wastewaters have been used to estimate illicit drug use in the investigated area by means of the sewage epidemiology approach. Despite the fact that estimations performed with this methodology are subject to uncertainty, this approach represents a very useful tool to evaluate drug use trends and detect possible hot-spots, which can help to allocate and develop more effective drug prevention programs.

BE, METH, and EDDP were also detected in finished drinking waters derived from the Llobregat River. Despite the fact that the levels found are too low to produce psychoactive effects through water ingestion, other health effects associated to these substances or the disinfection by-products formed during the drinking water process (halogenated derivatives) cannot be discarded and deserve investigation.

Acknowledgments This work has been supported by the Spanish Ministry of Science and Innovation (projects CGL2007-64551/HID and Consolider-Ingenio 2010 CSD2009-00065) and it reflects only the authors' view. Nicola Mastroianni acknowledges the CSIC for the JAE predoctoral grant.

References

1. Jones-Lepp TL, Alvarez DA, Petty JD, Huckins JN (2004) Polar organic chemical integrative sampling and liquid chromatography–electrospray/ion-trap mass spectrometry for assessing selected prescription and illicit drugs in treated sewage effluents. Arch Environ Contam Toxicol 47:427–439
2. Postigo C, López de Alda MJ, Barceló D (2008) Analysis of drugs of abuse and their human metabolic byproducts in water by LC-MS/MS: a non-intrusive tool for drug abuse estimation at the community level. Trends Anal Chem 27(11):1053–1069

3. Fatta-Kassinos D, Meric S, Nikolau A (2011) Pharmaceutical residues in environmental waters and wastewaters: current state of knowledge and future research. Anal Bioanal Chem 399:251–275
4. Boles TH, Wells MJM (2010) Analysis of amphetamine and methamphetamine as emerging pollutants in wastewater and wastewater-impacted streams. J Chromatogr A 1217(16): 2561–2568
5. Balducci C, Nervegna G, Cecinato A (2009) Evaluation of principal cannabinoids in airborne particulates. Anal Chim Acta 641(1–2):89–94
6. Cecinato A, Balducci C (2007) Detection of cocaine in the airborne particles of the Italian cities Rome and Taranto. J Sep Sci 30(12):1930–1935
7. Cecinato A, Balducci C, Budetta V, Pasini A (2010) Illicit psychotropic substance contents in the air of Italy. Atmos Environ 44(19):2358–2363
8. Cecinato A, Balducci C, Nervegna G (2009) Occurrence of cocaine in the air of the world's cities. An emerging problem? A new tool to investigate the social incidence of drugs? Sci Total Environ 407(5):1683–1690
9. Postigo C, López De Alda MJ, Viana M, Querol X, Alastuey A, Artiñano B, Barceló D (2009) Determination of drugs of abuse in airborne particles by pressurized liquid extracton and liquid chromatography-electrospray-tandem mass spectrometry. Anal Chem 81:4382–4388
10. Viana M, Postigo C, Querol X, Alastuey A, López De Alda MJ, Barceló D, Artiñano B, López-Mahia P, García Gacio D, Cots N (2011) Cocaine and other illicit drugs in airborne particulates in urban environments: a reflection of social conduct and population size. Environ Pollut 195 (5):1241–1247
11. Viana M, Querol X, Alastuey A, Postigo C, López de Alda MJ, Barceló D, Artiñano B (2010) Drugs of abuse in airborne particulates in urban environments. Environ Int 36:527–534
12. Jones-Lepp TL, Stevens R (2007) Pharmaceuticals and personal care products in biosolids/ sewage sludge: the interface between analytical chemistry and regulation. Anal Bioanal Chem 387(4):1173–1183
13. Kaleta A, Ferdig M, Buchberger W (2006) Semiquantitative determiantion of residues of amphetamien in sewage sludge samples. J Sep Sci 29:1662–1666
14. Petrovic M, De Alda MJL, Diaz-Cruz S, Postigo C, Radjenovic J, Gros M, Barcelo D (2009) Fate and removal of pharmaceuticals and illicit drugs in conventional and membrane bioreactor wastewater treatment plants and by riverbank filtration. Philos Trans R Soc A 367 (1904):3979–4003
15. Boleda MR, Galcerán MT, Ventura F (2009) Monitoring of opiates, cannabinoids and their metabolites in wastewater, surface water and finished water in Catalonia, Spain. Water Res 43(4):1126–1136
16. Boleda MR, Galcerán MT, Ventura F (2011) Behavior of pharmaceuticals and drugs of abuse in a drinking water treatment plant (DWTP) using combined conventional an ultrafiltration and reverse osmosis (UF/RO) treatments. Environ Pollut 159:1584–1591
17. Boleda MR, Galcerán MT, Ventura F (2007) Trace determination of cannabinoids and opiates in wastewater and surface waters by ultra-performance liquid chromatography-tandem mass spectrometry. J Chromatogr A 1175:38–48
18. Huerta-Fontela M, Galcerán MT, Martin-Alonso J, Ventura F (2008) Occurrence of psychoactive stimulatory drugs in wastewaters in North-eastern Spain. Sci Total Environ 397(1–3):31–40
19. Huerta-Fontela M, Galcerán MT, Ventura F (2007) Ultraperformance liquid chromatography-tandem mass spectrometry analysis of stimulatory drugs of abuse in wastewater and surface waters. Anal Chem 79:3821–3829
20. Huerta-Fontela M, Galcerán MT, Ventura F (2008) Stimulatory drugs of abuse in surface waters and their removal in a conventional drinking water treatment plant. Environ Sci Technol 42(18):6809–6816

21. Postigo C, López De Alda MJ, Barceló D (2008) Fully automated determination in the low nanogram per liter level of different classes of drugs of abuse in sewage water by on-line solid-phase extraction-liquid chromatography-electrospray-tandem mass spectrometry. Anal Chem 80(9):3123–3134
22. Daughton CG (2001) Illicit drugs in municipal sewage in pharmaceuticals and personal care products in the environment: scientific and regulatory issues. In: Daughton CG, Jones-Lepp TL (eds) Acs symposium series 791. The American Chemical Society, Washington DC, pp 116–139
23. Zuccato E, Chiabrando C, Castiglioni S, Calamari D, Bagnati R, Schiarea S, Fanelli R (2005) Cocaine in surface water: a new evidence-based tool to monitor community drug abuse. Environ Health 4:1–7
24. Castiglioni S, Zuccato E, Chiabrando C, Fanelli R, Bagnati R (2008) Mass spectrometric analysis of illicit drugs in wastewater and surface water. Mass Spectrom Rev 27(4):378–394
25. Commission Decision 2002/657/EC of 12 August 2002 implementing Council Directive 96/23/EC concerning the performance of analytical methods and the interpretation of results. Official Journal of the European Union L 221, 17 August 2002, pp 8–36
26. Bijlsma L, Sancho JV, Pitarch E, Ibáñez M, Hernández F (2009) Simultaneous ultra-high-pressure liquid chromatography-tandem mass spectrometry determination of amphetamine and amphetamine-like stimulants, cocaine and its metabolites, and a cannabis metabolite in surface water and urban wastewater. J Chromatogr A 1216(15):3078–3089
27. González-Mariño I, Quintana JB, Rodríguez I, Cela R (2010) Determination of drugs of abuse in water by solid-phase extraction, derivatisation and gas chromatography-ion trap-tandem mass spectrometry. J Chromatogr A 1217:1748–1760
28. González-Mariño I, Quintana JB, Rodríguez I, Rodil R, González-Peñas J, Cela R (2009) - Comparison of molecularly imprinted, mixed-mode and hydrophilic balance sorbents performance in the solid-phase extraction of amphetamine drugs from wastewater samples for liquid chromatography-tandem mass spectrometry determination. J Chromatogr A 1216 (48):8435–8441
29. Postigo C, López De Alda MJ, Barceló D (2010) Drugs of abuse and their metabolites in the ebro river basin: Occurrence in sewage and surface water, sewage treatment plants removal efficiency, and collective drug usage estimation. Environ Int 36:75–84
30. Martínez Bueno MJ, Uclés S, Hernando MD, Fernández-Alba AR (2011) Development of a solvent-free method for the simultaneous identification/quantification of drugs of abuse and their metabolites in environmental water by LC-MS/MS. Talanta 85(1):157–166
31. Berset JD, Brenneisen R, Mathieu C (2010) Analysis of llicit and illicit drugs in waste, surface and lake water samples using large volume direct injection high performance liquid chromatography - electrospray tandem mass spectrometry (HPLC-MS/MS). Chemosphere 81:859–866
32. Bones J, Thomas KV, Paull B (2007) Using environmental analytical data to estimate levels of community consumption of illicit drugs and abused pharmaceuticals. J Environ Monit 9 (7):701–707
33. Van Nuijs ALN, Pecceu B, Theunis L, Dubois N, Charlier C, Jorens PG, Bervoets L, Blust R, Meulemans H, Neels H, Covaci A (2009) Can cocaine use be evaluated through analysis of wastewater? A nation-wide approach conducted in Belgium. Addiction 104(5):734–741
34. Van Nuijs ALN, Pecceu B, Theunis L, Dubois N, Charlier C, Jorens PG, Bervoets L, Blust R, Neels H, Covaci A (2009) Cocaine and metabolites in waste and surface water across Belgium. Environ Pollut 157(1):123–129
35. Van Nuijs ALN, Pecceu B, Theunis L, Dubois N, Charlier C, Jorens PG, Bervoets L, Blust R, Neels H, Covaci A (2009) Spatial and temporal variations in the occurrence of cocaine and benzoylecgonine in waste- and surface water from Belgium and removal during wastewater treatment. Water Res 43(5):1341–1349

36. Van Nuijs ALN, Tarcomnicu I, Bervoets L, Blust R, Jorens PG, Neels H, Covaci A (2009) Analysis of drugs of abuse in wastewater by hydrophilic interaction liquid chromatography-tandem mass spectrometry. Anal Bioanal Chem 395(3):819–828
37. Castiglioni S, Zuccato E, Crisci E, Chiabrando C, Fanelly R, Bagnati R (2006) Identification and measurement of illicit drugs and their metabolites in urban wastewater by liquid chromatography-tandem mass spectrometry. Anal Chem 78:8421–8429
38. Mari F, Politi L, Biggeri A, Accetta G, Trignano C, Di Padua M, Bertol E (2009) Cocaine and heroin in waste water plants: a 1-year study in the city of Florence, Italy. Forensic Sci Int 189 (1–3):88–92
39. Hummel D, Löffler D, Fink G, Ternes TA (2006) Simultaneous determination of psychoactive drugs and their metabolites in aqueous matrices by liquid chromatography mass spectrometry. Environ Sci Technol 40:7321–7328
40. Terzic S, Senta I, Ahel M (2010) Illicit drugs in wastewater of the city of Zagreb (Croatia) - estimation of drug abuse in a transition country. Environ Pollut 158:2686–2693
41. Karolak S, Nefau T, Bailly E, Solgadi A, Levi Y (2010) Estimation of illicit drugs consumption by wastewater analysis in Paris area (France). Forensic Sci Int 200:153–160
42. Kasprzyk-Hordern B, Dinsdale RM, Guwy AJ (2008) Multiresidue methods for the analysis of pharmaceuticals, personal care products and illicit drugs in surface water and wastewater by solid-phase extraction and ultra performance liquid chromatography-electrospray tandem mass spectrometry. Anal Bioanal Chem 391(4):1293–1308
43. Kasprzyk-Hordern B, Dinsdale RM, Guwy AJ (2009) The removal of pharmaceuticals, personal care products, endocrine disruptors and illicit drugs during wastewater treatment and its impact on the quality of receiving waters. Water Res 43(2):363–380
44. Metcalfe C, Tindale K, Li H, Rodayan A, Yargeau V (2010) Illicit drugs in Canadian municipal wastewater and estimates of community drug use. Environ Pollut 158:3179–3185
45. Bartelt-Hunt SL, Snow DD, Damon T, Shockley J, Hoagland K (2009) The occurrence of illicit and therapeutic pharmaceuticals in wastewater effluent and surface waters in Nebraska. Environ Pollut 157(3):786–791
46. Boleda MR, Majamaa K, Aerts P, Gómez V, Galcerán MT, Ventura F (2010) Removal of drugs of abuse from municipal wastewater using reverse osmosis membranes. Desalinat Water Treat 21:122–130
47. Postigo C, Sirtori C, Oller I, Malato S, Maldonado MI, López De Alda MJ, Barceló D (2011) Solar transformation and photocatalytic treatment of cocaine in water: kinetics, characterization of major intermediate products and toxicity evaluation. Appl Catal B 104(1–2):37–48
48. Postigo C, Sirtori C, Oller I, Malato S, Maldonado MI, López De Alda MJ, Barceló D (2011) Photolytic and photocatalytic transformation of methadone in aqueous solutions under solar irradiation: kinetics, characterization of major intermediate products and toxicity evaluation. Water Res 45:4815–4826
49. Zuccato E, Chiabrando C, Castiglioni S, Bagnati R, Fanelli R (2008) Estimating community drug abuse by wastewater analysis. Environ Health Perspect 116:1027–1032
50. Kasprzyk-Hordern B, Dinsdale RM, Guwy AJ (2009) Illicit drugs and pharmaceuticals in the environment - forensic applications of environmental data. Part 1: estimation of the usage of drugs in local communities. Environ Pollut 157(6):1773–1777
51. OED (2009) Observatorio español sobre drogas. Informe 2009: Situación y tendencias de los problemas de drogas en España. 2009. Available at: http://www.pnsd.msc.es/categoria2/observa/pdf/oed-2009.pdf. Accessed in December, 2010
52. Van Nuijs ALN, Castiglioni S, Tarcomnicu I, Postigo C, López De Alda MJ, Neels H, Zucatto E, Barceló D, Covaci A (2011) Illicit drug consumption estimations derived from wastewater analysis: a critical review. Sci Total Environ 409:3564–3577
53. Kasprzyk-Hordern B, Kondakal VVR, Baker DR (2010) Enantiomeric analysis of drugs of abuse in wastewater by chiral liquid chromatography coupled with tandem mass spectrometry. J Chromatogr A 1217(27):4575–4586

54. Vazquez-Roig P, Andreu V, Blasco C, Picó Y (2010) SPE and LC-MS/MS determination of 14 illicit drugs in surface waters from the natural park of l'Albufera (Valencia, Spain). Anal Bioanal Chem 397:2851–2864
55. Gheorghe A, Van Nuijs A, Pecceu B, Bervoets L, Jorens PG, Blust R, Neels H, Covaci A (2008) Analysis of cocaine and its principal metabolites in waste and surface water using solid-phase extraction and liquid chromatography-ion trap tandem mass spectrometry. Anal Bioanal Chem 391(4):1309–1319
56. Huerta-Fontela M, Galcerán MT, Ventura F (2011) Presence and removal of illicit drugs in conventional drinking water treatment plants. In: Castiglioni S, Zuccato E, Fanelli R (eds) Illicit drugs in the environment: occurrence, analysis and fate using mass spectrometry. Wiley, Hoboken, pp 205–222

Risk Assessment of Pollutants in the Llobregat River Basin

Antoni Ginebreda, Carles Barata, and Damià Barceló

Abstract In this chapter, an overview of the risk assessment (RA) studies carried out in the Llobregat River is presented.

In the first section, general aspects of the RA process, such are occurrence (exposure) and harmful effects are reviewed. Particular emphasis is put on the assessment of mixture toxicity, since it is a key issue on real multichemical exposed scenarios.

The second part is focused on the description of the different RA studies carried out in the Llobregat River, covering the following aspects: general compound risk, pharmaceuticals, pesticides, metals in the whole basin, as well as the application of RA methodologies to water reuse.

Finally, the connection between RA and ecosystem status is discussed.

Keywords Emerging contaminants, Llobregat, Mixture toxicity, Risk assessment

Contents

A. Ginebreda (✉) and C. Barata
Institute of Environmental Assessment and Water Research (IDAEA-CSIC), C/ Jordi Girona 18-26, 08034 Barcelona, Spain
e-mail: agmqam@cid.csic.es

D. Barceló
Institute of Environmental Assessment and Water Research (IDAEA-CSIC), C/ Jordi Girona 18-26, 08034 Barcelona, Spain

Catalan Institute for Water Research (ICRA), c/Emili Grahit 101, 17003 Girona, Spain

S. Sabater et al. (eds.), *The Llobregat: The Story of a Polluted Mediterranean River*, Hdb Env Chem (2012) 21:263–296, DOI 10.1007/698_2012_142,

Abbreviations and Symbols

ANN	Artificial neural networks
CA	Concentration addition
DOC	Dissolved organic carbon
DSS	Decision support system
EC50	Effect concentration 50%
EEQ	Estradiol equivalent factor
EQS	Environmental quality standards
ES	Ecological status
GIS	Geographic information system
HQ	Hazard quotients
IA	Independent action
K_{oc}	Organic carbon partition coefficient
LC50	Lethal concentration 50%
MEC	Measured environmental concentration
msPAF	Multispecies potentially affected fraction
NOEC	Nonobserved effect concentration
PAF	Potentially affected fraction
PAH	Polyaromatic hydrocarbons
PBDE	Polybromodiphenyl ethers
PBT	Persistence bioaccumulation and toxicity
PCB	Polychlorobiphenyls
PCDD	Polychlorodibenzodioxins
PCDF	Polychlorodibenzofurans
PCN	Polychloronaphthalenes
PEC	Predicted environmental concentration
PNEC	Predicted no effect concentration
PRISW-1	Short-term pesticide risk index for the surface water system
RA	Risk assessment
SSD	Species sensitivity distribution
TER	Toxicological exposure ratio

TU	Toxic units
TUS	Toxic unit summation
WFD	Water framework directive
WWTP	Wastewater treatment plant

1 Introduction

Our technological society makes extensive and intensive use of chemicals (most of them organics) and this number is continuously growing. Thus, for instance the European Inventory of Commercial Chemical Substances (EINECS) reports up today 100,204 commercially available substances [1] and similar figures hold for the USA [2, 3].

Hence, depending on their properties, mode, and extent (volume) of use, some of these chemicals can potentially reach the environment, being their environmental and health effects unpredictable in long term. This has become a matter of major concern and constitutes the reason for new regulations related to the safety of chemicals are being promoted. Among them, and owing to its obvious relevance for the aquatic environment, it is worth noting the list of the so-called priority and dangerous priority substances included in the Directive 2008/105/EC ("daughter" of the WFD).

As it has been exposed in the foregoing chapters, the Llobregat River basin is subjected to heavy anthropogenic pressures, including urban (more than three million inhabitants are living in the basin), industrial and agricultural that translate into both point and diffuse pollution. The former two are especially relevant in the lower stretch of the river, in the neighborhood of Barcelona town. As a consequence of the application of Directive 91/271/EEC, most of the urban wastewater generated in the basin is subjected to treatment. Furthermore, the river shows a pronounced Mediterranean hydrological pattern, characterized by acute flow fluctuations (flow may vary between ca. 1 and more than 100 m^3/s), in such a way that during drought periods WWTP effluents may account for a very important part of the circulating water.

Under such circumstances, it is not strange that both emerging and classical pollutants are present in the river in variable amounts, being its relation to the ecosystem difficult to ascertain.

Bridging the gap between chemical exposure and ecosystem status is one of the current challenges of environmental sciences. One of the approaches used is the so-called risk assessment (RA), which essentially aims to overcome the said gap using ecotoxicology principles (understood in their broadest sense) as a mean to connect chemical multiexposure and ecosystem effects.

In this chapter, we like to present an overview of the existing RA studies that have been conducted up to now in the Llobregat River basin.

2 Outline of the RA Approach

Risk is broadly defined as the combination (i.e., product) of a probability of occurrence of some event by its associated hazard effects:

$$\text{Risk} = \text{Occurrence} \times \text{Hazard effects}$$

Correspondingly, the RA process may be defined as the set of procedures aiming to identify hazards and to quantify the associated risk (in our case, related to chemicals) concerning human health and/or ecosystems impairment.

In the case of the environmental risk posed by chemicals, "hazard effects" are related to the intrinsic properties of each compound [4], whereas "occurrence" is associated with its environmental exposure, usually expressed in terms of environmental concentration.

Different RA approaches have been developed in order to identify and rank compounds of environmental concern for both regulatory and monitoring purposes. Whereas most of all the existing schemes share the basic underlying RA paradigm, they basically differ on how both factors, i.e., occurrence and effects, are defined and hence quantified.

2.1 *Exposure: Assessing Environmental Concentrations*

There are two basic approaches for establishing environmental concentrations, namely, measurement or modeling, being the derived respective concentrations usually referred to as MECs (measured environmental concentrations) or the so-called PECs (predicted environmental concentrations). The most obvious and classical is through analytical chemistry. However, the development of environmental modeling provides an interesting alternative. Johnson et al. [5] have discussed in detail their respective pros and cons, which are summarized in Table 1. Essentially, the main drawback of measuring is that it is focussed on certain pre-selected analytes, and there is low chance of finding new targets. Other aspects of concern are their limited possibilities regarding time and space coverage, which can lead to missing certain events. Sampling issues become thus important. Analytical campaigns are usually expensive and time consuming. Conversely, on the positive side the main advantage of measurement is that it provides reliable results that are quite independent of the laboratories (at least, it is true for those that have an adequate Quality Assurance/Quality Control System). In turn, modeling is fast and relatively affordable and has very good time and space coverage possibilities, being, on the other hand, its main weakness the strong dependence on the model chosen and its inherent uncertainty because of the lack of reliable data (physical–chemical properties, emission sources, etc.) on experimental analytical information required.

Table 1 Comparative overview of strengths and weaknesses of analytical measurement vs. modeling in environmental risk assessment studies (adapted from [1])

Analytical measurement	Strengths	– Results obtained reflect well reality – Repeatability and reproducibility of results – Measurements are independent of information/data sources – Multipurpose analytical methods can cover many compounds on a single run – Even the best model will ultimately require experimental confirmation – Discovery of new emerging contaminants is possible
	Drawbacks	– Determination of compounds at very low quantities may be difficult – Time and space coverage require expensive monitoring campaigns – Sampling campaigns may miss crucial episodes. – Analytical measurements give a snapshot, rather than a continuous picture – Expensive analytical equipment and method development – Target monitoring may miss pollutants: "you only find what you are looking for"
Predicting (modeling)	Strengths	– Very good coverage capabilities of time and space – Computation equipment is affordable – Possibility of application to hypothetical scenarios: "What if?" – Useful for extrapolations to future (predictions on space and time, even for products not yet in the market) – Simultaneous modeling of many compounds is possible – Once the model is developed and validated is fast and cheap to use
	Drawbacks	– Different models may render very different results – Models are strongly dependent on parameter and data input – Diffuse sources of pollution may be very difficult to model – Experimental validation is often necessary

2.2 *Adverse Effects*

The second part of the RA process is related to the hazard associated with the compound considered. According to the nature and use of each compound, they may exert different deleterious effects against the ecosystem or human health. Such effects have been classically expressed by the so-called PBT approach, where the acronym stands for "Persistence, Bioaccumulation and Toxicity," each term characterized by means of the corresponding measurable properties. The first lists of priority organics were actually based on such PBT triad, and, in fact, still many recently proposed prioritization schemes still rely on it. Later on, other effects were joined to the foregoing PBT set, being particularly the "Toxicity" dimension expanded to related aspects such as carcinogenicity, mutagenicity, teratogenicity, or endocrine disruption, among others.

As far as environmental toxicity is concerned, there are many ecotoxicity biotests that have been utilized to characterize the ecotoxicological effects of chemicals in the aquatic environment, covering different trophic levels (fishes, benthic organisms, algae, etc.), endpoints and exposure times (chronic, subchronic, acute). An example of some of such tests is depicted in Table 2. They are conveniently collected in existing databases and scientific literature thus providing information regarding the ecotoxicological effects for thousands of chemical compounds. We just like to stress that ecotoxicology is a complex discipline, and the data used must be carefully managed in order to get reliable and comparable results: Chronic or acute toxicity (exposure time), species tested, intake routes, selected endpoints are issues to be taken into consideration on a properly done RA exercise. On the other hand, owing to the variable sensitivity displayed by the different biotests toward the existing chemicals, it is always advisable to refer RA studies to more than one, particularly if the results are to be extrapolated to the whole ecosystem.

3 Mixture Ecotoxicology and Related Concepts

In real-world scenarios, contaminants rarely occur alone. Instead, they usually appear as mixtures of many compounds being their combined effects difficult to predict (i.e., synergies or antagonistic effects may take place). Therefore, toxicological effects caused by mixtures must be taken into consideration in RA studies.

Mixture toxicity is a complex question and is a topic of active research. Interested readers are referred to recently published reviews [6, 7]. Up to now, most studies and regulatory protocols for conducting mixture hazard assessment focus on the toxicity mechanisms of individual chemicals in selecting models for predicting toxic effects. Specifically, in aquatic toxicology, two different conceptual models, respectively, termed concentration addition (CA) [8] and *independent action* (IA) [9], are considered to describe general relationships between the effects of single substances and their corresponding mixtures, for similarly and dissimilarly acting chemicals, respectively. Use of these models allows calculation of the toxicity of a mixture on the basis of known toxicities of the mixture's individual components [10]. The CA model is founded on the assumption that mixture components each possesses a similar pharmacological mode of action, and thus is most applicable for toxic substances that have the same molecular target site. The alternative model of IA or response addition assumes that mixture components possess dissimilar modes of action, interacting with different target sites, leading to a common toxicological endpoint via distinct chains or reactions within an organism. Under these premises, the effects of individual constituents are expected to be independent in a strictly probabilistic sense.

The mixture effects for both CA and IA modes of action are, respectively, given below [6].

Table 2 Examples of most used standardized aquatic ecotoxicology test

Type	Test species	Duration	Test system	Feeding	Endpoints	Parameter
Acute tests						
Bacteria bioluminescence test	*Vibrio fischeri*	5–15 min	Static	Salt-enriched	Luminescence	EC50
Acute immobilization tests with daphnids	*Daphnia magna*, *D. pulex*	48 h	Static test in tubes or beakers	No food	Immobility	EC50
Fish acute test	Guppy (*Poecilia reticulata*), zebrafish (*Danio rerio*), fathead minnow (*Pimephales promelas*), raimow trout (*Oncorhynchus mykiss*)	96 h	Static or renewal test	No food	Mortality	LC50
Algal growth inhibition test	Unicellular algae (*Selenastrum capricornutum, Chlorella vulgaris, Desmodesmus subspicatus*)	72–96 h	Static	Nutrient-enriched	Population growth/biomass	EC50
Chronic tests						
Daphnia reproduction test	Neonates *D. magna, D. pulex, Ceriodaphnia dubia*	7–21 days	Static renewal test	Algae	Mortality and reproduction (total number of young)	LC50, EC50, NOEC
Sediment Chironomus test	*Chironomus riparius*, *C. tentans*	10–21 days	renewal test	Commercial fish food	Mortality, larcal emergence, growth	LC50, EC50, NOEC
Fish early-life stage test	Zebrafish (*D. rerio*), fathed minnow (*P. promelas*), rainbow trout (*O. mykiss*)	5–90 days	Renewal or flow-through test	Commercial fish food	Survival, growth, development, behavioral responses, etc.	LC50, EC50, LOEC, NOEC

3.1 Concentration Addition

$$\mathrm{EC50}_{\mathrm{Mix}} = \sum_i \frac{c_i}{\mathrm{EC50}_i}, \tag{1}$$

where c_i is the concentration of component i; $\mathrm{EC50}_i$ is the toxicity of compound i expressed as EC50 and $\mathrm{EC50}_{\mathrm{Mix}}$ indicates the toxicity of the whole mixture expressed as EC50. $\mathrm{EC50}_{\mathrm{Mix}}$ is often referred to as "toxic unit summation" (TUS) and the individual terms "toxic units" (TU) [11], (1) thus becoming (see Sect. 4.3):

$$\mathrm{TUS} = \sum_i \mathrm{TU}_i. \tag{2}$$

3.2 Independent Action

$$E(c_{\mathrm{Mix}}) = 1 - \prod_{i=1}^{n} [1 - E(c_i)], \tag{3}$$

where $E(c_{\mathrm{Mix}})$ indicates the effect of a mixture of n-compounds; c_i is the concentration of the ith compound, and $E(c_i)$ is the effect of that concentration if the compound is applied singly.

Effects considered under IA mode of action can be measured using different variables and (3) is applicable providing that $E(c_i)$ 0–1 scaled. In the examples shown below, both hazard quotients (HQ) [12] (Sect. 4.2) and multispecies Potentially affected fraction (msPAF) [13] (Sects. 4.1 and 4.4.2) are used.

The applicability of both concepts in RA is still under debate. Whereas both of them have been proved acceptable if the corresponding mechanistic assumptions are fulfilled, since exact modes of action are often unknown for many compounds, both CA and IA must be regarded as two special extreme cases [14, 15]. In practice, both models have been more or less successfully applied, being the results obtained with both of them not very different, with CA tending to overestimate and IA to underestimate toxicity in controlled experiments [16, 17].

4 RA Studies in the Llobregat River Basin

In the following sections, we are briefly reviewing the RA studies published up to now in which the Llobregat River is included.

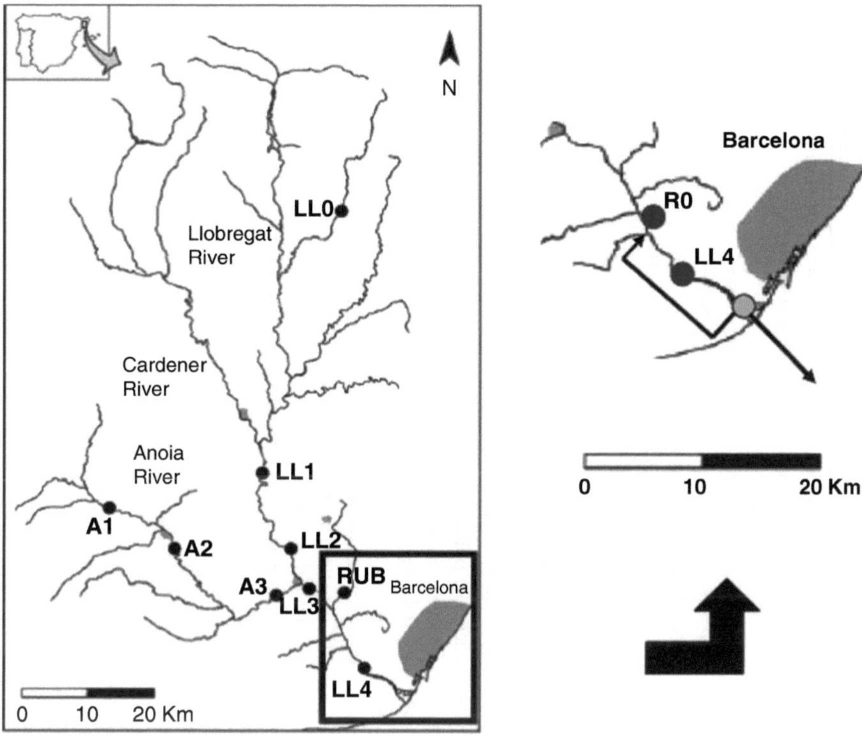

Fig. 1 Lobregat River Basin, indicating the sampling sites (*LL* Llobregat, *A* Anoia River, *RUB* Rubí)

Three types of studies are considered. First of all, general RA exercises in which the Llobregat is included as part of a broader scope at regional (Catalonia) or European scale. These studies are usually based on regulatory data gathered by the water authorities. A second class of RA studies specifically focused on the Llobregat are also presented. They may cover the whole basin (Fig. 1) or only one part, as well as one or more families of pollutants, both emerging (metals, pesticides, pharmaceuticals, and personal care products, illicit drugs, detergents, etc.). Finally, some of these studies were specifically addressed to assess specific problems, such as WWTP effluent reuse. They are summarized in Table 3.

4.1 *General RA Studies Based on Regulatory Purposes Generated Data*

Monitoring campaigns associated with the implementation of the WFD in Europe has been revealed as an invaluable source of data on chemical pollution. Even though they are usually restricted to the list of regulated priority and dangerous

Table 3 Summary of the published risk assessment studies carried out in the Llobregat River basin

Pollutants considered	Area/data source	Scope/remarks	Reference
Classical and priority contaminants	• Llobregat River basin • Data from the Catalan Water Agency collected for regulatory purposes	• Risk assessment methodology: toxic units (TUS) based on *Daphnia* toxicity. TU values were compared with measured effects in transplanted *D. magna* individuals • Compounds assessed: Seven types of contaminant assessed (copper, zinc, triazines, polycyclic aromatic hydrocarbons, organochlorine compounds, alkylphenols, organophosphorus pesticides) only the last group were likely to affect aquatic arthropods having similar sensitivities as *D. magna*	Damasio et al. [18]
Classical and priority contaminants	• Catalonian river basins • Data from the Catalan Water Agency collected for regulatory purposes (1997–2004)	• Risk assessment methodology: COMMPS (combined monitoring-based and modeling-based priority setting scheme) • A locally adapted list of priority pollutants at a regional scale and proposed a new site pollution risk index for the relative comparison of the chemical pollution status of the investigated sites in the region are proposed	Teixidó et al. [19]
Classical and priority contaminants	• Catalonian river basins • Data from the Catalan Water Agency collected for regulatory purposes (2007–2008) (WFD)	• Risk assessment methodology: exposure assessment with species sensitivity distribution (SSD) and mixture toxicity rules (CA and IA) were used to compute the multisubstances potentially affected fraction (msPAF) • The total dataset of chemical monitoring carried out in Catalonia (Llobregat is included) between 2007 and 2008 (232 sampling stations and 60 pollutants) has been analyzed using sequential advanced modeling techniques. Data on concentrations of contaminants in water were pretreated in order to calculate the bioavailable fraction, depending on substance properties and local environmental conditions. The resulting values were used to predict the potential impact on aquatic biota of toxic substances in complex mixtures and to identify hot spots	Carafa et al. [21]

Classical and priority contaminants	• Llobregat River basin • Data from the Catalan Water Agency collected for regulatory purposes (2001–2004) (WFD)	• Risk assessment methodology: integrated RA methodology for the classification of the ecological status (ES) based on the weight of evidence approach. It implements a fuzzy inference system that hierarchically aggregates a set of environmental indicators grouped into five lines of evidence, namely, biology, chemistry, ecotoxicology, physical-chemistry, and hydromorphology. The ES is expressed as the membership degree to one or two contiguous WFD status classes. The method is implemented within a freeware GIS (geographic information system)-based decision support system (DSS) developed as part of the MODELKEY project	Gottardo et al. [23, 24]
Classical, priority and emerging organic contaminants	• Water monitoring data collected at European level (Elbe, Scheldt, Danube and Llobregat rivers) • Data from water authorities collected for regulatory purposes (WFD)	Five hundred compounds are classified into categories according to the type of assessment required. This allows water managers to focus on distinct actions according to the classification of a substance To decide which compounds have the highest priority within each category two indicators are proposed: (a) The *frequency of exceedance* (b) The *extent of exceedance* of the lowest predicted no effect concentration (PNEC). For (a) maximum observed concentrations at each sampling site (MEC_{site}) are compared to the lowest PNEC, whereas for (b) the 95th percentile of all MEC_{site} were compared to the lowest PNEC values	Von der Ohe et al. [27]
Emerging contaminants (pharmaceuticals, pesticides, alkylphenols, and heavy metals)	• Water monitoring data collected in the Llobregat River middle and low basin • Cemagua, aquatoxigen, MODELKEY and keybioeffects research projects	• Risk assessment methodology: HQ based on Independent Action mode for invertebrates • RA are compared to responses to field collected and transplantated invertebrate species (*Hydropsyche exocellata*, *Echinogammarus longisetosus*, and *Daphnia magna*) using up to ten different endpoints including enzyme activities related to detoxication mechanisms (i.e., glutathione S transferase, catalase, esterases), the oxidative stress	Damasio et al. [29]

(continued)

Table 3 (continued)

Pollutants considered	Area/data source	Scope/remarks	Reference
		damage marker (lipid peroxidation), and individual responses (mortality, postexposure feeding rates) • Estimated hazard indexes of measured pollutants indicated that pesticides and metals accounted for most of the predicted toxicity (>95%) in the most contaminated site and that the predicted toxicity of pharmaceuticals was marginal (<5%)	
Pharmaceutical compounds	• Water monitoring data collected in the Llobregat River basin • MODELKEY research project	• Risk assessment methodology: HQ based on concentration addition mode for fish, *Daphnia*, and Algae • Survey was carried out along three campaigns in seven sampling points, located in the main river and in one of its tributaries (Anoia River). In each sample, 29 commonly used pharmaceuticals, belonging to different therapeutical classes (analgesics and nonsteroidal anti-inflammatories (NSAIDs), lipid regulators, psychiatric drugs, antihistamines, antiulcer agents, antibiotics and β-blockers) have been determined • HQ shows inverse correlation with Shannon–Wiener biodiversity index, being *Daphnia* the best one • For the fish-based bioassay, major contribution is due to gemfibrozil, followed by ibuprofen and diclofenac. Other compounds with significative effect are propyphenazone and bezafibrate. For *Daphnia*, major contributions are attributable to erythromycin, ibuprofen, and clofibric acid and, to a less extent, to diclofenac, acetaminophen, and sufamethoxazole. Algae appears to be mostly dependent of sulfamethoxazole, followed by ibuprofen and gemfibrozil	Ginebreda et al. [31]

Pesticides	• Water monitoring data collected in the Llobregat River middle and low basin • VIECO research project	• Risk assessment methodology: Pesticide risk index for the surface water system (PRISW-1), based on the pesticides concentrations and their overall toxicity (estimated as TUS) against algae, *Daphnia*, and fish • It investigates the occurrence of 16 selected pesticides belonging to the classes of triazines, phenylureas, 30 organophosphates, chloroacetanilides and thiocarbamates in surface waters from the Llobregat River and some tributaries (Anoia and Rubí) • Application of the PRISW-1 index indicated that, although pesticides levels fulfilled the European Union Environmental Quality Standards (EQS) for surface waters, the existing pesticide contamination poses a low to high ecotoxicological risk for aquatic organisms • The organophosphates diazinon and malathion and the phenylurea diuron are the major contributors to the overall toxicity	Köck-Schümeyer et al. [42]
Emerging contaminants (pharmaceuticals, illicit drugs, and estrogens)	• Water monitoring data collected in the Llobregat River and low basin • Data from the Catalan Water Agency (2008–2009)	• Risk assessment methodology: HQ based on concentration addition mode for fish, *Daphnia* and Algae • A total of 103 emerging contaminants belonging to the groups of pharmaceuticals (74), illicit drugs (17), and estrogens (12), were determined in river water samples during the water reuse campaign carried out in 2009 in the low Llobregat during a water reuse experiment • Differences between river upstream and downstream to the discharge point were perceivable, but not very significant, having pharmaceuticals higher contribution than illicit drugs. No relevant risks were identified	López-Serna et al. [51]

(continued)

Table 3 (continued)

Pollutants considered	Area/data source	Scope/remarks	Reference
Classical and emerging contaminants	• Sediment monitoring data collected at European level (Elbe, Scheldt, and Llobregat rivers) • MODELKEY research project	• Risk assessment methodology: Toxic units (TU) on the basis of acute toxicity to *Daphnia magna* and *Pimephales promelas*, and multisubstance Potentially affected fractions of species (msPAF) • The toxicity of four polluted sediments and their corresponding reference sediments were investigated using a battery of six sediment contact tests representing three different trophic levels. The tests included were chronic tests with the oligochaete *Lumbriculus variegatus*, the nematode *Caenorhabditis elegans* and the mudsnail *Potamopyrgus antipodarum*, a subchronic test with the midge *Chironomus riparius*, an early life stage test with the zebra fish Danio rerio, and an acute test with the luminescent bacterium *Vibrio fischeri* • The test battery could clearly detect toxicity of the polluted sediments. The msPAF and TU-based toxicity estimations confirmed the results of the biotests by predicting a higher toxic risk for the polluted sediments compared to the corresponding reference sediments, but partly having a different emphasis	Tuikka et al. [52]

priority substances, their coverage in time and space of the basins concerned is fairly good. Beyond the simple checking of their mere compliance with environmental quality standards (EQS), they can be further interpreted in terms of RA.

Most of the studies performed were based on the large data set resulting from the monitoring campaigns carried by the Catalan Water Agency (water authority responsible for the Catalonian river basins of which the Llobregat River forms part).

Damásio et al. [18] first translated measured levels of pollutants in water obtained from the Agencia Catalana de l'Aigua into Toxic Units [see (2), Sect. 3] using existing toxicological information, namely, $EC50_i$ the 48 h acute median lethal concentration (LC) in *D. magna* reported for the analyzed compounds. Then TU values were compared with measured effects in transplanted *D. magna* individuals. It was found that predicted toxic effects of organophosphorus pesticides estimated as TU were in average within three orders of magnitude those impairing survival and correlated quite well with the measured effects on their target (cholinesterase activity). Thus indicating that in the Llobregat River from the seven types of contaminant assessed (copper, zinc, triazines, polycyclic aromatic hydrocarbons, organochlorine compounds, alkylphenols, and organophosphorus pesticides) only the last group was likely to affect aquatic arthropods having similar sensitivities as *D. magna*.

Teixidó et al. [19] using the monitoring data gathered throughout all the Catalonian river basins between 1997 and 2004, applied the RA methodology developed by the Fraunhofer Institut (Schmallenberg, Germany) known as COMMPS (Combined Monitoring-based and Modelling-based Priority Setting Scheme) [20] that served to identify the list of WFD priority pollutants based on their exposure and effects. The authors obtained a locally adapted list of priority pollutants at a regional scale and proposed a new site pollution risk index for the relative comparison of the chemical pollution status of the investigated geographical region.

Carafa et al. [21] analyzed and interpreted the surveillance monitoring data obtained by the Catalan Water Agency in the period 2007–2008 using sequential advanced modeling techniques. Concentration data of 60 pollutants were pretreated in order to establish the bioavailable fraction depending on the substance properties (K_{oc}) and environment conditions (water hardness and DOC). The risk to aquatic biota associated with the mixture of so obtained concentration values was estimated using either CA or IA when appropriate. Subsequent exposure assessment with species sensitivity distribution (SSD) and mixture toxicity were used to compute the multispecies potentially affected fraction (msPAF) according to De Zwart and Postuma [13]. Visualization of the spatial distribution of the toxic risk was facilitated by means of the use self-organizing maps (SOM) based on Kohonen's artificial neural networks (ANN) [22] algorithm and other data representation techniques as well. The results obtained in the study show that the most impacted sites are located in the urban and industrial areas surrounding Barcelona, corresponding to the lower parts of Llobregat and Besòs Rivers in which ca. five million people are concentrated. The Llobregat River is suffering not only the toxic impacts of chemical pollution, but also those of metals, nutrients and salinity, the later resulting from both natural and mining activities located in the medium course of the basin. Among the chemicals responsible for toxicity, nonylphenol and octylphenol (degradation products from surfactants) are worth to be mentioned.

Gottardo et al. [23, 24] developed in the context of the WFD an integrated RA methodology for the classification of the ecological status (ES) based on the weight of evidence approach. It implements a fuzzy inference system that hierarchically aggregates a set of environmental indicators grouped into five lines of evidence, namely, biology, chemistry, ecotoxicology, physical-chemistry, and hydromorphology. The ES is then expressed as the membership degree to one or two contiguous WFD status classes. The method is implemented within a freeware GIS (geographic information system)-based decision support system (DSS) developed as part of the MODELKEY project [25]. The tool was tested in the Llobregat River basin using the monitoring database of the Catalan Water Agency corresponding to the period 2001–2004 including biological data (macroinvertebrate species abundances), but unfortunately the lack of some required ecotoxicological and hydromorphological data, were a limitation to the full application of the methodology.

Monitoring databases gathered by the responsible water authorities of several European rivers (Elbe, Scheldt, Danube, and Llobregat) were deeply examined by Von der Ohe et al. [26, 27] aiming to prioritize chemical pollutants in terms of their risk effects. The authors assessed the risk associated with ca. 500 organic compounds for which observations were recorded. Whereas most of the data available referred to WFD priority substances, emerging contaminants were not included in the prioritization exercise since they were mostly absent from the databases, (i.e., they are rarely analyzed by water authorities on routine monitoring). Depending on the information, available compounds were classified into six categories for which different management actions were proposed. Within each category, priority was established on the ground of two indicators, namely the *frequency of exceedance* and *extent of exceedance* of measured concentrations (MEC) respect to PNEC. Two different PNECs were compared, respectively, based on acute or chronic toxicity data. It is worth noting that ca. 75% of the 44 substances showing higher MEC/PNEC ratio corresponded to pesticides.

The study was designed having in mind its practical utility for water managers in the context of the WFD implementation, fact that certainly provided an added value.

4.2 Combined (Multichemical) RA Studies

Many of the rivers located in highly populated areas suffer from many different pressures. As a result, the water of those rivers is contaminated by many different substances coming from different sources. In relation to this, it is important to identify which of these substances are likely to affect aquatic organisms to a greater extent. The use of mixture toxicity assessment modeling tools may allow estimating environmental hazards of complex mixtures of chemicals if their concentration in water is known and toxicological information exist. Underlying principles of such an approach has been outlined in Sect. 3 of this chapter. One of the most used approaches is to consider that all pollutants act similarly and hence assume that their combined toxicity is proportional to the sum of their estimated hazards.

RA studies described in the previous section (Sect. 4.1.) were limited only to the contaminants routinely analyzed by the local water environmental agency. However, water catchments are highly susceptible to be at risk for potential contamination by various pharmaceutical products due to increasing human population density and intensive animal farming techniques. A wide range of human pharmaceuticals including analgesics, antibiotics, steroids, cardiovascular drugs, and various drugs used to treat mental illness are present in effluents from sewage treatment plants that continuously enter freshwater systems [28, 29]. Although concentrations of these chemicals in the aquatic environment are generally in the ng/L, these compounds show high reactivity with biological systems; and are often highly stable. Contrary to pharmaceuticals, the toxicity of many metals, pesticides and detergents is known for a vast number of organisms. Therefore, it is possible to estimate with reasonable accuracy predicted no effect concentration (PNEC) levels from SSDs using chronic data and hence to estimate environmental hazards in the field [30]. For some pharmaceuticals, it is also possible to estimate PNEC from SSD from available information. The uncertainty of PNEC estimated from SSD is from 10- to 1,000-fold lower than those obtained from other methods [30]. For the above-mentioned reasons, there is an urgent need to evaluate rigorously the environmental hazards of different contaminant substances on aquatic organisms and predict their combined effect.

Damasio et al. [28] reanalyzed data from a previous study [18] to address until which extent predicted combined hazards of complex mixtures in real field situations differed using either the concentration or response addition concepts and if such predictions agree with observed detrimental effects measured on field toxicity tests. The study was performed in three locations of the middle–lower part of the Llobregat River (Fig. 1) that included a relative clean upstream location situated in a mid-mountain stream (Merles stream, LL0) and two sites, located in the main river channel in Castellbell (LL1) and Sant Joan Despí (LL4). The three sites showed an increasing gradient of pollution, and a decreasing ecological status. Pollutants considered included dissolved metals (Al, Cr, Co, Cu, Zn, As, and Pb) and up to 91 different organic contaminant residues (target analytes included 22 pesticides, and 64 pharmaceuticals belonging to different classes and alkylphenols).

4.2.1 Determination of hazard quotients

Environmental risks caused by certain contaminants to aquatic invertebrates were estimated using hazard quotient indexes (HQ, (4)) associated with the most representative pollutants found in the Llobregat River [31]. For substances with limited toxicological information, PNEC can be estimated using just a few chronic NOEC or acute LC50s considering application factors (AF) ranging from 10 to 1,000. For known substances, most accurate PNEC estimates can be obtained from large data sets using SSDs of no effect concentrations (NOECs) [30]. In this study, PNECs were estimated considering SSDs, NOECs, or LC50s and using AF ranging from 1 to 1,000 according to (4). Toxicological information was obtained from different sources and the number of species considered.

$$\mathrm{HQ} = \frac{C_i}{\mathrm{PNEC}_i} \tag{4}$$

$$\mathrm{PNEC}_i = \frac{\mathrm{ENPOINT}_i}{\mathrm{AF}} \tag{5}$$

where C_i and PNEC are the measured concentration and PNEC of contaminant i and AF the application factor.

Total hazard quotients $\mathrm{HQ}_{\mathrm{Total}}$ considering all metals, alkylphenols, pesticides, and pharmaceuticals were determined using either the CA (6) or the response addition concept (7) assuming that compounds act similarly or independently, respectively, and that their effects were uncorrelated. These results are shown in Table 4.

$$\mathrm{HQ}_{\mathrm{Total}} = \sum_{i=1}^{n} \mathrm{HQ}_i \tag{6}$$

Table 4 Summed hazard estimates (HQ × 100) determined from Damasio et al. [28] of metals, alkylphenols, pesticides, and pharmaceuticals in each site considering the concentration or response addition concepts

Compounds	HQ × 100			
	LL0	LL1	LL4	
Metals	6.1 (6.1)	10.8 (10.4)	10.1 (9.7)	
Alkylphenols	0.08 (0.10)	0.20 (0.20)	0.56 (0.60)	
Pesticides	0.44 (0.40)	3.67 (3.60)	13.97 (13.92)	
Pharmaceuticals	0.55 (0.60)	0.57 (0.60)	0.95 (0.91)	
Total	7.25(7.11)	15.28 (14.40)	25.61 (23.40)	
	% Contribution to the total predicted toxicity			
Metals	85.2 (84.9)	70.9 (70.4)	39.6 (39)	Zn (28%), Pb (27%)
Alkylphenols	1.1 (1.1)	1.3 (1.4)	2.2 (2.4)	NP (32%)
Pesticides	6.0 (6.2)	24.0 (24.2)	54.5 (54.6)	Diazinon (91%)
Pharmaceuticals	7.6 (7.8)	3.7 (4.0)	3.7 (4)	Clorofibric acid (8%), paracetamol (7%), carbamazepine (7%)

The percentage contribution of each chemical class to the total predicted hazard and compounds accounting for the greatest effects is also included. Response addition estimates are in *parentheses*

$$HQ_{Total} = 1 - \prod_{i=1}^{n} [1 - HQ_i] \tag{7}$$

where HQ_i is the hazard quotient of compound i.

4.2.2 Environmental RA

Based on hazard indexes for each measured compound (for more details, see supplementary information, [29]), the total hazard quotient HQ_{Total} considering all metals, alkylphenols, pesticides, and pharmaceuticals were determined using the CA or response addition concepts ((6) and (7), respectively). These results are shown in Table 4. For clarity, HQ has been multiplied by 100. The closest HQ values to 100 the greatest is the hazard to invertebrates. Both concepts CA and IA predicted almost the same combined toxicity. The overall hazard index was increasing when going downstream being, respectively, those of LL0 and LL4 (Fig. 1) 2.1- and 3.5-fold higher than that of LL0 (see Table 4, under "Total"). Metals and pesticides accounted for 39% and 54% of the estimate total hazard in LL4, respectively, and pharmaceuticals to only 4%. Within each chemical class, Zn, Pb, nonylphenol, and diazinon accounted for most of the predicted toxicity of metals, alkylphenols and pesticides, respectively. Eight compounds explained the greatest predicted toxicity of pharmaceuticals (7–13%): diclofenac, roxithromycin, fenofibrate, clarithromicyn, propyphenazone, clorofibric acid, paracetamol, and carbamazepine, but note that only for the last three there was empirical toxicity data available. These results indicate that pharmaceuticals contributed only marginally to the predicted overall toxicity of water at the studied sites.

To test the predicted responses, Damasio et al. [29] studied how transplanted invertebrate species such as caddisfly larvae, amphipods, and *D. magna* responded to river water after 4 days. Measured responses included survival, feeding rates, and several biomarkers of general stress. Observed responses of deployed organisms correlated quite well with estimated combine environmental hazards but were not conclusive since other water parameters such as salinity or levels of ammonia also correlated with biological effects. Again predicted environmental hazards of organophosphorus pesticides showed a high correlation with inhibitory effects observed in its target (cholinesterase). Tentatively [29], it was concluded that pesticides were likely to have major effects in river aquatic invertebrate species within the Llobregat River. Nevertheless, the use of general stress markers instead of those specific for pharmaceuticals prevented to properly test the toxicity of pharmaceuticals on aquatic invertebrates. Future research, thus, should be focused on developing and using specific markers for pharmaceutical research.

4.3 RA Studies Focused on Specific Pollutant Families

4.3.1 Pharmaceuticals

Occurrence of pharmaceuticals in the aquatic environment is nowadays a well-established issue [32]. A plethora of different classes of pharmaceuticals reach the environment at variable degrees, after their use and excretion through wastewater and sewage treatment systems in which, depending on the compound's properties and treatment conditions, they are only partially degraded [33, 34].

While their occurrence in the aquatic environment is well established, their long-term effects and environmental consequences are still matter of active research [35–37]. However, there is a general consensus on the main features of the problem which can be summarized on the following facts:

(a) Pharmaceuticals are intrinsically bioactive compounds; therefore, they are able to cause potential effects on living systems.
(b) There is a continuous increase on their use and, thus, on their subsequent input into the environment. Indeed, they do not need to be particularly persistent to be present since they are continuously feed.
(c) Hundreds of different pharmaceuticals are currently and regularly used simultaneously, which are susceptible of interaction and synergistic effects, that are basically unknown [38–40].
(d) Data regarding effects on the aquatic ecosystems resulting from long-term low-dose exposure to pharmaceuticals are almost lacking.

Direct estimation of effects caused by environmental pharmaceuticals on ecosystems is not a straightforward task and RA procedures based on ecotoxicological data may be used as an indirect alternative. One of the most common is through the estimation of the aforementioned hazard quotients HQ (see Sect. 4.2.1).

Ginebreda et al. [31] estimated the risk associated with pharmaceuticals in every sampling point expressing it in terms of hazard quotients using data gathered along three monitoring campaigns in seven sampling points distributed throughout the Llobregat River basin (four located in the main river and three in the Anoia River tributary; see Fig. 1). Twenty-nine pharmaceutical compounds, belonging to different therapeutical classes (analgesics, nonsteroidal anti-inflammatories, beta-blockers, lipid lowering agents, antibiotics, anti-histaminics, psychiatric drugs, and antiulcers) were analyzed.

The HQ_i index associated with a given compound i in each sampling point was calculated according to (4) and (5) (with $\mathrm{ENDPOINT}_i = \mathrm{EC50}_i$, and $\mathrm{AF} = 1{,}000$). Overall, HQ_{Total} were calculated for every sampling point through aggregation of compounds' single HQ_i (6) under the CA mode of action assumption for three different bioassays, namely fish, *Daphnia*, and algae. Under this framework, if HQ_{Total} equals to unity there is a potential environmental risky situation (i.e., $HQ_{\mathrm{Total}} \geq 1$ is tentatively interpreted as environmental risk, while $HQ_{\mathrm{Total}} < 1$ should indicate a no risk). Results are summarized in Table 5.

Table 5 Biodiversity Shannon–Wiener Index (calculated from individuals density and biomass) and total hazard quotients based on *Daphnia*, fish and algae for the Llobregat River basin (adapted from [31])

Site		Shannon index (density)	Shannon index (biomass)	Hazard quotients		
				Fish	*Daphnia*	Algae
A1	Mean	2.219	1.960	0.112	0.044	0.514
	SD	0.264	0.249	0.005	0.005	0.569
A2	Mean	0.517	0.416	2.302	2.848	8.687
	SD	0.111	0.238	0.997	1.198	12.246
A3	Mean	1.456	1.322	2.910	2.351	5.328
	SD	0.711	0.804	1.440	1.117	5.284
LL1	Mean	2.139	1.784	0.623	0.648	1.813
	SD	0.915	0.561	0.150	0.520	1.788
LL2	Mean	1.690	1.350	0.535	0.540	1.893
	SD	1.530	1.447	0.208	0.360	1.735
LL3	Mean	1.928	1.652	1.210	0.888	2.385
	SD	0.595	0.591	0.561	0.698	1.467
LL4	Mean	0.887	0.802	9.389	45.622	39.030
	SD	0.472	0.530	1.635	27.998	55.599

Hazard Quotient Indexes

As expected, since the respective toxicities are different for the three bioassays, each one shows its own sensitivity with respect to certain compounds. Thus, for the fish-based bioassay, major contribution is due to gemfibrozil, followed by ibuprofen and diclofenac. Other compounds with significant effect are propyphenazone and bezafibrate. For *Daphnia*, major contributions should be attributable to erythromycin, ibuprofen, and clofibric acid in point LL4 and, in less extent, to diclofenac, acetaminophen, and sulfamethoxazole. Finally, because of its high toxicity reported toward algae [41], this bioassay appears to be mostly dependent on sulfamethoxazole, followed by ibuprofen and gemfibrozil.

Relationship (Correlations) Between HQ and Biodiversity Indexes

The key point of the study was addressed to ascertain the existence of possible relationships between biodiversity and ecotoxicity, respectively, expressed, as explained above, in terms of Shannon–Wiener diversity indexes and hazard quotients (Table 5).

After long time exposure to stressor factors, ecosystems are supposed to get integrative responses adapted (until certain extent) to them. Therefore, since we are assessing long-term situations, for a given point we considered more realistic to reflect its diversity status as the average of the results obtained along the three sampling campaigns. The same rationale holds for hazard quotients.

Roughly speaking, both types of indicators are mutually consistent and evidence a decrease of quality when going downstream (see Table 5 and Fig. 2). This can be observed in either the Llobregat and in its tributary Anoia. Diversity indexes are

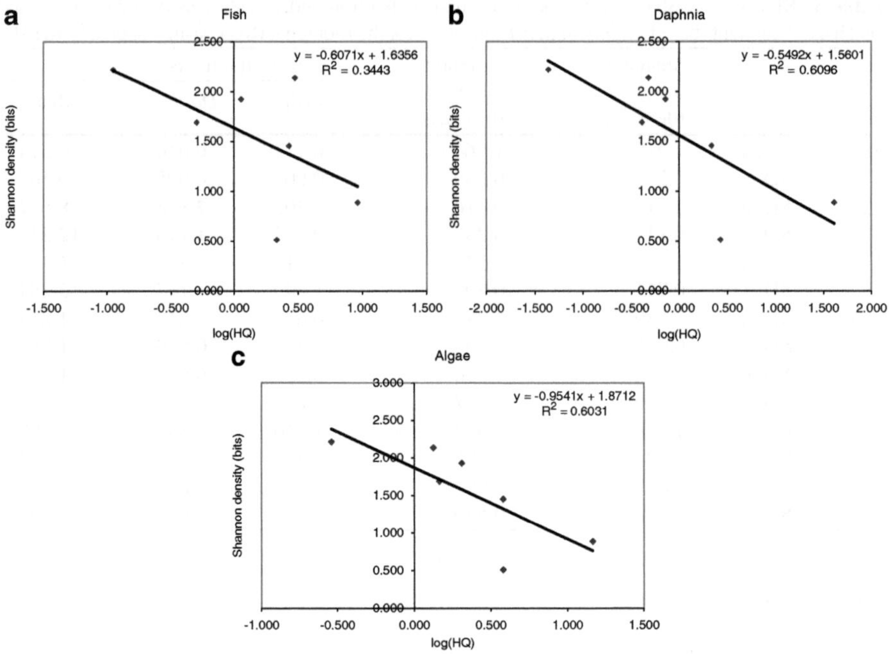

Fig. 2 Correlations between log-transformed total hazard indexes based on different reference species indicators and Shannon diversity. (**a**) Fish vs. Shannon diversity; (**b**) Daphnia toxicity vs. Shannon diversity, and (**c**) algae toxicity vs. Shannon diversity (adapted from [31])

comprised between ca. 3 and 0.4 bits, but the ranges are highly variable, depending on the sampling point. For instance, point A2 shows a fairly constant value of about 0.5 bits, while point LL1 shows a wide variation of ca. one order of magnitude (0.36–3.36 bits).

4.3.2 Pesticides

Agricultural activities are still important in the Llobregat River basin. Therefore, contamination of surface waters by pesticides is a relevant concern. On the other hand, it is well proved that agriculture is not the only origin of pesticides and effluent discharges from WWTPs are too a nonnegligible source, especially for herbicides. Owing to the intrinsic bioactive characteristics of pesticides and their mode of use, their potential deleterious effects on the aquatic ecosystems is obvious.

In this context, Köck–Schulmeyer et al. [42] have investigated the occurrence and potential effects of 16 polar pesticides belonging to different chemical and functional families, namely triazines, phenylureas, organophosphates, chloroacetanilides, and thiocarbamates in the Llobregat River (NE Spain) and some of its tributaries (Anoia and Rubí) using the "Short-Term Pesticide Risk Index for the Surface Water System (PRISW-1)" [43, 44], a recently developed index which takes into account the pesticides concentrations and their overall toxicity against three aquatic organisms (algae, *Daphnia*, and fish).

Briefly, Toxic Units for each organism and sample (TUS) were calculated using the simple additive approach as the sum of the individual pesticide toxic units (2).

The PRISW-1 index was originally developed not only for individual chemicals, but it has also been applied to mixtures of pesticides. This index is based on, first, the calculation of TERs (toxicological exposure ratios) for each sample and the three nontarget organisms algae, *Daphnia*, and fish as the ratio between the critical toxicological "end point" (EC50 for algae and *Daphnia* and LC50 for fish) and the PEC, which in this study is substituted for the real pesticides concentrations measured in the samples, and, second, the application of the following equation (8) that assigns to each TER a score and a weight related to the type of organism:

$$\text{PRISW-1} = (A \times 3) + (B \times 4) + (C \times 5.5) \tag{8}$$

where A, B, and C are the scores assigned to the TERs for algae, *Daphnia*, and fish, respectively, and 3, 4 and 5.5 are the corresponding weights. The scores (for all three organisms) are assigned as follows: 0 for TERs $> 10{,}000$, 1 for TERs in the range 10,000–1,000, 2 for TERs in the range 1,000–100, 4 for TERs in the range 100–10, 6 for TERs in the range 10–2, and 8 for TERs < 2. As it can be deduced from the explanations above the TERs are the reciprocal of the corresponding TU (i.e., TER $= 1/\text{TU}$)

Finally, the environmental risks caused by the pesticides present in the samples are classified, according to the resulting PRISW-1 values, into negligible (PRISW-1 ≤ 5), low ($5 <$ PRISW-1 ≤ 15), medium ($15 <$ PRISW-1 ≤ 40), high ($40 <$ PRISW-1 ≤ 80), and very high (PRISW-1 > 80).

The study showed that pesticide contamination increased downstream of the river (Fig. 3). As far as risk is concerned, application of PRISW-1 indicated that in spite that legal EQS levels are mostly fulfilled the existing pesticide contamination still involves a certain risk for aquatic organisms ranging from low to high depending on the location considered. The risk appears to be higher for *Daphnia* and algae than for fishes and the organophosphates diazinon and malathion and the phenylurea diuron are the major contributors to the overall toxicity and therefore the most likely compounds to cause ecosystem impairment.

4.3.3 Environmental Hazards of Metals to Aquatic Organisms

Many metals are known to be persistent pollutants that are accumulated by aquatic organisms. Indeed, most aquatic environments and in particular rivers are polluted by metals coming from natural and anthropogenic activities. Metals can be essential for life such as copper and zinc or nonessential such as cadmium and mercury among many others. The toxicity of metals is extremely difficult to predict since it

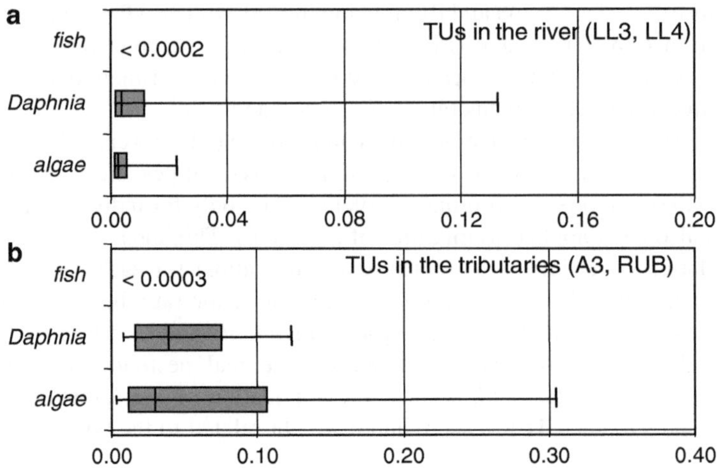

Fig. 3 Boxplot showing the TUS ranges calculated in the samples from the Llobregat main river and its tributaries Anoia and Rubí for the different species indicators (fish, *Daphnia*, algae) (adapted from [42])

varies dramatically across species and is affected by speciation. In water in general it is considered that the free hydrated form of a given metal is bioavailable to organisms [45], whereas those complexed with other ions or into organic material are much less bioavailable. For the above-mentioned reasons, it is difficult to assess the risk of metals with only measurements of their total ion concentration in water. Good examples of dramatic changes of metal toxicity across metal species are those of mercury, whose methylated form is 90% more bioavailable to aquatic organisms than the free hydrated dissolved form Hg^{2+} [46]. Copper also has a complex speciation in freshwater being easily completed with CO_3, NO_3, SO_4, and OH. Indeed only <5% of copper occurs in its bioavailable form Cu^{2+} in waters [47] having an ionic composition similar to that of the Llobregat River (hard waters with high levels of salt). Furthermore, the toxicity of many metals and in particular of essential ones like copper may have a very steep curve changing from being essential to toxic within a narrow range. This means that uncertainty extrapolation factors usually apply to convert acute data to predicted no effect concentration are not applicable. For the above-mentioned reasons, it is difficult to predict toxicity and/or associate correctly a specific hazard to a metal level measured in water without knowing its speciation neither its toxico-kinetics. An example of such difficulties can be found in laboratory studies performed with different essential and nonessential metals using different water types [35]. In this study, it was reported tenfold differences in copper toxicity across waters differing in hardness and two orders of magnitude differences in toxicity of cadmium among different clones of *Daphnia magna*. Within the Llobregat River [48, 49] studying metal body burdens and biomarkers in aquatic larvae of the caddisfly *Hydropsyche exocellata* reported accumulated levels of metals orders of magnitude lower than those causing toxicity in this specie. In the earlier study [35] identify some metal body burdens that correlated with measured tissue damage markers but in the latter study of [37] it was found that such correlations were related to accumulated hazardous persistent organic pollutants such as PCBs, endosulfan, alkylphenols and polycyclic aromatic hydrocarbons.

4.4 Purpose-Specific RA Studies

4.4.1 Water Reuse in the Low Llobregat

During the years 2007 and 2008, the Llobregat River basin suffered from a severe drought, which affected the supply of drinking water facilities that rely on the exploitation of the river water. This extreme situation affected the town of Barcelona and its heavily populated surrounding area. Among other measures, the Catalan Water Agency as the responsible water authority in the basin, implemented a water reuse experiment with the objective of maintaining the river flow rate at sufficiently high level such as to ensure the supply of raw water Barcelona's major drinking water facility. Part of the secondary effluent of Barcelona's major WWTP (El Prat) was subject to a suitable tertiary treatment prior its discharge eight kilometers upstream of the Barcelona's drinking water facility (Sant Joan Despi). Beyond the compliance with the current regulations, it was worth considered the occurrence and effects of emerging contaminants. For that purpose, some studies were undertaken in 2008 [50] and 2009 [51]. In the latter, a RA comparing river upstream and downstream to the discharge point was performed for pharmaceuticals and illicit drugs. Hazard quotients for the concentration ranges found and ecotoxicological data reported in the literature for three reference organisms belonging to different trophic levels (fish, daphnids and algae), as it is recommended by the WFD were considered. Estrogenicity was similarly evaluated using estradiol equivalent factors (EEQs) for the estrogenic compounds identified.

Hazard quotients HQ_i were calculated using (4) and (5) (with $ENDPOINT_i$ = $EC50_i$, and AF = 1,000) for every compound aggregated by simple addition on an overall hazard quotient HQ_{Total} (6) under the assumption of the CA mode of action, so the sum of single hazard quotients must be interpreted as a first rough approach.

In the referred work, hazard quotients were calculated for the concentration ranges (maximum–minimum) found in river sites upstream and downstream to the point of discharge (Fig. 1, sites R0 and LL4, respectively) for pharmaceuticals and illicit drugs. Experimental EC50 values extracted from the literature were preferred, but when they were not available EC50 estimated with ECOSAR were used. When more than one EC50 value was available, the lowest was taken into consideration [41]. Therefore, the risk in the most extreme situation was in any case considered. The results are summarized in Table 6.

Differences between river upstream and downstream to the discharge point were perceivable, but not very significant. Overall, no relevant risks were identified, with

Table 6 Data on sediment biotests and toxicity estimates of msPAF and TU (adapted from Tuikka et al. [52])

Biotest	Endpoint	Site		
		A1	A3	L14
Caenorhabditis elegans	Growth (% of c.)	66	72	47
	Reproduction (% of c.)	56	97	18
Danio rerio	Development (% of c.)	93	61	17
Lumbriculus variegatus	Reproduction (% of s.)	200	173	89
	Biomass (% of s.)	91	91	46
	Individual mass (% of s.)	46	52	52
Chironomus riparius	Survival (% of c.)	124	118	124
	Biomass (% of s.)	82	217	158
	3rd instar stage (% of c.)	65	115	112
Potamopyrgu antipodarum	Reproduction (% of c.)	96	96	103
Ibrio fischeri	EC50 (g, ww/L)	4	3	n.a.
Risk assessment		A1	A3	L14
Number of chemicals analyzed		157	157	39
msPAF (%)		2.73	2.58	1.03
log10 TU_{sum} *Daphnia magna*		−2.4	−1.87	−1.78
log10 TU_{max} *Daphnia magna*		−2.61	−2.18	−2.25
log10 TU_{max} *D. magna* (most toxic compound)		Phenitrothion	Tributyltin	Phenitrothion
log10 TU_{sum} *Pimephelas promelas*		−2.52	−1.77	−1.71
log10 TU_{max} *Pimephelas promelas*		−2.87	−1.96	−2.00
log10 TU_{max} *P. promelas* (most toxic compound)		4-Nonylphenol	Tributyltin	4-Nonylphenol

% of c percentage of corresponding control, *% of s* percentage of the start of the experiment, *EC50* concentration of sediment causing 50% inhibition in light production (g, wet sediment/L, suspension), *msPAF* multisubstance potentially affected fraction of the species, *log10 TU_{sum} values* logarithmical toxic unit of the sum toxicity of compounds, based on their LC50 values for *D. magna* or *P. promelas*, *log10 TU_{max} values* logarithmical toxic unit of the most toxic compound, based on its LC50 values for *D. magna* or *P. promelas*, *n.a.* not analyzed

the only exception of pharmaceuticals vs. algae that were likely to occur both in river upstream and downstream. Such algae toxicity was mostly attributable to the extremely low EC50 reported for two compounds, i.e., ciprofloxacin and sulfamethoxazole. The importance of having reliable and comparable ecotoxicological data is once again stressed; otherwise, RA studies such as the present one may lose part of its meaning.

As far as estrogenicity was concerned, the upstream site showed higher estrogenicity than the downstream one, attributable to the occurrence of estrone and diethylstilbestrol.

4.4.2 Sediments RA

In a recent work, Tuikka et al. [52] investigated the toxicity of sediments from three polluted European river basins (Elbe, Scheldt, and Llobregat) as well as their corresponding references sediments taken from the upper respective rivers. In the Llobregat case, the selected points (Fig. 1) were LL4 (Llobregat in Sant Joan Despí), A3 (Anoia in Martorell), and A1 (Anoia in Jorba) as reference. A battery of six different toxicity contact test representing three different trophic levels were assayed, including the chronic tests with the oligochaete *Lumbriculus variegatus*, the nematode *Caenorhabditis elegans*, and the mudsnail *Potamopyrgus antipodarum*, a subchronic test with the midge *Chironomus riparius*, an early life stage test with the zebra fish *Danio rerio*, and an acute test with the luminescent bacterium *Vibrio fischeri*. All of them are characterized by different endpoints (survival, growth, reproduction, embryo development, and light inhibition, respectively).

Interestingly, the toxic effects obtained from the foregoing tests in each sample were compared to the calculated ecotoxicity of the sediment contamination translated into toxic units (TU) (2) [11] on the basis of acute toxicity to *Daphnia magna* and *Pimephales promelas*, and msPAFs of species [13, 53] as an estimate for expected community effects under the respective assumption of CA in the former case and combined CA and IA depending on the compounds mode of action in the second.

Chemicals considered included persistent polychlorinated compounds (PCDDs, PCDFs, PCBs, PBDEs, PCNs, organochlorine pesticides), perfluorinated compounds (carboxylates and sulfonates), methyl mercury, organotins, nonorganochlorine pesticides (herbicides and insecticides), alkylphenols and their ethoxylate derivatives, PAHs and heavy metals.

Results are summarized in Fig. 4. In general, the toxicity tests were able to discriminate the more polluted sites A3 and LL4 from the cleaner one (A1) used as reference (even though it was proved to be also partially polluted). However, the RA toxicities based on TU and msPAF only partially explained the biotest results. For instance, TUS (*D. magna*) estimated sublethal effects on microinvertebrates on A3 and LL4, whereas TUS (*P. pomelas*) and msPAF ranked the sediments

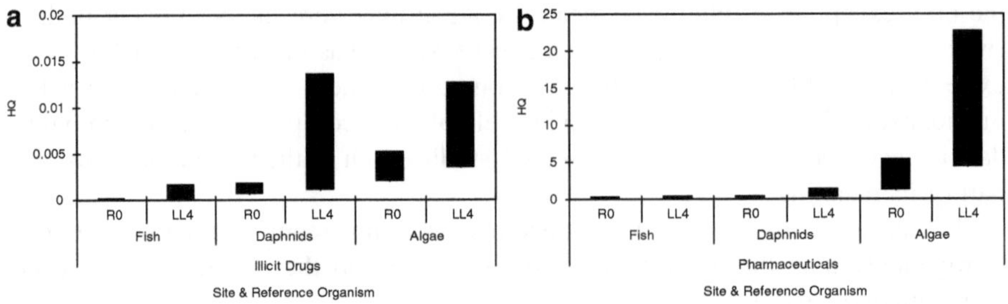

Fig. 4 Total hazard quotient ranges calculated for different reference species indicators at R0 (control site *upstream* to the reclaimed water discharge) and LL4 (control site located *downstream* to reclaimed water discharge) (adapted from [52])

similarly. The most toxic identified compounds using TUs were the organophosphate insecticide fenitrothion, tributyltin, and 4-nonylphenol.

As far as biotests are concerned, the results showed differences in their respective sensitivities, thus emphasizing the convenience of using several toxicity tests in order to have a more comprehensive estimation of the effects of pollution on the aquatic ecosystem.

5 From RA to Ecosystem Effects

Probably, the greatest problem of risk estimates is the uncertainty attendant when extrapolating from lab to field situations and from individuals to the whole community [54]. In relation to these risk assessors often use two approaches to derive PNEC from laboratory toxicological endpoints (LC50, EC10, EC50, NOEC, etc.): "extrapolation or application factors" or the species sensitivity distribution method [55–57]. The first one consists in dividing selected endpoints by a factor that range from 10 to 1,000. Low and high factors are used when chronic (NOEC) or acute data (LC50) are used [42], respectively. The other approach is to compile existing information of toxicological endpoints across species and estimate the concentration levels that would protect 95% of a hypothetical community composed of those species [43]. When NOEC from many species are used, estimates based on species sensitivity distributions did not require extrapolation factors. Here in this chapter we used both approaches. The application factor is more conservative than the species sensitivity one but both approaches are not based on a robust scientific knowledge. Factors such as the lack of relevance of laboratory populations and laboratory exposure scenarios make difficult to apply correctly safety margins to both components of risk, the PNEC and PEC, respectively [40]. For example from the thousands or hundreds of species that coexist in a community only a few of them can be reared in the lab. Often species with long lifecycles, low reproductive output or sensitive to lab environment are underrepresented. Simulation and empirical studies have shown that when considering populations in a steady stage, which are common stage in the filed (populations neither grow nor decrease, they maintain a constant number of individuals in time) toxic effects become less apparent due to indirect compensatory effects (Tolerant individuals had more food and hence performed better under toxic stress since density-dependent effects were relaxed) [43, 58]. According to those studies, extrapolation factors to estimate PNEC from toxicological endpoints should be relaxed. On the other hand, lab exposure scenarios are often constant, whereas in real field situation the environment changes and is heterogeneous. This can make contaminants more or less bioavailable in the filed than in the lab. This is critical for contaminants that speciate in water like many metals, those that bound to particles or food and hence are bioconcentrated by organisms. All these factors affect PEC and should be considered when deriving measured levels in water to PEC. Finally when considering communities affected by many contaminants, risk estimates have to be implemented taking into account combined effects. As explained in a previous section the most conservative approach is to consider that all substances act similarly and just summing estimate hazards for each measured substance. The complementary approach is to consider that measured compounds act differently. In that case, their combined effect should be proportional to the product of their probability effects, which require to know the complete concentration-response curves, which are impossible to obtain for the whole community. For these reasons, often the CA combine effect model is used. However, this model suffers from an important statistical problem that is the more substances are considered the most likely to find a measurable combined hazard irrespective of the toxicity and measured levels of the individual mixture constituents. On the other hand, the response addition model assumes that there is a threshold concentration level below which there was no effect. Thus, those substances whose measured levels are below such threshold are not computed in the overall combined effect. In relation to this, thus, risk assessors working with complex mixtures of many chemicals at low levels should consider the fact that at low levels measured contaminants may not have any biological activity. Indeed, these problems may have overestimated the risks predicted in previous sections [29, 31].

Finally, estimated environmental risk should be validated with experimental data. There are several approaches to do that, one is to do it in the lab dosing a water sample with the same contaminant levels as those measured in the filed and assayed its toxicity using lab toxicity tests. The other approach is complementing risk estimates with filed studies that measured changes in the whole community or responses of caged organisms deployed at the studied locations. Both field approaches have been used and described in previous sections but probably need refinement. A possible one should be to combine community studies with omic techniques to include specific responses able to detect the many targets of pollutants that occur in the Llobregat. For example in aquatic organisms most specific targets of pharmaceuticals are not known. At the moment, such targets are limited to estrogenic, dioxin like and anticholinesterase compounds. The use of omic methods such as microarrays in model species may allow to measure and quantified simultaneously effects on thousands of potential targets [59].

6 Conclusions

One of the premises (and hence the challenges) posed by the WFD is the relationship between chemical and ecological status. Progressing on the understanding of the effects caused by chemical pollution on ecosystem status can be approached on different ways, being one of the most promising the interpretation of the effects caused by chemical exposure on the grounds of ecotoxicological principles. This is the objective of environmental RA.

The availability of plenty of chemical and ecological data relative to the Llobregat River basin gathered, respectively, by both water authorities and researchers as a consequence of routine regularly performed monitoring (WFD implementation) and scientific research projects has turned this river into an interesting and illustrative case study, thus becoming one of the Mediterranean rivers better known from the environmental point of view [28].

In the preceding lines, we have summarized the most representative RA studies carried out in the Llobregat River basin. They have served not only to evidence the extent and effects attributable to several families of pollutants (heavy metals, pharmaceuticals, pesticides, etc.), thus generating valuable scientific knowledge, but also to provide a simple methodology useful for management purposes.

As a whole, RA has revealed as a promising tool and further research in that field can be recommended. Among other issues, the still lacking of reliable ecotoxicological information concerning many chemical compounds is perhaps one of its main drawbacks in which more effort would be desirable.

Acknowledgment This research study was funded by the Spanish Ministry of Science and Innovation through the project SCARCE (Consolider Ingenio 2010 CSD2009-00065).

References

1. European Inventory of Existing Commercial Chemical Substances of 15 June 1990. OJ C/146A. Corrigendum published in OJ C54, 2002
2. Arnot JA et al (2006) Screening level risk assessment model for chemical fate and effects in the environment. Environ Sci Technol 40(7):2316–2323
3. Muir DCG, Howard PH (2006) Are there other persistent organic pollutants? A challenge for environmental chemists. Environ Sci Technol 40(23):7157–7166
4. Arnot JA, Mackay D (2008) Policies for chemical hazard and risk priority setting: can persistence, bioaccumulation, toxicity, and quantity information be combined? Environ Sci Technol 42(13):4648–4654
5. Johnson AC et al (2008) Assessing the concentrations of polar organic microcontaminants from point sources in the aquatic environment: measure or model? Environ Sci Technol 42(15):5390–5399
6. Kortenkamp A, Backhaus T, Faust M (2009) State of the art report on mixture toxicity. Final Report. Contract No. 070307/2007/485103/ETU/D.1. http://ec.europa.eu/environment/chemicals/effects.htm. Accessed 11 Aug 2011

7. Van Gestel CAM et al (eds) (2010) Mixture toxicity: linking approaches from ecological and human toxicology. CRC, Boca Raton. ISBN: 9781439830086
8. Loewe S, Muischnek H (1926) Über Kombinationswirkungen. Naunyn Schmiedebergs Arch Pharmacol 114(5):313–326
9. Bliss C (1939) The toxicity of poisons applied jointly. Ann Appl Biol 26:585–615
10. Barata C, Baird DJ, Nogueira AJA, Soares AMV, Riva MC (2006) Toxicity of binary mixtures of metals and pyrethroid insecticides to *Daphnia* magna Straus. Implications for multi-substance risks assessment. Aquat Toxicol 78:1–14
11. Sprague JB (1970) Measurement of pollutant toxicity to fish. II. Utilizing and applying bioassay results. Water Res 4:3–32
12. Teuschler LK, Herzberg RC (1995) Current and future risk of assessment guidelines, policy and methods development for chemical mixtures. Toxicology 105:137–144
13. De Zwart D (2002) Observed regularities in SSDs for aquatic species. In: Posthuma L, Suter GW II, Traas TP (eds) Species sensitivity distributions in ecotoxicology. Lewis, Boca Raton, pp 133–154
14. Backhaus T et al (2003) The BEAM-project: prediction and assessment of mixture toxicities in the aquatic environment. Cont Shelf Res 23(17–19):1757–1769
15. Vighi M et al (2003) Water quality objectives for mixtures of toxic chemicals: problems and perspectives. Ecotoxicol Environ Saf 54(2):139–150
16. Junghans M et al (2006) Application and validation of approaches for the predictive hazard assessment of realistic pesticide mixtures. Aquat Toxicol 76(2):93–110
17. Altenburger R, Walter H, Grote M (2004) What contributes to the combined effect of a complex mixture? Environ Sci Technol 38(23):6353–6362
18. Damasio J, Tauler R, Teixido E, Rieradevall M, Prat N, Riva MC, Soares A, Barataa C (2008) Combined use of Daphnia magna in situ bioassays, biomarkers and biological indices to diagnose and identify environmental pressures on invertebrate communities in two Mediterranean urbanized and industrialized rivers (NE Spain). Aquat Toxicol 87:310–320
19. Teixidó E, Terrado M, Ginebreda A, Tauler R (2010) Quality assessment of river waters using risk indexes for substances and sites, based on the COMMPS procedure. J Environ Monit 12:2120–2127
20. CEC (1999) Revised Proposal for a List of Priority Substances in the Context of the Water Framework Directive (COMMPS Procedure). Fraunhofer-Institut Umweltchemie und Oekotoxicologie, 98/788/3040/DEB/E1
21. Carafa R, Faggiano L, Real M, Munné A, Ginebreda A, Guasch H, Flo M, Tirapu L (2011) Water toxicity assessment and spatial pollution patterns identification in a Mediterranean River Basin District. Tools for water management and risk analysis. Sci Total Environ. doi:10.10116/j.scitotenv.2011.06.053
22. Kohonen T (2001) Self-organizing maps. Vol. 30. of Springer Series in Information Sciences, 3rd edn. Springer, Berlin
23. Gottardo S, Semenzin E, Giove S, Zabeo A, Critto A, de Zwart D, Ginebreda A, Marcomini A (2011) Integrated risk assessment for WFD ecological status classification applied to Llobregat river basin (Spain). Part I – Fuzzy approach to aggregate biological indicators. Sci Tot Environ 409:4701–4712
24. Gottardo S, Semenzin E, Giove S, Zabeo A, Critto A, de Zwart D, Ginebreda A, Marcomini A (2011) Integrated risk assessment for WFD ecological status classification applied to Llobregat river basin (Spain). Part II – Evaluation process applied to five environmental lines of evidence. Sci Tot Environ 409:4681–4692
25. MODELKEY research project (Models for Assessing and Forecasting the Impact of Environmental Key Pollutants on Marine and Freshwater Ecosystems and Biodiversity) funded by the European Commission within the Sixth Framework Programme (SSPI-CT-2003-511237-2)
26. Von der Ohe PC, de Deckere E, Prüß A, Munoz I, Wolfram G, Villagrasa M et al (2009) Towards an integrated assessment of the ecological and chemical status of European river basins. Integr Environ Assess Manag 5:50–61

27. Von der Ohe PC, Dulio V, Slobodnik J, De Deckere E, Kühne R, Ebert R-U, Ginebreda A, De Coomann W, Schüürmann G, Brack W (2011) A new risk assessment approach for the prioritization of 500 classical and emerging organic microcontaminants as potential river basin specific pollutants under the European Water Framework Directive. Sci Tot Environ 409:2064–2077
28. González S, López-Roldán R, Cortina JL (2012) Presence and biological effects of emerging contaminants in Llobregat River basin: a review. Environ Pollut 161:83–92
29. Damásio J, Barceló D, Brix R, Postigo C, Gros M, Petrovic M, Sabater S, Guasch H, de Alda ML, Barata C (2011) Are pharmaceuticals more harmful than other pollutants to aquatic invertebrate species: a hypothesis tested using multi-biomarker and multi-species responses in field collected and transplanted organisms. Chemosphere 85:1548–1554
30. ECHA (2008) Guidance on information requirements and chemical safety assessment: characterisation of dose [concentration]-response for environment. European Chemicals Agency (Chapter R.10)
31. Ginebreda A, Munoz I, de Alda ML, Brix R, Lopez-Doval J, Barcelo D (2010) Environmental risk assessment of pharmaceuticals in rivers: relationships between hazard indexes and aquatic macroinvertebrate diversity indexes in the Llobregat River (NE Spain). Environ Int 36:153–162
32. Daughton CG, Ternes TA (1999) Pharmaceuticals and personal care products in the environment: agents of subtle change? Environ Health Perspect 107:907–938
33. Gros M, Petrovic M, Barceló D (2007) Wastewater treatment plants as a pathway for aquatic contamination by pharmaceuticals in the Ebro river basin (northeast Spain). Environ Toxicol Chem 26:1553–1562
34. Gros M, Petrović M, Ginebreda A, Barceló D (2010) Removal of pharmaceuticals during wastewater treatment and environmental risk assessment using hazard indexes. Environ Int 36:12–26
35. Sanderson H, Johnson DJ, Reitsma T, Brain RA, Wilson CJ, Solomon KR (2004) Ranking and prioritization of environmental risks of pharmaceuticals in surface waters. Regul Toxicol Pharmacol 39:158–183
36. Fent K, Weston AA, Caminada D (2006) Ecotoxicology of human pharmaceuticals. Aquat Toxicol 76:122–159
37. Nunes B, Carvalho F, Guilhermino L (2005) Acute toxicity of widely used pharmaceuticals in aquatic species: *Gambusia holbrooki*, *Artemia parthenogenetica* and *Tetraselmis chuii*. Ecotoxicol Environ Saf 61:413–419
38. Pomati F, Orlandi C, Clerici M, Luciani F, Zuccato E (2008) Effects and interactions in an environmentally relevant mixture of pharmaceuticals. Toxicol Sci 102:129–137
39. Cleuvers M (2003) Aquatic ecotoxicity of pharmaceuticals including the assessment of combination effects. Toxicol Lett 142:185–194
40. Cleuvers M (2004) Mixture toxicity of the anti-inflammatory drugs diclofenac, ibuprofen, naproxen and acetylsalicylic acid. Ecotoxicol Environ Saf 59:309–315
41. Grung M, Kallqvist T, Sakshaug S, Skurtveit S, Thomas KV (2008) Environmental assessment of Norwegian priority pharmaceuticals based on the EMEA guideline. Ecotoxicol Environ Saf 71:328–340
42. Köck-Schulmeyer M, Ginebreda A, González S, Cortina JL, de López Alda M, Barceló D (2011) Analysis of the occurrence and risk assessment of polar pesticides in the Llobregat River basin (NE Spain). Chemosphere 82:670–678
43. Finizio A, Calliera M, Vighi M (2001) Rating systems for pesticide risk classification on different ecosystems. Ecotoxicol Environ Saf 49:262–274
44. Finizio A, Calliera M, Villa S, Vighi M (1999) Predictive approach to ecotoxicological properties for use in a rating system in the hazard assessment of pesticides. In: Del Re AA et al (eds) Human and environmental exposure to xenobiotics: proceedings of the XI symposium pesticide chemistry. La Goliardica Pavese, Pavia, pp 533–541

45. Campbell PGC (1995) Interactions between trace metals and aquatic organisms: a critique of the free-ion activity model. In: Tessier A, Turner DR (eds) Metal speciation and bioavailability in aquatic systems. Wiley, Chichester, pp 45–102
46. Diez S, Barata C, Raldúa D (2008) Exposure to mercury: a critical assessment of adverse ecological and human health effects. In: Prasad MNV (ed) Trace elements as contaminants and nutrients: consequences in ecosystems and human health. Wiley, New York, pp 345–373
47. Barata C, Markich SJ, Baird DJ (1998) Influence of genetic and environmental factors on the tolerance of *Daphnia magna* Straus to essential–non essential metals. Aquat Toxicol 42:115–137
48. Barata C, Lekumberri I, Vila-Escalé M, Prat N, Porte C (2005) Trace metal concentration, antioxidant enzyme activities and susceptibility to oxidative stress in the tricoptera larvae *Hydropsyche exocellata* from the Llobregat riven basin (NE Spain). Aquat Toxicol 74:3–19
49. Damásio J, Fernández-Sanjuan M, Sánchez-Avila J, Lacorte S, Prat N, Rieradevall M, Soares AMVM, Barata C (2011) Multi-biochemical responses of benthic macroinvertebrate species as a complementary tool to diagnose the cause of community impairment in polluted rivers. Water Res 45:3599–3613
50. Köck-Schulmeyer M, Ginebreda A, Postigo C, López-Serna R, Pérez S, Brix R, Llorca M, Alda MLD, Petrović M, Munné A, Tirapu L, Barceló D (2010) Wastewater reuse in Mediterranean semi-arid areas: the impact of discharges of tertiary treated sewage on the load of polar micro pollutants in the Llobregat river (NE Spain). Chemosphere 82:670–678
51. López-Serna R, Postigo C, Blanco J, Pérez S, Ginebreda A, López de Alda M, Petrović M, Munné A, Barceló A (2012) Assessing the effects of tertiary treated wastewater reuse on a Mediterranean river (Llobregat, NE Spain). Emerging contaminants. Environ Sci Pollut Res (in press)
52. Tuikka AI, Schmitt C, Höss S, Bandow N, von der Ohe PC, de Zwart D, de Deckere E, Streck G, Mothes S, van Hattum B, Kocan A, Brix R, Brack W, Barceló D, Sormunen AJ, Kukkonen JVK (2011) Toxicity assessment of sediments from three European river basins using a sediment contact test battery. Ecotoxicol Environ Saf 74:123–131
53. De Zwart D, Posthuma L (2005) Complex mixture toxicity for single and multiple species: proposed methodologies. Environ Toxicol Chem 24:2665–2676
54. Parker ED, Forbes VE, Nielsen SL, Ritter C, Barata C, Baird DJ, Admiraal W, Levin L, Loesshke V, Lyytikäinen-Saarenmaa P, Hokenstrom JC, Calow P (1999) Stress in ecological systems. Oikos 86:179–184
55. Baird DJ (1993) Testing vs research in ecotoxicology: a response to Baird and Calow. Funct Ecol 7:509–512
56. Barata C, Baird DJ, Soares AMVM (2002) Determining genetic variability in the distribution of sensitivities to toxic stress among and within field populations of *Daphnia magna*. Environ Sci Technol 36:3045–3049
57. Forbes VE, Calow P, Sibly RM (2001) Are current species extrapolation models a good basis for ecological risk assessment. Environ Toxicol Chem 20:442–447
58. Barata C, Baird DJ, Soares AMVM (2002) Food supply on density-dependent effects on demographic responses of the cladoceran Moinodaphnia macleayi to heavy metal exposure. Ecol Appl 12:552–564
59. Piña B, Barata C (2011) A genomic and ecotoxicological perspective of DNA array studies in aquatic environmental risk assessment. Aquat Toxicol 105:40–49

Human Pressure and Its Effects on Water Quality and Biota in the Llobregat River

Antoni Munné, Carolina Solà, Lluís Tirapu, Carlos Barata, Maria Rieradevall, and Narcís Prat

Abstract The Llobregat River has severely been impacted by anthropogenic pressures since long time ago. The mid and lower Llobregat basin holds an important concentration of industries, agricultural activities, and urban areas, with high associated water demand and wastewater discharge. Salt mine activities, hydropower water diversion, and flow regime alteration by dams affect both the Llobregat headwaters and middle reaches. These impacts have historically caused the degradation of riparian biological communities and the loss of habitats along the river. The high amount of information available on water quality and biological community composition allows establishing a suitable monitoring program aimed to improve its ecological status. Some measures have been applied to mitigate the impacts, and Llobregat's biological quality status has progressively improved. The biological communities, mainly diatoms and macroinvertebrates, have recovered even those inhabiting the river mouth, but mostly during wet periods. However, some anthropogenic pressures still remain and Llobregat's biological status is not completely restored. The high amount of small weirs and hydropower water diversion along the Llobregat and Cardener Rivers, together with flow regime regulation by dams, riparian degradation, and point nutrient discharges (from water sewage plants) and salt debris due to mine activities, result in a poor biological quality status in the mid and lower Llobregat River. Fish fauna is the most altered community, with a high number of nonnative species present.

A. Munné (✉), C. Solà, and L. Tirapu
Department of Monitoring and Aquatic Ecosystem Improvement, Catalan Water Agency, c/ Provença 204-208, 08036 Barcelona, Spain
e-mail: anmunne@gencat.cat

C. Barata
Department of Environmental Chemistry, IDÆA-CSIC, c/ Jordi Girona 18-26, 08034 Barcelona, Spain

M. Rieradevall, and N. Prat
Department of Ecology, University of Barcelona, Av. Diagonal 654, 08028 Barcelona, Spain

S. Sabater et al. (eds.), *The Llobregat: The Story of a Polluted Mediterranean River*, Hdb Env Chem (2012) 21:297–326, DOI 10.1007/698_2012_149,

The occurrence of some priority substances and emergent pollutants (e.g., endocrine disruptors, heavy metals, pesticides, flame retardants, drugs, and pharmaceuticals), even at low concentrations, further alter the biological quality. The changes in the biological community structure in the middle part of the river can be detected by using biomarkers, and these should additionally be considered as biological monitoring tools necessary for an integral ecological status diagnosis.

Keywords Biological indices, Biomarkers, Chemical status, Ecological status, Human pressure, Llobregat basin, Monitoring program, Water Framework Directive

Contents

1 Introduction

The Llobregat River is the second longest in the Catalan River Basin District (170 km long). The Llobregat basin has a total drainage area of 4,957 km^2, with an average total annual discharge of 700 Hm^3, which can range between 219 (i.e., 1944–1945) and 2,040 Hm^3 (i.e., 1971–1972) depending on year's climatology. Water flow regime also varies along the year because of the Mediterranean climate influence. High flow values are usually present in spring and autumn (between 30 and 50 m^3/s), while water flow may be scarce in summer (between 2 and 20 m^3/s). Some tributaries can even dry up in summer periods. Also, sudden floods eventually appear in spring and autumn due to heavy rain events, and water flow can rise up to 170 m^3/s (once every 2–10 years), or occasionally up to 800 m^3/s (once every 10–50 years) in the lower Llobregat River. The Llobregat ends in the Mediterranean sea close to Barcelona, where large catastrophic floods have been recorded in its vicinity (e.g., 3,080 m^3/s in 1971) [1].

The Llobregat River is heavily impacted by anthropogenic pressures as a result of its geographical location. The mid and lower Llobregat basin area holds an important concentration of industries, agricultural activities and urban areas, with an important water demand [2, 3]. River flow is regulated by three large dams in the Llobregat and Cardener headwaters which are mainly used to provide water to the lower basin (Fig. 1). Moreover, several small weirs and hydropower stations are located along Llobregat River, which, together with salt mine activities mainly located in the mid-Llobregat basin and in the headwaters of Cardener River, result in a heavy human pressure on the river ecosystem.

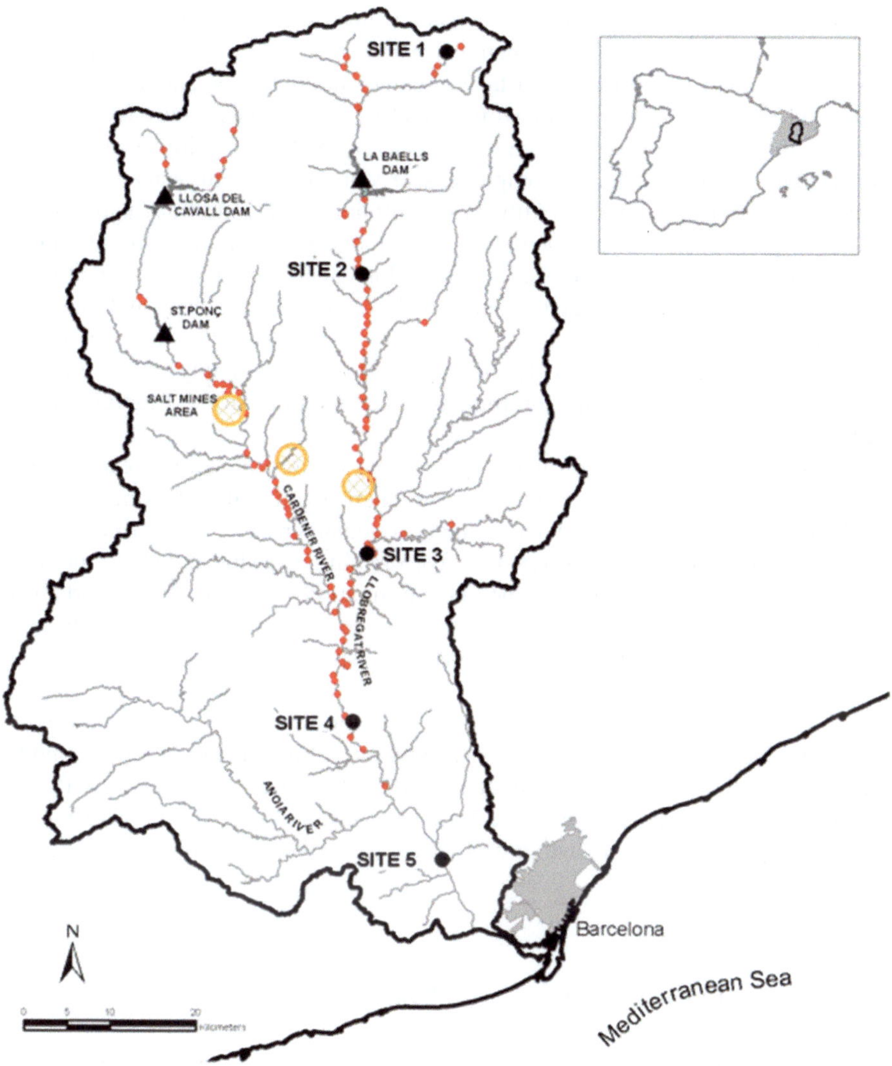

Fig. 1 The main human pressures in the Llobregat River. Major dams (*dark triangles*) and weirs (small hydropower stations highlighted using *red dots*) are located along Llobregat, Cardener, and Anoia Rivers. Also salt mine activities are shown in the basin (*orange circles*). Five sampling sites (Sites 1–5) have been selected from Llobregat headwater to the river mouth (close to Barcelona), in order to analyze quality changes along river and over time

Llobregat's water quality and ecosystem damages have been monitored for a long time by water authorities and research centers in Catalonia. The biological quality of the Llobregat River is monitored since the 1970s by using macroinvertebrate indices [4–6]. Since then, several studies and monitoring programs have been carried out to assess the river ecosystem health (e.g., [7–9]).

Furthermore, water authorities have been analyzing the chemical quality since mid-1940s. Firstly focusing on basic chemical parameters (nutrients, COD, etc.) [10], and later by analyzing priority substances, emergent pollutants, and the ecological status [11, 12]. The Water Framework Directive (2000/60/CE) introduced a new concept for water ecosystem monitoring that set apart the traditional concept of water quality, and launched a comprehensive water ecosystem analysis by using biological elements, together with hydromorphological condition analysis and chemical parameters. Macroinvertebrates, diatoms, macrophytes, and fish fauna have been more recently used to assess the biological quality in the Llobregat River and in the rest of Catalan water watersheds [13–16]. Also, priority and emergent pollutants have been monitored [17, 18], and new tools for ecosystem risk assessment have been implemented [11, 19]. Recent studies have analyzed the effect of certain emerging contaminants on biological communities in the Llobregat River [20, 21].

There is a large dataset on water quality and ecological status measurements related with the Llobregat River, and this chapter aims at analyzing the river ecological status, as well as the main effects of human pressure on the river quality. This analysis has been mostly based on five sampling sites located along the main Llobregat watercourse (Fig. 1), and the information on those sites mostly was originated from monitoring programs currently carried out (in particular those of the Catalan Water Agency and the FEM research group from the University of Barcelona). Site 1 is located upstream of La Baells reservoir in a rather natural area. Site 2 is located downstream of La Baells reservoir and is therefore submitted to flow regime alteration, but with low urban pressures. Site 3 is located in the middle part of the river, downstream of the inputs of the salt mines, and in an area with high urban discharges. Finally, Sites 4 and 5 are located in the lower Llobregat River, in the vicinity of the main industrial and urban areas, which are highly affected by sewage discharges.

2 Human Pressure and Water Management

The Llobregat River flows throughout one of the most industrialized and inhabited areas of Catalonia in its middle and lower course [22]. Waste and accidental discharges from industrial and urban activities as well as from mining activities (salt brines) are the main impacts [3, 18, 19]. High nutrient loads, salt concentrations and conductivity, and the presence of some priority and hazardous chemical substances co-occur in the lower Llobregat River [17, 18] resulting in a moderate and poor ecological and chemical status [3, 20, 23, 24]. In the same way, hydromorphological alterations by water withdrawals, flow regime alteration, and riparian occupation produce several ecosystem damages from headwaters to the river mouth.

The Llobregat River has been severely disturbed by human uses for long time ago. In the mid-nineteenth century, Llobregat River was mainly used as an energy source for the textile industry. Many weirs were built along the river, and most of them remained in place long after the mills were closed. Nowadays, most weirs still provide hydropower and they are impeding the natural flow regime and sediment transport downstream. A total of 106 hydropower stations plus 96 water withdrawal points for agricultural and urban uses are currently located along the Llobregat, Cardener, and Anoia Rivers, throughout 200 km of river watercourse (one almost every kilometer) (Fig. 1). Several weirs are placed one following the other and continuously derive a high water percentage from the river. Water is completely stored in weirs and later flowing alongside the river through canals or pipes (with an average of 1–2 km length) until the downstream hydropower station, where water is dumped back into the river. This pressure is repeated downstream again and again leaving long river reaches with poor flow or, in some cases, completely dried up. The derivation flow exceeds the natural flow regime in a total of 81 out of 106 hydropower stations (76%), which results in a scarce or null water flow in the affected river reaches.

The middle sections of the Llobregat and Cardener Rivers are also impacted by salt mine activities (Fig. 1). Saline wastes from the mining activities increase substantially the salinity of the Llobregat and Cardener Rivers. Some small tributaries like the riera Salada and the riera d'Hortons in the mid-Cardener basin, or the riera de Saldes and the Gavarresa in the mid-Llobregat are usually affected by natural salt springs resulting in high water conductivity. However, the high salt concentration along the Llobregat and Cardener Rivers appeared when salt mine industry started in Suria (1925), Cardona (1931), Salient (1932), and Balsareny areas (1954). Mine salt activities produce big saline waste dumping mountains, from the separation processes of sodium (Na) and potassium (K). Potassium is marketed, whereas sodium and other wastes have been deposited continuously near the mines and close to the river course. Saline rubbles result in a high concentration of sodium, chlorides, and bromides downstream the river due to lixiviation, in particular after rains. In natural conditions, the Cardener and Llobregat Rivers should have chloride concentrations from 40 to 100 mg/L (measured in 1915), whereas current measures are in the range of 200–500 mg/L (Sites 3–5) (Fig. 2a). Chloride concentration suddenly increases in Site 3 just downstream salt mine activities, and values remain high downstream until the river mouth, with a high variability along the year. Occasionally, high concentrations are detected in Sites 3–5, with values up to 700 mg/L of chloride. Bromide concentration increases up to 0.5–0.8 mg/L in the lower Llobregat. Hence, water conductivity usually ranges from 1,300 to 2,000 μS/cm and may rise up to 4,000 and 8,000 μS/cm. This high salt concentration affects the biological communities as well as the water uses. Moreover, water supply for urban uses needs to be properly treated to reduce chloride and bromide concentrations in order to ensure water quality, and some agricultural uses are restricted downstream.

The increasing high population inhabiting the mid and lower Llobregat basin (from 600,000 inhabitants in 1900 to 2,700,000 in 2010) mainly located close to Barcelona and nearby cities causes a high urban and industrial wastewater discharge pressure. A total of 63 sewage plants have been built by the Catalan Water

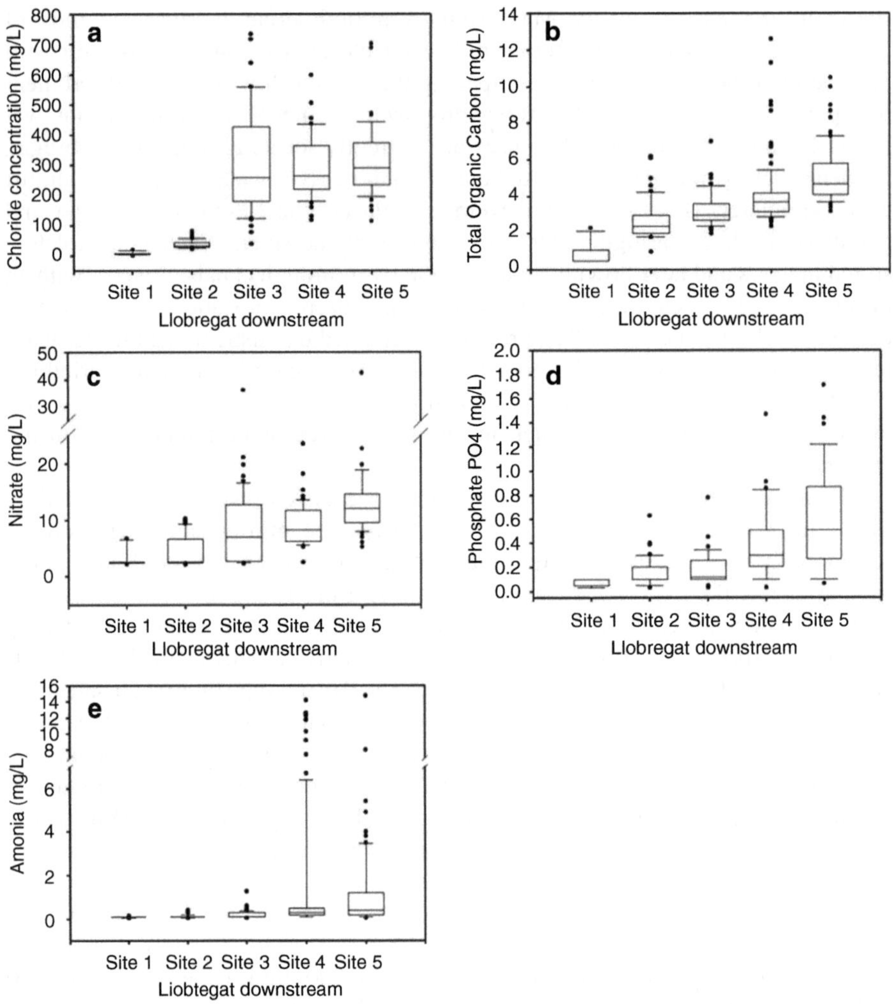

Fig. 2 Main chemical parameters analyzed along the Llobregat main watercourse (see Fig. 1). (**a**) Chloride, (**b**) total organic carbon, (**c**) nitrate, (**d**) orthophosphate, and (**e**) ammonia concentration. Values are calculated from 2007 to 2010 monthly. A total of 48 samples per site have been considered

Agency in the last 25 years in the Llobregat basin in order to mitigate wastewater effects on the river ecosystem. All sewage plants treat a total of 1,303,598 inhabitant equivalents, annually discharging up to 107 Hm^3 of treated wastewater into the river, which is equivalent to 15% of the total annual Llobregat River flow. Wastewater-treated plants discharge a total of 2,576 Mt. of N and 283 Mt. of P into the river. Therefore, a total of 44% of total nitrogen and 37% of total phosphorus are annually removed by urban and industrial sewage plans. However, the remaining nutrient load and the scarce dilution in the river cause water eutrophication. This is more evident during low flow and drought periods, when the wastewater discharge dilution is significantly reduced. Furthermore, a significant amount of treated urban wastewater (3.23 m^3/s) that comes from Barcelona urban area (1,034,297 inhabitant equivalents) is directly discharged into the Mediterranean sea through a marine outfall (2 km long).

As a result of these discharges, total organic carbon (TOC), nitrate and phosphate concentrations increase along the Llobregat River (Fig. 2b–d), and this increase is more evident from Site 2 to the river mouth (Site 5). High TOC values are mainly found in the mid-Llobregat, nearby main urban areas located close to the river (Site 3), and progressively increase downstream with values between 3 and 6 mg/L. Peak concentrations up to 10–12 mg/L are found in the lower Llobregat (Sites 4 and 5). Nutrient concentration (nitrate and phosphate), and organic matter (TOC), progressively increases from Site 2 to the river mouth (Sites 3–5) with peak values mainly found close to urban and industrial areas (Sites 4 and 5). In lower Llobregat River reaches (Sites 4 and 5), high concentrations of phosphate are also detected, from 0.3 to 0.8 mg/L, with peak concentrations up to 1.5 mg/L. Moderate nitrate values (from 5 to 15 mg/L) occur in Site 3 but quickly increase and maximum values ranging from 25 to 40 mg/L occur in Sites 3–5. On the other hand, ammonia values usually remain lower than 1 mg/L along the Llobregat River (Fig. 2e), even close to urban and industrial areas. The high impact of ammonia on river communities is due to peak concentrations nearby urban and industrial areas (Sites 4 and 5), when values are occasionally up to 15 mg/L, which may cause important detrimental effects on the biological communities [25].

Besides chemical impacts and water diversion by hydropower previously mentioned, the Llobregat River is also affected by water flow regime alteration. A significant amount of Llobregat's water withdrawal is used for urban and agricultural activities (205 Hm^3), especially in its lower water course. A large drinking water treatment plant is located close to Barcelona, which provides water from the Llobregat River to a large part of Barcelona city. To enhance the water resources of the basin, the Llobregat basin was regulated by three big dams located in Llobregat and Cardener headwaters (Fig. 1). "St Ponç" dam (24 Hm^3 of capacity), "la Llosa del Caball" dam (80 Hm^3 of capacity), both located in the upper Cardener River, and "la Baells" dam (109 Hm^3 of capacity) located at the headwaters of the Llobregat River. These dams eliminated flood events and regulated the flow along time. Flow regime attenuation can reduce habitat availability, affecting the biological community and its diversity [26], especially on fish communities [27]. Water flow stored in dams and weirs and flow regime regulation result in new habitats that can be colonized by nonnative species [28], which may have adverse impacts on Llobregat water ecosystems [29]. More than 100 nonnative species, considering riparian vegetation and aquatic flora and fauna, have been cited in the lower Llobregat River [30].

The Catalan Water Agency published the IMPRESS document (analysis of pressures and impacts on Catalan water bodies) in 2005 [22]. This document aims at highlighting and quantifying the main human pressures that could affect the good water body status according to the Water Framework Directive guidelines (http://circa.europa.eu/Public/irc/env/wfd/library?l=/framework_directive/guidance_documents/gds10srefcondspolicyssum/_EN_1.0_&a=d). Pressures from human activities were calculated and classified by combining their magnitude and possible effect on the aquatic ecosystem (sensitivity risk assessment analysis). Finally, a quality target was established for each pressure in order to quantify the risk of nonachievement of good status. The magnitude of each pressure, corrected by river sensitivity (e.g., flow regime dilution) and divided by the selected quality target, was calculated. Values near "1" (from 0.8 to 1.2) mean that the pressure corrected by the river sensitivity is close to the quality standard (low risk), values over "1" (between 1.2 and 2) mean that the pressure exceeds the quality threshold (quality target) and the risk is moderate, and values over "2" mean high risk. Besides, values under 0.8 were classified as without risk. Risk assessment was calculated for a total of ten human pressures identified in the Llobregat River (Table 1). Results showed that the main highlighted pressures were due to urban and industrial wastewater discharges into the river. A total of 29 out of 80 water bodies identified in the Llobregat basin (34%) are affected by wastewater discharges. Also, water diversion for hydropower affects a total of 16 water bodies (19%), salt mine wastes affect a total of 13 water bodies (15%), flow alteration by upstream dam regulation affect 10 water bodies (12%), and river channelization affect 10 water bodies (12%) mainly located downstream. Invasive species also produce a high pressure on the Llobregat River water bodies, affecting a total of 18 water bodies (21%),

Table 1 Main human pressures identified in the Llobregat basin. Number and percentage of affected water bodies have been classified according to the magnitude of pressures and risk assessment. River water bodies in the Llobregat basin have an average length of 15 km

Human pressures on Llobregat River	High risk		Moderate risk		Low risk		With some risk	
	No. of water bodies	%	No. of water bodies	%	No. of water bodies	%	No. of water bodies	%
Hydromorphological pressures								
Connectivity lost (weirs and dams)	3	3.6	3	3.6	2	2.4	8	9.5
River canalization	4	4.8	3	3.6	3	3.6	10	11.9
Water withdrawals	1	1.2	0	0.0	2	2.4	3	3.6
Flow regime alteration (dam regulation)	3	3.6	3	3.6	4	4.8	10	11.9
Water diversion by hydropower stations (flow reduction)	14	16.7	0	0.0	2	2.4	16	19.0
Chemical pressures								
Urban and industrial wastewater discharge	15	17.9	10	11.9	4	4.8	29	34.5
Rainwater sewage system discharges	5	6.0	0	0.0	0	0.0	5	6.0
Salt mine activities	3	3.6	6	7.1	4	4.8	13	15.5
Other pressures								
Invasive species	3	3.6	7	8.3	8	9.5	18	21.4
Accumulated pressure	29	34.5	16	19.0	6	7.1	51	60.7

although the main pressures were classified to have low and moderate levels of impacts. These pressures have continuously increased in the last years. The two main high pressures identified in the Llobregat basin were urban and industrial wastewater discharge, and water diversion by hydropower uses, mainly classified as pressures producing high risk. A total of 15 water bodies for urban and industrial discharge (18% of the total river water bodies) and 14 water bodies for water diversion by hydropower (17% of the total river water bodies) were classified as having a high risk (Table 1). Other detected pressures were poor river connectivity that affected 8 river water bodies (9%), rainwater discharges affecting a total of 5 water bodies (6%), and water withdrawals mainly due to agricultural activities and water supply to urban areas, affecting a total of 3 water bodies (4%). Pressures and their magnitude progressively increase downstream. Thus, water bodies located in the mid and lower Llobregat River show higher risk due to several pressures (water withdrawal, urban and industrial wastewater discharge, water diversion, flow regime alteration, invasive species, salt mine waste, etc.). However, the headwaters are mainly affected by hydropower water diversion and low connectivity because of the presence of high amount of weirs. A total of 51 out of 80 river water bodies (61%) have some risk to nonachieve a good ecologically status due to human pressures on the Llobregat basin.

3 Biological Communities and Quality Assessment

The Water Framework Directive (2000/60/EC) (WFD) and the Priority Substances Directive (105/2008/EC) provided significant changes for water quality assessment in the European aquatic ecosystems [31–33]. The application of these Directives requires using a new monitoring program, which EU Member States were bound to apply since 2007, according to the WFD requirements. Both, ecological status and priority substances must be taken into account in order to establish a comprehensive water status diagnosis. Chemical and biological elements must be combined to set the final water quality status. Ecological status assessed using biological elements, together with the chemical status, have been implemented in the Llobregat basin by the Catalan Water Agency since the new Monitoring Program started in 2007 [24]. Moreover, some Catalan research centers have also been analyzing the chemical and ecological status through research projects launched several years ago (e.g., ECOBILL, ModelKey, and KeyBioeffects projects).

In Europe, diverse biological indices and metrics have been developed for ecological status assessment in rivers using macroinvertebrate [34, 35], diatoms [36], macrophytes [37], and fish communities (e.g., [38]). Biological indices have also been applied in Catalan Rivers since long time ago, basically developed in research centers [3, 7, 9, 14], and more recently applied by Water Authorities following the WFD requirements [13, 16, 24, 39]. The assessment of biological quality in rivers has been developed and enhanced by contributions of several research centers and water authorities in order to achieve the normative definitions of WFD and their compliance with a gradient stressor.

3.1 Historical Perspective on Biomonitoring

The study of freshwater communities in the Llobregat River and their use as bioindicators has a long tradition [3]. Ramon Margalef (University of Barcelona) in 1951 made a first attempt to use algae as indicators in his monograph entitled "Indicator organism in Limnology" [40]. Moreover, Margalef adapted the Saprobic system from central European countries to Spanish rivers [41]. Much of this work was done in the Llobregat basin, and in 1978, he helped to start a survey of the rivers Llobregat and Besos, which was the seed for the future monitoring programs currently being carried out. A first biotic index using macroinvertebrates called "BILL" was later defined in the Llobregat and Besos Rivers [4]. Data were presented in a series of papers [5, 9, 42]. A regular biomonitoring program for the Llobregat basin was definitively set up in 1994 after a series of large floods by the FEM research group (University of Barcelona). Since then, a Llobregat monitoring program has been established and continued mainly using macroinvertebrate communities and chemical parameters [7, 8]. The sampling site network carried out by the ECOBILL monitoring program consists of 25 sampling sites covering the main stream and permanent and temporary tributaries in the Llobregat basin. The IBMWP index uses macroinvertebrates at family level, which requires a multihabitat sampling and an extensive searching of macroinvertebrate families (GUADALMED protocol: [43]). Each sampling site is visited twice a year (spring and summer), and data are available on http://ecobill.diba.cat.

The evolution of biological water quality by using macroinvertebrate communities (IBMWP index) in 25 sites of the basin has been analyzed in the Llobregat River through the FEM research group data (Fig. 3). The IBMWP index is a useful quality index based on macroinvertebrate community composition at the family level [44], which has been commonly used by Spanish Water Authorities. The percentage of sampling sites classified in each of the five quality classes can be compared from 1997 to 2010 (Fig. 3a). From 1997 until 2006, close to 50% of sites were classified below a good quality level with some changes along years and high variability between dry and wet years. The worst quality values coincide with dry years (e.g., 2004, 2006, and 2008), when close to 60–70% of sites were classified below good quality status. The importance of dry years in the biological quality assessment of Mediterranean rivers was studied by Munné and Prat [39], who concluded that reference condition values should be lowered for dry periods in order to properly interpret quality status and compare it along time. In wetter years the biological quality of Llobregat's water improves, also because of the urban and industrial sewage treatments implemented in its basin.

Furthermore, when the IBMWP evolution is analyzed over 17 years (from 1994 to 2010) in the five sites considered in this study (Fig. 3b) is evidenced that the

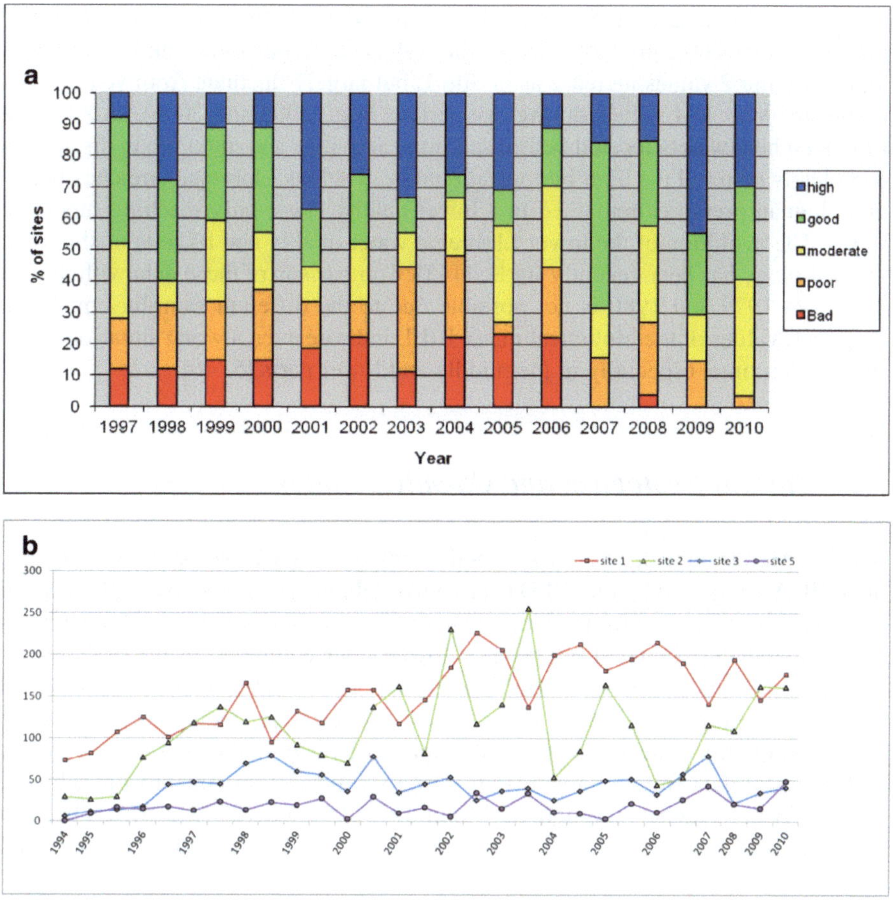

Fig. 3 Temporal evolution of the biological quality in the Llobregat basin using the IBMWP index. (**a**) The percentage of sampling sites classified in five quality classes is shown from 1997 to 2010. A total of 25 sampling sites covering the main Llobregat watercourses and tributaries have been selected from the ECOBILL database (FEM research group, University of Barcelona). (**b**) Changes along time (from 1994 to 2010) of the IBMWP index are shown for selected sites located along the Llobregat River from headwaters to the river mouth (see Fig. 1)

water quality of the upper part of the basin has continuously been improved (Sites 1 and 2), while the lower part (Sites 4 and 5) is still remaining in a poor or moderate quality status. All sites show low IBMWP values at the beginning of the analyzed period due to the large floods occurred in 1994. High floods dragged most macroinvertebrate communities, and this event provided a good basis to analyze the capacity of macroinvertebrate colonization and biological quality improvement along the Llobregat River. While quality values continuously increased along time in headwaters (Sites 1 and 2), with some interannual fluctuations, water quality values did not recover so much along time in downstream locations (Sites 3 and 5).

In the mid-Llobregat part (Site 3), values never achieved the good status (IBMWP values upper to 60), probably due to the influence of salt mines and local urban inputs. In Site 2 values increase as in Site 1, but largely fluctuate from year to year with some very low values during dry periods (e.g., 2004 and 2006). In 2002 the Llobregat headwaters located before salt mine activities and main urban discharges completely restored the high biological quality, whereas Llobregat stretches below main human pressures remained in a lower quality class, only reaching moderate status. The evolution of the lower Llobregat water quality and its relationship with the hydrology has been recently studied [45]. Comparison of these data with the old data from 1979 and 1980 is not possible due to the different sampling methods. However, values of the biological index BILL indicate even a worst situation in the river at that time, especially in the middle and lower parts [5, 46].

3.2 Current Ecological and Chemical Status

Suitable Mediterranean river type-specific indices for each biological quality element (BQE), required by the WFD (macroinvertebrates, diatoms, macrophytes, and fish), have been applied for the biological quality assessment in the Llobregat basin (Table 2 and Fig. 4). Quality classes must be later combined using BQE values according to "one out, all out" criteria [47] in order to establish the final biological quality class (Table 2 and Fig. 5a). This is a restrictive procedure since the worst biological quality item is used to set the final biological quality status.

Biological data were obtained from spring samples (from April to June) by the Catalan Water Agency through its Monitoring Program (2007–2012). The data available up to now were obtained during 2007–2010. Macroinvertebrates and

Table 2 Number and percentage (in parentheses) of river water bodies classified in five biological quality classes for each biological quality element (macroinvertebrates, diatoms, macrophytes, and fish fauna) in the Llobregat basin. Chemical status is also shown using two quality classes according to the Water Framework Directive. Data are provided by the Catalan Water Agency monitoring program carried out from 2007 to 2010. A total of 80 river water bodies have been established in the Llobregat basin

Biological elements	Quality classes					Water bodies without data
	High	Good	Moderate	Poor	Bad	
Macroinvertebrates (IBMWP index)	41 (51%)	16 (20%)	11 (14%)	7 (9%)	0 (0%)	5 (6%)
Diatoms (IPS index)	24 (30%)	17 (21%)	3 (4%)	16 (20%)	8 (10%)	12 (15%)
Macrophytes (IBMR index)	0 (0%)	2 (3%)	1 (1%)	5 (6%)	2 (3%)	70 (87%)
Fish (IBICAT index)	5 (6%)	3 (4%)	20 (25%)	3 (4%)	18 (22%)	31 (39%)
Biological quality	9 (11%)	14 (18%)	6 (7%)	33 (41%)	15 (19%)	3 (4%)
	Good		Bad			Water bodies without data
Chemical status	66 (83%)		10 (12%)			4 (5%)

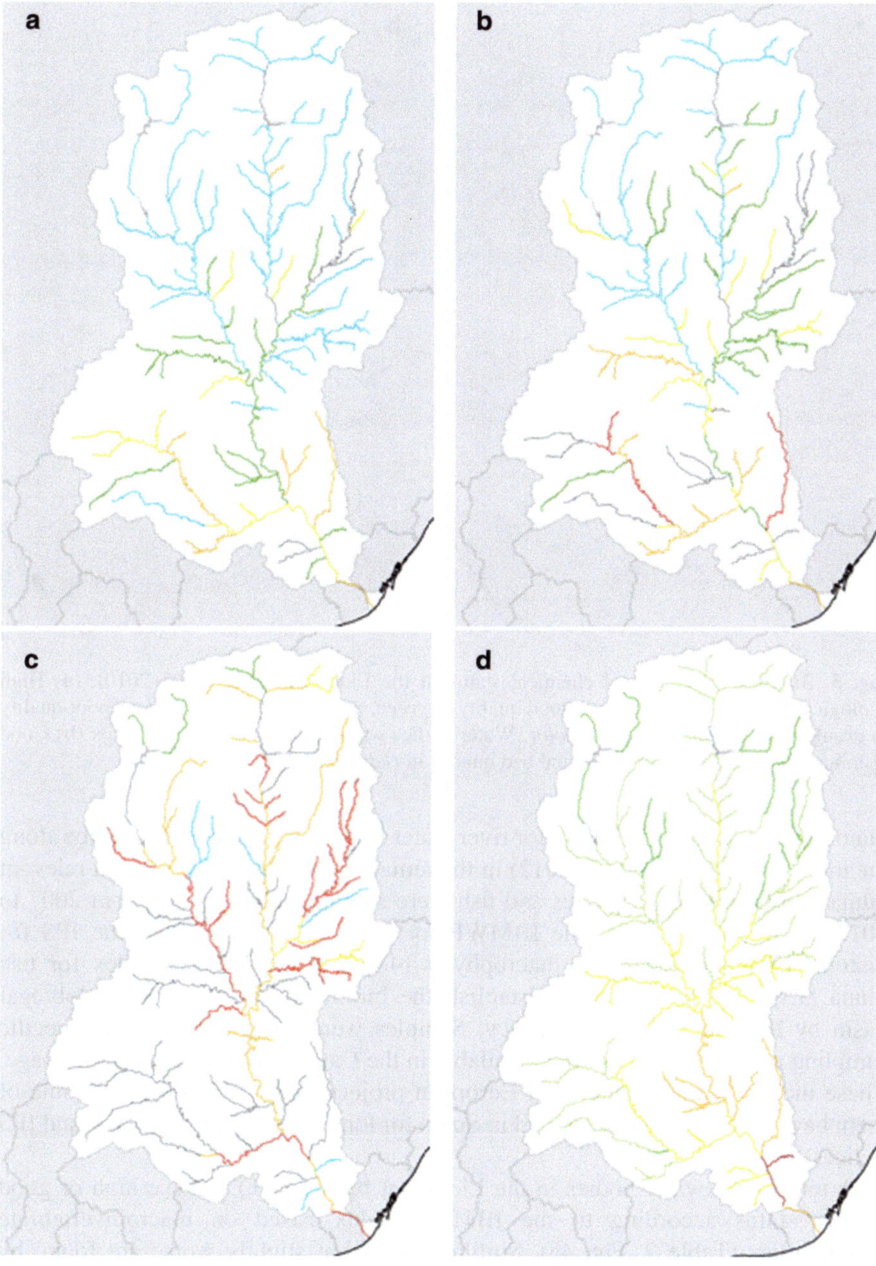

Fig. 4 Biological quality in the Llobregat basin (2007–2010). High quality class is shown in *blue*, good quality in *green*, moderate quality in *yellow*, poor quality in *orange*, and bad quality in *red color*. Water bodies without data are shown in *gray*. Water quality is measured by using: (**a**) macroinvertebrates (IBMWP index), (**b**) diatoms (IPS index), and (**c**) fish (IBICAT index). (**d**) The number of total nonnative aquatic species is shown for each river water body. Water bodies with a total nonnative species from 1 to 30 (considering riparian vegetation and all aquatic organisms) are shown in *dark green color*, from 30 to 60 with *light green*, from 60 to 80 with *yellow*, from 80 to 100 with *orange*, and more than 100 are shown in *red color*

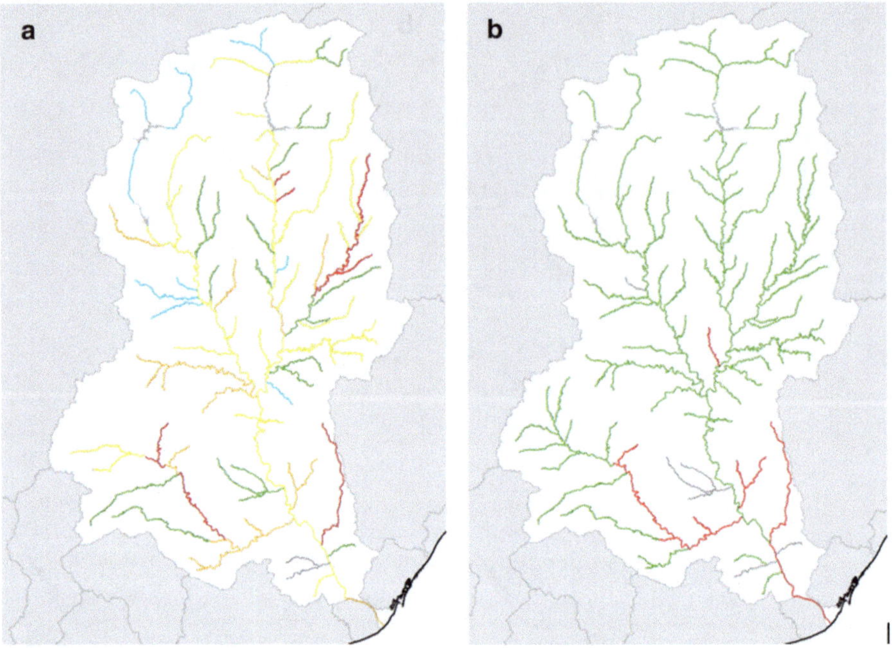

Fig. 5 Biological quality and chemical status in the Llobregat basin (2007–2010). (**a**) High ecological status is shown in *blue*, good quality in *green*, moderate quality in *yellow*, poor quality in *orange*, and bad quality in *red color*. Water bodies without data are shown in *gray*. (**b**) Good chemical status is shown in *green* and bad quality in *red color*

diatoms were sampled annually for river water body at risk, and at least twice along the monitoring period (2007–2012) in the remaining water bodies without relevant human pressures. Macrophytes and fish were sampled at least once from 2007 to 2012 in all water bodies. The IBMWP for macroinvertebrate [44], the IPS for diatoms [36], the IBMR for macrophytes [48], and the IBICAT index for fish fauna [15] have been used to establish the biological quality in the Llobregat basin by the Catalan Water Agency. Samples were gathered following specific sampling protocols, and data are available in the Catalan Water Agency web page. These indices mainly came from European projects (e.g., [38, 49]), and some of them have commonly been applied in Spain for long time (mainly IBMWP and IPS indices).

A total of 57 water bodies in the Llobregat basin (71%) achieve high or good quality status according to the IBMWP index based on macroinvertebrate communities (Table 2, Fig. 4a). Similar results but slightly worse are found by using diatoms as a BQE (Table 2, Fig. 4b). Quality objectives (good and high quality) are achieved in 51% of river water bodies (41 out of 80 water bodies established in the Llobregat basin) when diatom index is used (IPS index). Diatoms are very sensitive organisms for organic pollution, and they have a rapid response to chemical disturbances [36]. Low-quality values are found in more river water bodies than using macroinvertebrates specially in water bodies classified as high quality with macroinvertebrate index only obtain a good status with diatom indexes. Macroinvertebrates are mainly sensitive to organic pollution, but they can have a slower and sustained response to eutrophication than diatoms [50]. However, in most cases the IBMWP index values were quite consistent with diatoms results, which are low in mid-Anoia and lower Llobregat Rivers. Low IPS indexes are found near urban and industrial areas receiving high nutrient loads, and mainly in the Anoia, low Cardener, and mid and low Llobregat watercourses. A total of 23% river water bodies for macroinvertebrates (18 out of 80 water bodies), and 34% for diatoms (27 out of 80 water bodies), do not achieve good biological quality in the Llobregat basin. Moreover, 6% of river water bodies for macroinvertebrates (5 out of 80 water bodies), and 15% for diatoms (12 out of 80 water bodies), have not been analyzed yet. These water bodies will be further sampled along the monitoring program period (2007–2012).

Macrophytes have scarcely been used in the Llobregat River for quality purposes so far. Quality indices based on macrophytes are currently tested and still under discussion [51]. Several macrophyte quality indices are currently available (IM, IVAM, etc.), and some of them have been applied in the Llobregat basin [52]. More recently the IBMR index [48], which was intercalibrated for the Mediterranean rivers according to the WFD requirements, has been applied by the Catalan Water Agency in the Catalan Rivers, but only data of ten river water bodies are nowadays available in the Llobregat basin (Table 2). From these ten sampled river reaches, a few show good quality and the majority (eight out of ten) are classified in a moderate, poor, and bad quality status. Macrophyte-based indexes are heavily dependent on the substrata type and on the hydro-morphological conditions. Also inappropriate sampling protocol can negatively affect the output. Results should be therefore carefully considered because of insufficient knowledge on its application in Mediterranean rivers. Macrophyte indices, and its interpretation in terms of water quality, need to be improved to be properly used for ecological status assessment in the Llobregat River.

Fish fauna shows the worst quality scenario from any of the measured BQEs. Only 10% of water bodies (8 out of 80 water bodies) show fish communities dominated by native species (Table 2, Fig. 4c). Fish fauna are sensitive not only to water pollution, but also to morphological and flow alterations, river discontinuity, and habitat loss [27, 53]. Low-quality values using fish fauna denote lack of suitable habitat conditions, and hence hydrological alteration and water abstractions can considerably affect native fish community composition. Moreover, additional threats such as exotic alien species invasions, basically introduced for fishing, negatively affect fish communities and prevent them to achieve a good biological quality. The main Llobregat watercourse, and mainly its lower part, hosts a high number of nonnative fish species (Table 3, Fig. 4d). Close to 100% of fish species in the mid and lower Llobregat main watercourse are nonnative. A total of three to five nonnative fish species are usually present in the mid and lower Llobregat. Also, over 100 nonnative species (considering all riparian and aquatic flora and fauna) have sometime been cited in the lower Llobregat River [30]. Nonnative species are

Table 3 Quality indices and main metrics of the biological quality elements along the main Llobregat watercourse (see Fig. 1) using the biological quality elements (macroinvertebrates, diatoms, macrophytes, and fish fauna) required by the Water Framework Directive to assess the ecological status in rivers

Biological quality elements	Quality indexes or biological metrics	Site 1	Site 2	Site 3	Site 4	Site 5
Macroinvertebrates	IBMWP	155 (high)	138 (high)	82 (good)	70 (good)	44 (moderate)
	EPT	16	11	6	4	3
	S (num. fam.)	34	31	28	21	14
Diatoms	IPS	19.6 (high)	15.3 (high)	11.8 (moderate)	11.7 (high)	6.7 (moderate)
	TDI	28.8	89.8	90.8	81.1	96.2
	H′	1.45	2.52	3.94	4.15	3.51
	S (num. sp.)	15	24	34	38	34
Macrophytes	IBMR	13 (good)	3 (bad)	nd	9.25 (poor)	nd
	S (num. sp.)	18	3	nd	19	nd
Fish	IBICAT2010[a]	4.0 (moderate)	7.5 (poor)	3.7 (bad)	9.9 (poor)	8.4 (poor)
	S (num. sp.)	3	3	5	5	4
	S (num. alien sp.)	1	1	5	5	3
Alien sp. (all aquatic communities)[b]	S (num. sp.)	37	66	78	97	141

[a]Results for each site are calculated according to its reference condition and type-specific metrics. That is the reason why values are not comparable. Quality classes can be compared between sites

[b]The total amount of alien species has been calculated trough EXOAQUA project (supported by the Catalan Water Agency and carried out by the CREAF). All nonnative aquatic species cited in several studies and projects were collected in a large database (EXOAQUA). Riparian vegetation, vertebrate and invertebrate fauna, and also algae and macrophytes were considered

basically distributed in the lower basin, close to the river mouth and near urban areas. Headwaters near nonimpacted areas held communities with less than 30 aquatic nonnative species. Nonnative species invasions are considered a high risk for aquatic ecosystems due to their usually high dispersal capacity [54]. A database with nonnative and invader species cited in the Catalan aquatic ecosystems can be consulted online through the Catalan Water Agency web page (http://aca-web.gencat.cat).

Finally, high and good biological quality, combining macroinvertebrate, diatoms, and fish biological elements, are achieved in 29% of the Llobregat water bodies (23 out of 80 water bodies) (Table 2, Fig. 5a). Biological quality has been set out using the worst three BQEs measured (macroinvertebrate, diatoms, and fish fauna). Due to scarcity of current data, macrophytes are not considered yet. Results show high and good quality in headwaters, basically rivers upstream the major human pressures (salt mine activities and urban discharges), and small streams far from urban and agricultural areas, mostly tributaries. The middle and lower Llobregat River, in areas below salt mine and nearby urban and industrial areas, show many water bodies with moderate, poor, and bad biological quality. The main Llobregat watercourse below the dam shows a moderate biological quality mainly due to fish community alterations, since macroinvertebrate and diatoms provide a good quality. The worst biological quality is at the lowest Llobregat main watercourse, close to Barcelona, and in some tributaries flowing throughout high-industrialized areas, like as the Anoia and Rubi Rivers, both located in the lower Llobregat basin.

By analyzing the evolution of biological patterns and quality indices along the Llobregat main watercourse (Table 3), we can observe how the number of macroinvertebrate and diatom taxa decreases from the Llobregat headwater (Site 1) to the river mouth (Site 5). From a total of 34 families of macroinvertebrates that are found in the Llobregat headwaters, only 14 families are sampled at the lower part. Also, the number of families of Ephemeroptera, Plecoptera, and Trichoptera (EPT) diminish throughout the main Llobregat watercourse, from 16 families sampled in headwater to only three families close to the river mouth (Baetidae, Caenidae, and Hydropsychidae). Besides, a clear change of macroinvertebrate taxonomic composition along river is detected. Some taxa progressively disappear downstream, while other ones appear. Macroinvertebrate families that disappear are more sensitive to water pollution, and prefer well oxygenated and cold water. Some families of Plecoptera, Trichoptera, Ephemeroptera, and Diptera are commonly found in headwaters, over the salt mine activity and main water flow alteration by dams (Site 1) (e.g., Chloroperlidae, Perlidae, Perlodidae, Nemouridae, Leuctridae, Ephemerellidae, Leptophlebiidae, Rhyacophilidae, Limnephilidae, Athericidae, and Glossossomatidae), but some of them disappear downstream of dams (Site 2) (e.g., Chloroperlidae, Perlidae, Perlodidae, Nemouridae, Athericidae, and Glossossomatidae). Also, downstream salt mine activities (Site 3) more tolerant families are found (e.g., Baetidae, Hydropsychidae, and Simulidae), and major taxonomic groups disappear (e.g., Leuctridae, Ephemerellidae, Leptophlebiidae, and Rhyacophilidae). On the other hand, some other taxa, mainly Trichoptera

(Brachycentridae and Leptoceridae), are not found in the Llobregat headwaters (Site 1), but they are often sampled in the mid-Llobregat, even downstream of salt mine areas (Sites 2 and 3). Those taxa are also sensitive to water pollution and human pressures and, therefore, they indicate good biological quality. Sites 1 and 2, respectively, located upstream and downstream of the flow regulation caused by La Baells dam show high-quality status by using the macroinvertebrate index. Evenly Site 3, located downstream major salt mine activities and small urban discharges, shows a good quality status by using macroinvertebrates (IBMWP). On the other hand, only few macroinvertebrate taxa inhabit the lower Llobregat River, heavily affected by urban and industrial discharges (Sites 4 and 5). In the lower Llobregat River, the macroinvertebrate fauna is basically composed by Ephemeroptera (Baetidae, Caenidae), Trichoptera (Hydropsychidae), Diptera (Chironomidae), some Mollusca (Physidae), and Oligochaeta, taxa resistant to pollution. Nevertheless, the richness of macroinvertebrates is very low nearby the river mouth (Site 5), and biological quality by using macroinvertebrates is at most moderate (just under the quality target). In summary, macroinvertebrate community alteration slightly change under salt mine activities, and most quality damages are detected downstream urban and industrial discharges (Sites 4 and 5), where biological indices are not achieving the quality objectives by using macroinvertebrate community (IBMWP index).

The number of taxa (species) increases from headwaters (15 species sampled in Site 1) to the river mouth (34 species found in Site 5). Most of the diatoms species inhabiting in the lower Llobregat River are highly tolerant to water pollution and nutrient loads (e.g., *Cocconeis placentula* and *Nitzschia inconspicua*), whereas less tolerant species are found upstream (Sites 1 and 2) (e.g., *Achnanthidium biasolettianum*). Water quality measured by using diatoms (IPS) and macroinvertebrates (IBMWP) is quite similar. Quality values decrease downstream, although major changes are downstream of salt mine activities, where quality class shifts from good to moderate. Also, trophic index (TDI) increases downstream, from 28.8 in Site 1 to 96.2 in Site 5 (Table 3), which coincides with an increasing nutrient load. The fish community is the most restrictive BQE applied in the Llobregat River. Quality values using the IBICAT index only show near good quality status upstream dams (Site 1), where the fish quality class is moderate, and fish communities are dominated by native species (e.g., *Salmo trutta*, *Barbus haasi*). In the mid and lower Llobregat main watercourse, below dam flow regime regulation, and nearby the high urban areas, quality status is poor or bad due to the poor hydromorphological quality, habitat loss, riparian degradation, and high abundance of nonnative fish species found (e.g., *Lepomis gibbosus*, *Cyprinus carpio*, *Alburnus alburnus*, and *Phoxinus* sp.).

Regarding the chemical status, a total of 97 priority substances and group of substances (isomers, metabolites, etc.) have been analyzed in the Llobregat basin according to the 105/2008/EC Directive. Atomic fluorescence spectroscopy for mercury, inductively coupled plasma mass spectrometry for heavy metals, headspace extraction procedure for solvent substances, solvent extraction with simultaneous derivatization for pentachlorophenol, and solid-phase stirred bar extraction for the rest of organic compounds [55–57] are used in order to analyze priority substances and to set out the chemical status. All chemicals were also analyzed or confirmed using GC–MS according to the 2009/90/EC Directive. From these 97 substances, 42 are included in the Annex I of the 105/2008/EC Directive, while 55 remaining substances are required by Spanish-national laws, or are likely to be found in Catalan Rivers due to industrial or agricultural activities. Only substances with thresholds provided by the 105/2008 Directive (EQSs) were used for chemical status assessment. Values of heavy metals (lead, cadmium, mercury, and nickel), chlorinated solvents, pesticides (chlorine, phosphorus, and triazine), polycyclic aromatic hydrocarbons, and endocrine disruptors (nonylphenols, octilphenols, and brominated diphenyl ether compounds) are analyzed according to the EQS provided by the 105/2008/EC Directive.

The good chemical status is achieved in 83% of river water bodies (66 out of 80 water bodies) in the Llobregat basin (Table 2, Fig. 5b). A total of ten river water bodies do not achieve good chemical status. Unfulfilled environmental quality standards (EQS) are located close to industrial areas, in the lower part of Llobregat, Rubi, and Anoia Rivers. Also, additional unfulfilled quality standards exist in some small streams located in the mid-Llobregat basin, close to the Manresa city industrial area (riu d'Or). Pesticides and endocrine disruptors are the main substances responsible of quality standard failures. Endocrine disruptors are mainly found in the lower main Llobregat watercourse. A total of nine river water bodies do not achieve the quality standards in the Llobregat basin, from which four are located in the mid and lower Llobregat below Cardener River. The remaining high endocrine disruptor concentrations are found in the lower Anoia River, the Rubí stream, and riu d'Or, all located downstream of big industrialized zones. Nonylphenols (EQS: 0.3 μg/L) and octilphenols (EQS: 0.1 μg/L) are present with an average concentration of 0.4–0.8 μg/L. Similar endocrine disruptors concentrations have been found by other authors close to industrial areas, where similar compounds are mainly used in industrial processes [58]. Moreover, chlorinated pesticides, triazines, organophosphates, and some miscellaneous compounds are also found in the lower Llobregat basin close to industrial and agricultural areas. Pesticides do not achieve quality standards in five water bodies (6% of the total Llobregat River water bodies), and they are mainly located in the mid and lower Llobregat, the lower Anoia River, riu d'Or, and Rubí stream. Mostly trifluralin and also hexachlorocyclohexanes (lindane) are the main hazardous substances found over their EQS. Trifluralin is found in tributaries of lower Llobregat (Rubi stream) with an average concentration of 0.05 μg/L (slightly over its EQS value: 0.03 μg/L). Lindane is found in riu d'Or, in the mid-Llobregat River with an average value of 0.03 μg/L (EQS: 0.02 μg/L). Regarding triazines, organophosphates, and miscellaneous compounds, they do not achieve quality standards in two river water bodies, basically located in the lower Anoia River, and in Rubí stream, nearby industrial areas. Chlorpyrifos and chlorfenvinfos are the most detected compounds, though mainly found at low concentrations. Measured chlorpyrifos show an average concentration of 0.03–0.09 μg/L (EQS: 0.03 μg/L), whereas chlorfenvinfos are detected at 0.1–0.15 μg/L (EQS: 0.1 μg/L). Most pesticides have been detected slightly over the EQS values and their threshold detection. Therefore, they must be tentatively considered and later evaluated along time to be confirmed. Heavy metals are also found near industrialized areas in the lower Llobregat and Anoia Rivers. Nickel values are detected over the quality standard (EQS: 20 μg/L) in three water bodies (in lower Anoia and Llobregat Rivers), with an average value of 30 μg/L.

4 New Tools for Ecosystem Risk Assessment

Community-based monitoring approaches such as those described above, provide useful information on composition and structure, but do not provide clear insight into the impact of pressures of ecosystem functioning neither on specific pollution effects on river biota. Community-based indexes can only detect relatively strong effects that usually involve the eradication of one or several species from a particular site. Thus, they cannot diagnose low levels of ecological impairment caused by sublethal physiological effects. In relation to this, the development of innovative ecological assessment methods such as *i*n *s*itu *b*ioassays (ISBs), that employ caged species deployed at sites of concern, offer great potential for use in a tiered assessment scheme since they show good abilities to discriminate among chemical pressures [59]. The development of biomarkers in macroinvertebrate species has allowed identifying major pollutants with detrimental effects on river biota [23, 59]. ISBs per se can provide valuable information on functional processes such as changes in food processing and hence on trophic food webs. For example, the use of postexposure feeding inhibition responses of *Daphnia magna* transplanted across different locations along the Llobregat River basin has allowed characterizing "hot" sites whose water pollutants impair grazing rates [59]. Such detrimental effects may resolve in the elimination of filter feeders like tricopters. The use of a large set of biomarkers that involve several metabolic and detoxication paths may also be useful to indentify particular pollutants that are affecting river biota. Biomarkers can be measured in field collected organisms and then may inform us about the physiological state of organisms inhabiting those sites [19, 23]. If biomarkers are combined with the analysis of pollutants, it is possible to correlate specific effects with putative stressors [19, 21]. Nevertheless, one of the greatest advantages of the above-mentioned tools is its combination. When ISBs, biomarkers, and chemical analyses are combined and related with community-based indices, it is possible to discriminate and identify stressors that are affecting communities from those that are impairing key physiological functions of benthic species. Such an approach has been used in the Llobregat River to characterize the effect of habitat quality and water pollutants in river biota [19, 23, 59]. Furthermore, the use of multivariate tools, such as Principal Component (PCA) and Multivariate Partial Least Square Projections to Latent Structures regression analyses (PLS), allowed identifying specific stressors affecting the studied organisms. In the study of Barata et al. [23] latter reevaluated by Damásio et al. [19], caddisfly larvae of *Hydropsyche exocellata* were sampled from seven

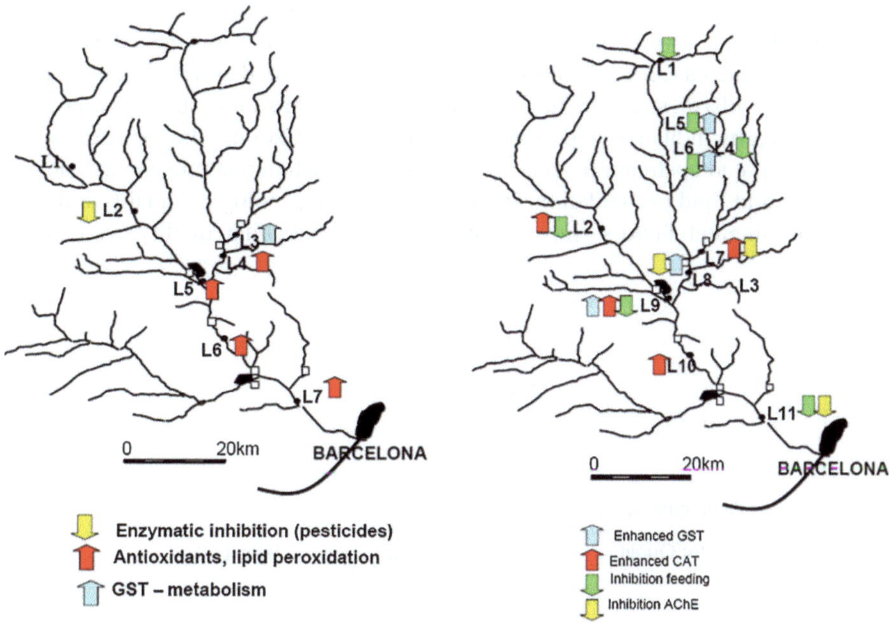

Fig. 6 Biomarker analysis in the Llobregat River. Samples were collected downstream of the main urban and industrial wastewater discharges. (**a**) Main results of *Hydropsyche exocellata* responses are superimposed to sites as *arrows* that show the direction of response compared to reference sites. Additional information can be found in [19, 23]. (**b**) Results of transplanted *Daphnia magna* individuals are superimposed to sites as *arrows* that show the direction of response compared to reference sites. Additional information can be found in [59]

locations in spring and summer receiving increasing levels of urban and industrial waste water discharges along the Llobregat River during spring and summer 2003 (Fig. 6a). Locations were selected to include aquatic communities in poor and good ecological state according to measured physicochemical water parameters and the analysis of benthic macroinvertebrate communities. Contaminant levels in water were determined in conjunction with antioxidant enzymes (superoxide dismutase, SOD; catalasa, CAT; glutathione peroxidase, GPx), a phase II enzyme (glutathione-*S*-transferase, GST) and lipid peroxide levels measured as thiobarbituric reactive species (TBARs), with the aim of investigating whether resident macroinvertebrate benthic species were responsive to changes in water quality. Metals such as Cu may also produce reactive oxygen species by redox cycling. The increment of reactive oxygen species alter the redox status of cells and antioxidant defenses: the enzyme SOD aid cells converting superoxide ions (O^{2-}) into hydrogen peroxide, catalasa, and glutathione peroxidase convert superoxides of hydrogen to water. Another set of enzymes, such as the GST, aid cells to eliminate contaminants by conjugating glutathione with contaminant metabolites generated by phase I enzymes. If production of reactive species is greater than their removal, tissue damage occurs. One marker of such tissue damage is the presence of lipid peroxidation levels.

The results of this study showed increased levels of pollutants downstream and enhanced activity levels of two (CAT and GST) out of the four tested enzymes, coupled with increased levels of lipid peroxidation measured as TBARs, indicating increasing levels of stress in the studied species toward downstream reaches or locations nearby industrial and urban areas (Fig. 6a). PCA on biomarker responses (Fig. 7) separated upstream from downstream sites, the latter having elevated levels of catalasa and lipid peroxidation. PCA on biomarkers, ecological quality indexes, and environmental factors showed that salinity and habitat quality were those

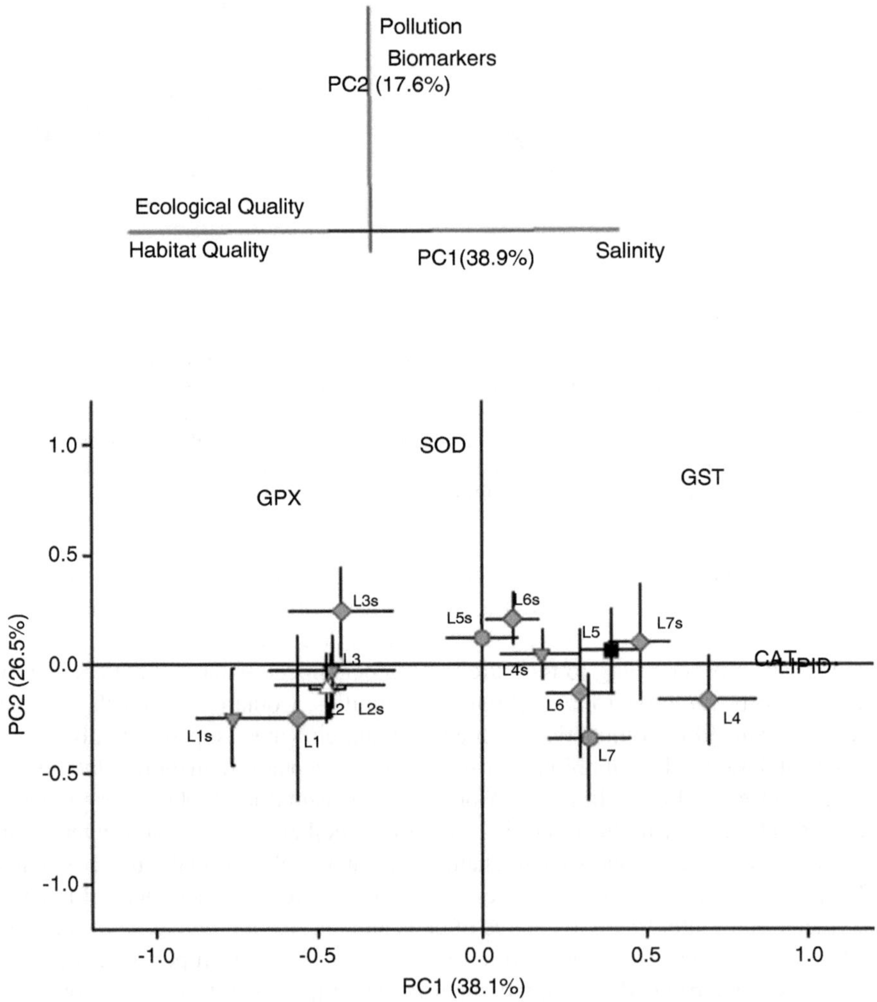

Fig. 7 PCA analysis results performed on biomarker responses. The means and 95% CI of site loadings are shown (*bottom graph*) together with the PCA results of environmental and biological variables (*top graph*). Additional information can be found in [19, 23]

factors affecting most macroinvertebrate water quality indexes, whereas pollutants did so to biomarker responses of *H. exocellata*. In resume these data indicate that the use of biomarker responses in *H. exocellata* allowed a complementary classification of water quality based on physiological stage of representative organisms rather than in whole community composition. These tools thus could be used in the future to complement currently used WFD biological indices.

In Fig. 6b we present data on a second study aimed to characterize environmental hazards of pollutants using transplanted *D. magna* individuals and determining individual (survival, feeding inhibition effects) and biomarkers [59]. This time biomarkers included the antioxidant enzyme catalasa, the metabolizing one GST and cholinesterasa, which is specifically inhibited by organophosphorous pesticides. The use of transplants using lab species allow to discard historical effects of adaptation to pollutants as well as to increase the number of sites since the presence of a particular species for sampling is not relevant. The results obtained also showed a clear deterioration of their ecological water quality parameters and benthic communities toward downstream reaches. In all but one location (L3) studied, transplanted organisms evidenced effects in at least one of the five measured responses in spring or summer. In 7 out of the 11 studied sites, significant effects were detected in at least two traits.

In two other studies, transplants and biomarkers were used to study specific effects of pollutants. *Daphnia magna* transplants and *H. exocellata* biomarkers were used to investigate potential effects of the application of the herbicide Herbolex (Aragonesas Agro, S.A., Madrid, Spain), which has glyphosate as active ingredient, to control giant reed (*Arundo donax*) [60]. Glyphosate is intended to be used in the Llobregat to eliminate foreign riparian vegetation and hence to restore autochthonous riparian vegetation. Just before and after application of glyphosate, *D. magna* individuals were deployed upstream and downstream the treated location and caddisfly larvae collected. The treated location was situated close to L10 according to Fig. 6b. Effects on benthic macroinvertebrate structure assemblages were also monitored. Measured glyphosate levels in river water following herbicide application were quite high (20–60 μg/L), with peak values of 137 μg/L after 3 days. After 12 days of its application, leaching of glyphosate from sprayed riverbanks was quite high in pore water (20–85 μg/L), but not in the river. Closely linked with the measured poor habitat and water physicochemical conditions, macroinvertebrate communities were dominated by taxa tolerant to pollution and herbicide application did not affect the abundance or number of taxa in any location. Nevertheless, only significant specific toxic effects on transplanted *D. magna* and field collected *H. exocellata* were observed. Effects included *D. magna* feeding inhibition and oxidative stress-related responses such as increased antioxidant enzyme activities related with the metabolism of glutathione and increased levels of lipid peroxidation.

In a more recent study founded by the Catalan Water Agency [61] biomarker responses of the caddisfly larvae *H. exocellata* were used to evaluate the effects of reclaimed wastewater on the ecological status of the lower part of Llobregat River, assessing if the introduction of reclaimed water during a period of low water flow (reclaimed water up to 50% of the final river flow) has any additional effect on physiological responses of the caddisfly larvae *H. exocellata*. The study was conducted close to site L11 of Fig. 6b. Again the discharge of reclaimed water did not affect the composition and abundance or the dominant taxa, consequently the ecological status measured using estructural indicators was similar before and after the introduction of treated water. Nevertheless, significant specific toxic effects on field collected *H. exocellata* larvae were observed using biomarkers probably due to the increase of chlorine levels in water together with ammonia and phosphate.

5 Conclusions

A great effort to enhance the quality status in the Llobregat basin has been carried out through sewage plant construction and habitat restoration by the Catalan Water Agency and several local institutions. That has been possible due to the high amount of information available on water quality and biological community composition, and chemical and bioassessment studies mainly provided by research centers and water authorities which have analyzed the quality status and biota in the Llobregat basin since long ago [3–5, 7, 9, 20, 23, 41]. The quality and abundance of such information have been a key element to fulfill the challenge to improve the Llobregat ecological status, and to establish a suitable monitoring program. The Llobregat River suffers a considerable ecological impact basically due to human pressures throughout the river basin [3, 6, 20]. The most important anthropogenic impacts within the Llobregat basin included salt mine activities, hydropower water diversion, and flow regime alteration by dams in headwaters and mid-Llobregat basin, together with urban and industrial sewage discharges mainly located downstream [22]. Some programs of measures have been progressively applied along time in order to mitigate such impacts. A total of 63 sewage treatment plants have been built in the last 25 years in order to reduce urban and industrial discharge impacts, and also salt runoff control has been set out around mine activities. Quality status has progressively enhanced and some chemical parameters have been reduced downstream. Ammonia concentration and, in general, nutrient loads decreased during the last decade in mid and lower Llobregat River. Such amelioration has allowed restoring biological communities, mainly diatoms and macroinvertebrates downstream, even close to the river mouth in wet periods. However, some anthropogenic pressures are still remaining and the Llobregat's biological status is not completely restored along the river. The high amount of small weirs and hydropower water diversion along Llobregat and Cardener Rivers, together with flow regime regulation by dams, riparian degradation, and eventual peak concentrations of nutrients and salts due to mining activities, result in a poor biological quality status in mid and lower Llobregat basin, where fish communities show the highest community alteration, with a high number of nonnative species appearing. Moreover, the high industrial concentration and urban discharges in mid and lower Llobregat River cause the detection of some priority substances and emergent pollutants (e.g., endocrine disruptors, heavy metals, pesticides, brominated flame retardants, drugs, and pharmaceuticals) [17–19, 23, 59], which all together increase the ecological threats.

Quality elements (macroinvertebrate, diatoms, and fish) reveal differences among human pressure impacts along the main Llobregat watercourse, from headwaters to the river mouth. Diatoms and macroinvertebrate communities show good quality in the headwaters and in the mid-Llobregat River. Downstream salt mines the macroinvertebrate community is altered, and non-tolerant taxa disappear downstream, but the biological quality status remain between moderate to good in wet periods when higher minimum flows reduce salt concentrations. The low condition of the fish fauna is related to the high habitat disturbances downstream the dams, that basically cause hydromorphological alterations, low environmental flows, habitat loss, riparian degradation, and the abundance of nonnative species. While the Llobregat headwaters and middle reaches have a good biological quality according to macroinvertebrate and diatom communities, fish quality indices indicate a poor and bad quantity status. Downstream the salt mines and in the area or large concentration of urban and industrial activities, close to the river mouth, all indices show low values. Consequently, additional analytical and biological studies have been implemented in the middle and lower Llobregat course to identify specific pressures. Chemical risk assessment studies have established the impacts of emerging contaminants, such as surfactants, pharmaceuticals, and pesticides, on local biota [17–19, 23, 59]. Biomarkers also inform us of the actual ecological status when used together with community indices, though their results are sometimes difficult to interpret [62]. It is problematic to determine whether a single biomarker response is indicator of impairment or is part of the homeostatic response, indicating that an organism is successfully dealing with the exposure [62]. However, the use of large set of biomarkers representing several metabolic paths overcome problems of interpretation and characterize the physiological effects of pollutants. The results presented herein also demonstrate the usefulness of biomarkers in detecting subtle changes of water quality in locations with deteriorated benthic communities. This is mainly due to the resilience of tolerant species assemblages to change and the great phenotypic plasticity of tolerance species such as *H. exocellata* to cope with stress. Indeed our results showed that *H. exocellata* is able to adjust quite rapidly its physiological mechanisms of defense to tolerate chemical inputs, such as glyphosate, salinity, and water flow changes. On the other hand, the use of transplants of laboratory sensitive species such as those of *D. magna* may also allow standardizing field assays. Such field assays are experimentally robust and reliable. Biomarkers should be considered in the future as tools for implementation of the WFD, in addition to community indices [63, 64]. By 2020, EU member states will have to improve the quality of their surface waters and report those changes to the WFD. In this sense, the use of markers sensitive to water pollution may provide useful information on small changes in ecological quality especially in the threshold between moderate and good.

Acknowledgments Main information on quality status, biological data, and chemical values has been obtained from the Catalan Water Agency through its monitoring program. Additional and useful historical data and biomonitoring results over time (since 1994) have been provided by the FEM research group (Department of Ecology, University of Barcelona). Historical surveillance data since 1994 were possible due to financial support provided by the Diputació de Barcelona.

References

1. Llasat MC, Barriendos M, Rodriguez R, Martín-Vide J (1999) Evolución de las inundaciones en Cataluña en los últimos quinientos años. Ingeniería Agua 6(4):353–362
2. Kuster M, López de Alda MJ, Hernando MD, Petrovic M, Alonso JM, Barceló D (2008) Analysis and occurrence of pharmaceuticals, estrogens, progestogens and polar pesticides in sewage treatment plant effluents, river water and drinking water in the Llobregat River basin (Barcelona, Spain). J Hydrol 358(1–2):112–123
3. Prat N, Rieradevall M (2006) 25-Years of biomonioring in two mediterranean streams (Llobregat and Besòs basins, NE Spain). Limnetica 25(1–2):541–550
4. Prat N, Puig MA, González G (1983) Predicció i control de la qualitat de les aigües dels rius Besós i Lobregat. II. El poblament faunístic i la seva relació qualitat-aigües. Col. Monografies., vol 9. Diputació de Barcelona, Barcelona, 164 pp
5. Prat N, Puig MA, González G, Tort MJ, Estrada M (1984) The Llobregat: a Mediterranean river fed by the Pyrenees. In: Whitton BA (ed) Ecology of European rivers. Blackwell, Oxford, pp 527–552
6. Prat N (1991) Present trends in river studies. Oecol Aquat 10:1–12
7. Prat N, Rieredevall M, Munne A, Chacon G (1996) La qualitat ecològica de les aigües del Besòs i el Llobregat. Diputació de Barcelona. Servei de Medi Ambient. Col. Estudis de la qualitat ecològica dels rius, vol 1, 102 pp
8. Prat N, Vila-Escale M, Bonada M, Casanovas-Berenguer R, Punti T, Sola C, Jubany J, Miralles M, Ordeix M, Acosta R, Rios B, Andreu R, Rieradevall M (2005) La qualitat ecològica del Llobregat, el Besòs i el Foix. Informe 2003. Barcelona: Diputació de Barcelona. Servei de Medi Ambient. Col. Estudis de qualitat ecològica dels rius, vol 13. Edición CD-Rom
9. Muñoz I, Prat N (1994) A comparison between different biological water quality indexes in the Llobregat basin (NE Spain). Verh Int Verein Limnol 1(25):1945–1949
10. Queralt R (1982) La calidad de las aguas de los ríos. Tecnol Agua 4:49–57
11. Carafa R, Fanggiano L, Real M, Munné A, Ginebreda A, Guasch H, Flo M, Tirapu L, Carsten von der Ohe P (2011) Water toxicity assessment in Catalan rivers (NE Spain) using Species Sensitivity Distribution and Artificial Neural Networks. Sci Total Environ 409:4269–4279
12. Munné A, Tirapu L, Solà C, Olivella L, Vilanova M, Ginebreda A, Prat N (2012). Comparing Chemical and Ecological Status in Catalan rivers. Analysis of river quality status following the Water Framework Directive. In: The handbook of Environmental Chemistry. Emerging and Priority Pollutants in Rivers: Bringing science into River Management Plans. H. Guasch et al. (eds.), 19:243–266
13. Munné A, Prat N (2009) Use of macroinvertebrate-based multimetric indices for water quality evaluation in Spanish Mediterranean rivers: an intercalibration approach with the IBMWP index. Hydrobiologia 628:203–225
14. Sabater S, Tornés E, Leira M, Trobajo R (2003) Anàlisi de viabilitat i proposta d'indicadors fitobentònics de la qualitat de l'aigua per als cursos fluvials de Catalunya (Muga, Fluvià, Ter i Daró). Documents tècnics de l'Agència Catalana de l'Aigua, 113 pp
15. Sostoa A, Caiola N, Casals F (2004) A new IBI (IBICAT) for local application of the E.U. Water Framework Directive. In: Fifth ecohydraulics conference, Madrid, September 2004

16. Munné A, Solà C, Prat N (2006) Estado ecológico de los ríos en Cataluña. Diagnosis del riesgo de incumplimiento de los objetivos de la Directiva Marco del Agua. Tecnol Agua 273:30–46
17. Ginebreda A, Muñoz I, López de Alda M, Brix R, López-Doval J, Barceló D (2010) Environmental risk assessment of pharmaceuticals in rivers. Relationships between hazard indexes and aquatic macroinvertebrate diversity indexes in the Llobregat River (NE Spain). Environ Int 36:153–162
18. Ricart M, Guasch H, Barcelo D, Brix R, Conceicao MH, Geiszinger A, de Alda MJL, Lopez-Doval JC, Munoz I, Postigo C, Romani AM, Villagrasa M, Sabater S (2010) Primary and complex stressors in polluted Mediterranean rivers: pesticide effects on biological communities. J Hydrol 383:52–61
19. Damásio J, Fernández-Sanjuan M, Sánchez-Avila J, Lacorte S, Prat N, Rieradevall M, Soares AMVM, Barata C (2011) Multi-biochemical responses of benthic macroinvertebrate species as a complementary tool to diagnose the cause of community impairment in polluted rivers. Water Res 45:3599–3613
20. Muñoz I, López-Doval JC, Ricart M, Villagrasa M, Brix R, Geszinger A, Ginebreda A, Guasch H, López de Alda M, Romaní AM, Sabater S, Barceló D (2009) Bridging levels of pharmaceuticals in river water with biological community structure in the Llobregat river basin (NE Spain). Environ Toxicol Chem 28:2706
21. Damasio J, Navarro-Ortega A, Tauler R, Lacorte S, Barcelo D, Soares A, Lopez MA, Riva MC, Barata C (2010) Identifying major pesticides affecting bivalve species exposed to agricultural pollution using multi-biomarker and multivariate methods. Ecotoxicology 19: 1084–1094
22. ACA – Agència Catalana de l'Aigua (2005) Caracterització de les masses d'aigua i anàlisi del risc d'incompliment dels objectius de la Directiva Marc de l'Aigua (2000/60/CE) a Catalunya. Agència Catalana de l'Aigua. Departament de Medi Ambient i Habitatge de la Generalitat de Catalunya, Octubre 2005, 860 pp
23. Barata C, Lekumberri I, Vila-Escalé M, Prat N, Porte C (2005) Trace metal concentration, antioxidant enzyme activities and susceptibility to oxidative stress in the tricoptera larvae *Hydropsyche exocellata* from the Llobregat riven basin (NE Spain). Aquat Toxicol 74:3–19
24. ACA – Agència Catalana de l'Aigua (2011) Estat de les Msses d'Aigua a Catalunya. Resultats del Programa de Seguiment i Control (Dades 2007-2010). Departament de Territori i Sostenibilitat, Generalitat de Catalunya, Octubre 2011, 63 pp
25. Prat N, Munne A (2003) Water use and quality and stream flow in a Mediterranean stream. Water Res 34(15):3876–3881
26. Boix D, García-Berthou E, Gascón S, Benejam L, Tornés E, Sala J, Benito J, Munné A, Solà C, Sabater S (2010) Response of community structure to sustained drought in Mediterranean rivers. J Hydrol 383:135–146
27. Benejam L, Angermeier PL, Munné A, García-Berthou E (2010) Assessing effects of water abstraction on fish assemblages in Mediterranean streams. Freshw Biol 55:628–642
28. Aparicio E, Vargas MJ, Olmo JM, de Sostoa A (2000) Decline of native freshwater fishes in a Mediterranean watershed on the Iberian Peninsula: a quantitative assessment. Environ Biol Fishes 59:11–19
29. Harrison IJ, Stiassny MLJ (1999) The quiet crisis: a preliminary listing of the freshwater fishes of the World that are extinct or "missing in action". In: MacPhee (ed) Extinctions in near time. Kluwer Academic/Plenum, New York, pp 271–331
30. Andreu J, Pino J, Rodríguez-Labajos B, Munné A (2011) Avaluació de l'estat i el risc d'invasió per espècies exòtiques dels ecosistemes aquàtics de Catalunya. Agència Catalana de l'Aigua, Departament de Territori i Sostenibilitat, Generalitat de Catalunya, 97 pp
31. Bloch H (1999) European water policy facing the new millennium: the Water Framework Directive. In: Assessing the ecological integrity of running waters, Vienna, pp 9–11
32. Allan IJ, Vranaa B, Greenwooda R, Millsb GA, Knutssonc J, Holmbergd A, Guiguese N, Fouillace AM, Laschif S (2005) Strategic monitoring for the European Water Framework Directive. Trends Anal Chem 25(7):704–715

33. Coquery M, Morin A, Bécue A, Lepot B (2005) Priority substances of the European Water Framework Directive: analytical challenges in monitoring water quality. Trends Anal Chem 24(2):117–127
34. Sandin L, Hering D (2004) Comparing macroinvertebrate indices to detect organic pollution across Europe: a contribution to the EC Water Framework Directive intercalibration. Hydrobiologia 516(1–3):55–68
35. Bonada N, Prat N, Resh VH, Statzner B (2006) Developments in aquatic insect biomonitoring: a comparative analysis of recent approaches. Annu Rev Entomol 51:495–523
36. Kelly MG, Cazaubon A, Coring E, Dell'Uomo A, Ector L, Goldsmith B, Guasch H, Hürlimann J, Jarlman A, Kawecka B (1998) Recommendations for the routine sampling of diatoms for water quality assessments in Europe. J Appl Phycol 10(2):215–224
37. Szoszkiewicz K, Ferreira T, Korte T, Baattrup-Pedersen A, Davy-Bowker J, O'Hare M (2006) European river plant communities: the importance of organic pollution and the usefulness of existing macrophyte metrics. Hydrobiologia 566(1):211–234
38. Pont D, Hugueny B, Rogers C (2007) Development of a fish-based index for the assessment of river health in Europe: the European Fish Index. Fish Manag Ecol 14(6):427–439
39. Munné A, Prat N (2011) Effects of Mediterranean climate annual variability on stream biological quality assessment using macroinvertebrate communities. Ecol Indic 11:651–662
40. Margalef R (1955) Organismos indicadores en la Limnología. Instituto Forestal de Inv. Exper., 308 pp
41. Margalef R (1969) El concepto de polución en limnología y sus indicadores biológicos. Agua 7:105–133
42. Muñoz I, Prat N (1998) Effects of water abstraction and pollution on macroinvertebrate community in a mediterranean river. Limnetica 12(1):9–16
43. Jáimez-Cuéllar P, Vivas S, Bonada N, Robles S, Mellado A, Álvarez M, Avilés J, Casas J, Ortega M, Pardo I, Prat N, Rieradevall M, Sáinz-Cantero CE, Sánchez-Ortega A, Suárez ML, Toro M, Vidal-Albarca MR, Zamora-Muñoz C, Alba-Tercedor J (2002) Protocolo GUADALMED (PRECE). Limnetica 21(3–4):187–204
44. Alba-Tercedor J, Jáimez-Cuéllar P, Álvarez M, Avilés J, Bonada N, Casas J, Mellado A, Ortega M, Pardo I, Prat N, Rieradevall M, Robles S, Sáinz-Cantero CE, Sánchez-Ortega A, Suárez ML, Toro M, Vidal-Albarca MR, Vivas S, Zamora-Muñoz C (2002) Caracterización del estado ecológico de los ríos mediterráneos ibéricos mediante el índice IBMWP (antes BMWP'). Limnetica 21(3–4):175–185
45. Perrée I, Rieradevall M, Prat N, Martin J, Céspedes R (2010) Cambios en el estado ecológico de tres ríos producidos por el vertido de depuradoras. Tecnol Agua 320:21–29
46. Prat N, Ward JV (1994) The tamed river. In: Margalef R (ed) Lymnology now. Elsevier Science, London, pp 219–236
47. European Commission (2003) Common implementation strategy for the Water Framework Directive (2000/60/EC). Working Group REFCOND. Guidance document nº 10. Rivers and lakes – typology, reference conditions and classification systems
48. Haury J, Peltre MC, Trémolières M, Barbe J, Thiébaut G, Bernez I, Daniel H, Chatenet P, Haan-Archipof G, Muller S (2006) A new method to assess water trophy and organic pollution – the Macrophyte Biological Index for Rivers (IBMR): its application to different types of river and pollution. Macrophytes in aquatic ecosystems: from biology to management. Dev Hydrobiol 190(2):153–158
49. Buffagni A, Erba S, Cazzola M, Murria-Bligh J, Soszka H, Genomi P (2006) The Star common metrics approach to the WFD intercalibration process: full application for small, lowland rivers in three European countries. Hydrobiologia 566:379–399
50. Hering D, Johnson R, Kramm S, Schmutz S, Szoszkiewicz K, Verdonschot PFM (2006) Assessment of European streams with diatoms, macrophytes, macroinvertebrates and fish: a comparative metric-based analysis of organism response to stress. Freshw Biol 51(9): 1757–1785

51. Thiébaut G (2006) Aquatic macrophyte approach to assess the impact of disturbances on the diversity of the ecosystem and on river quality. Int Rev Hydrobiol 91(5):483–497
52. Moreno JL, De las Heras J, Prat N, Rieradevall M (2008) Evaluación del estado trófico de tres cuencas interiores de Cataluña (Foix, Besòs y Llobregat) mediante la vegetación acuática: aplicación de un índice trófico (IVAM-FBL). Limnetica 27(1):107–118
53. Segurado P, Santos JM, Pont D, Melcher AH, Jalon DG, Hughes RM, Ferreira MT (2011) Estimating species tolerance to human perturbation: expert judgment versus empirical approaches. Ecol Indic 11:1623–1635
54. Claudi R, Leach JH (eds) (1999) Nonindigenous freshwater organisms: vectors, biology, and impacts. Lewis, Boca Raton, 464 pp
55. UNE-EN ISO 10301 (1997) Water quality. Determination of highly volatile halogenated hydrocarbons. Gas-chromatographic methods
56. Lee H, Weng L, Chau AS (1984) Chemical derivatization analysis of pesticides residues. VIII. Analysis of 15 chlorophenols in natural water by in situ acetylation. J Assoc Off Anal Chem 67(4):789–794
57. León VM, Llorca-Pórcel J, Álvarez B, Cobollo MA, Muñoz S, Valor I (2006) Analysis of 35 semivolatile compounds in water by stir bar sorptive extraction-thermal desorption-gas chromatography-mass spectrometry. Part II: method validation. Anal Chim Acta 558:261–266
58. Barceló D, Petrovic M (2007) Under the analytical spotlight, contaminants emerge: report on the 2nd EMCO Workshop. Emerging contaminants in wastewaters: monitoring tools and treatment technologies. Belgrade (Serbia), 26 and 27 April 2007. Trends Anal Chem 26: 647–649
59. Damasio J, Tauler R, Teixido E, Rieradevall M, Prat N, Riva MC, Soares A, Barataa C (2008) Combined use of *Daphnia magna* in situ bioassays, biomarkers and biological indices to diagnose and identify environmental pressures on invertebrate communities in two Mediterranean urbanized and industrialized rivers (NE Spain). Aquat Toxicol 87:310–320
60. Puertolas L, Damasio J, Barata C, Soares A, Prat N (2011) Evaluation of side-effects of glyphosate mediated control of giant reed (*Arundo donax*) on the structure and function of a nearby Mediterranean river ecosystem. Environ Res 110:556–564
61. Prat N, Rieradeval M, Barata C, Munné A (2012) The combined use of metrics of biological quality and biomarkers as a tool to detect the effects of tertiary treated water on macroinvertebrate assemblages in the lower part of a polluted Mediterranean river (Llobregat, NE Spain) (submitted)
62. Forbes VE, Palmqvist A, Bach L (2006) The use and misuse of biomarkers in ecotoxicology. Environ Toxicol Chem 25:272–280
63. Damásio JB, Barata C, Munne A, Ginebreda A, Guasch H, Sabater S, Caixach J, Porte C (2007) Comparing the response of biochemical indicators (biomarkers) and biological indices to diagnose the ecological impact of an oil spillage in a Mediterranean river (NE Catalunya, Spain). Chemosphere 66:1206–1216
64. Mills GA, Greenwood R, Gonzalez C (2007) Environmental monitoring within the Water Framework Directive (WFD). Trends Anal Chem 26:450–453

Wastewater Reuse in the Llobregat: The Experience at the Prat de Llobregat Treatment Plant

Sandra Pérez, María-Eugenia Rubiano, Antoni Ginebreda, Cristina Postigo, Rebeca López-Serna, Juan Blanco, Victoria Osorio, Miren López de Alda, Mira Petrović, Jordi J. Pastor, Lleonard Matia, Jordi Martín-Alonso, Antoni Munné, Joan Jofre, Francisco Lucena, Míriam Agulló-Barceló, Narcís Prat, and Damià Barceló

S. Pérez (✉), A. Ginebreda, C. Postigo, R. López-Serna, J. Blanco, V. Osorio and M.L. de Alda
Department of Environmental Chemistry, IDAEA-CSIC, c/Jordi Girona 18-26, 08034 Barcelona, Spain
e-mail: spsqam@idaea.csic.es

M.-E. Rubiano, J. Jofre, F. Lucena and M. Agulló-Barceló
Faculty of Biology, Department of Microbiology, University of Barcelona, Avinguda Diagonal 643, 08028 Barcelona, Spain

M. Petrović
Catalan Institute for Water Research (ICRA), Edifici H2O, Parc Científic i Tecnològic de la Universitat de Girona, c/Emili Grahit 101, 17003 Girona, Spain

Catalan Institution for Research and Advanced Studies (ICREA), Passeig Lluis Companys 23, 08010 Barcelona, Spain

J.J. Pastor
Water Idea, Rambla Catalunya 123, 1-2, 08008 Barcelona, Spain

L. Matia and J. Martín-Alonso
Water Quality Department, Aigües de Barcelona, Avda. General Batet 5-7, 08028 Barcelona, Spain

A. Munné
Agència Catalana de l'Aigua (ACA), c/Provença 204-208, 08036 Barcelona, Spain

N. Prat
Department of Ecology, Grup de recerca F.E.M. (Freshwater Ecology and Management), University of Barcelona, Diagonal 645, 08028 Barcelona, Spain

D. Barceló
Department of Environmental Chemistry, IDAEA-CSIC, c/Jordi Girona 18-26, 08034 Barcelona, Spain

Catalan Institute for Water Research (ICRA), Edifici H2O, Parc Científic i Tecnològic de la Universitat de Girona, c/Emili Grahit 101, 17003 Girona, Spain

S. Sabater et al. (eds.), *The Llobregat: The Story of a Polluted Mediterranean River*, Hdb Env Chem (2012) 21:327–346, DOI 10.1007/698_2012_151,

Abstract Water scarcity is a consequence of both natural and anthropogenic factors including highly variable temporal and spatially heterogenic distribution of precipitations, growing populations, increasing water demand particularly for agriculture, and the widespread contamination of water resources by a plethora of organic and inorganic contaminants. In the countries bordering the Mediterranean Sea, water is a renewable resource relatively scarce during certain periods. Therefore, sustainable management of water resources is of utmost importance in these countries. In Catalonia a Mediterranean region located in the northeast of Spain, the reuse of tertiary treated wastewater is evaluated as alternative resource for water supply. On this basis, several campaigns on water reuse were conducted by the Catalan Water Agency in order to artificially enhance the flow along the river. In fall of 2009, the impact of tertiary treated effluent discharge on the ecology (pathogens and indicators, macroinvertebrate community assemblages, and biomarkers) and water quality of the river (priority and emerging pollutants: polar pesticides, illicit drugs, estrogens and pharmaceuticals as well as on the pathogens and microbial indicators of faecal contamination) and consequently to the drinking water supplies was evaluated in the lower stretch of the Llobregat River located in the vicinity of the town of Barcelona (NE Spain). The key findings of this study are reported in this book chapter. Chemical parameters were not significantly affected by the reclaimed water discharge with the exception of a slight increase in the ammonium concentration, conductivity, and TOC. Concerning priority substances, three pesticides increased in concentration but only diazinon exceed the quality threshold for human supply. Regarding the ecotoxicological assessment of illicit drugs and pharmaceuticals using algae for the toxicity value, differences between river upstream and downstream to the discharge of the treated tertiary wastewater were detected. For the other organisms the differences were imperceptibles. No perceptible effects were detected either in the ecological status of the river or in the load of pathogens and fecal indicators of the river water. In conclusion, the use of tertiary treated effluent for water reuse did not produce important alterations on river water quality downstream of the reclaimed water discharge.

Keywords Ecology, Emerging pollutants, Pathogens, Water quality, Water reuse

Contents

1 Introduction: Water Scarcity in the Mediterranean Region

Global hydrological change is a consequence of climate change including global warming and changes in rain events, increase of agricultural impacts, and also anthropogenic changes in the morphology of the rivers like water transfers and river engineering. One of the most critical consequences is caused by water scarcity in extended drought periods. This may result in changes in the hydrologic resources, in carbon and nutrients transfer as well as their storage affecting the biodiversity, water quality including pollution by chemicals, and river ecosystem functioning, sediment supply to ocean, and also economic sectors that use and depend on water such as agriculture, tourism, industry, energy, and transport. The lack of water is of growing concern, and it attracts attention worldwide particularly in the Mediterranean regions where water shortages are generally ascribable to a large extent to growing demand regardless of the limited renewable water resources of irregular and unequal quality. Mediterranean rivers are characterized by important fluctuations in flow rates, with typical low winter and summer discharges and periodical floods in spring and autumn. In addition, some rivers in the Mediterranean basin present heavy contamination pressures from extensive urban, industrial and agricultural activities which translate into generally higher contamination.

One of the regions included in the Mediterranean basin that suffers from periodic water scarcity and eventual droughts is Catalonia (NE Spain). The observed tendency in Catalonia indicates that water demands are close to available water resources. Moreover, available water resources are expected to decrease in the medium and longer run. Undoubtedly, demands during the last decade have been higher than they were during the historic droughts of the 1940s or 1970s. Nowadays, in Catalan basins, the total water demand, including domestic supply, industrial activities, and irrigation, amounts to 3,123 hm^3 per year. Under the pressure of water scarcity and the expected increasing occurrence of drought events in the Mediterranean area [1], measures like water reclaim and reuse, proposed to provide alternative resources, are gaining major relevance [2]. Such measures are doubtlessly necessary from the point of view of resource sustainability and are usually promoted by water authorities. The reutilization of treated wastewater aims to be introduced in the future Spanish National Plans for Water Management and must be a key component of the new environmental policy that promotes a new water culture, trying to enhance the exploitation of nonconventional resources while ensuring the physical, chemical, and bacteriological water quality. In Spain, the legal basis for the regulation of the quality standards for reclaimed water for different applications is established by the Real Decreto (Royal Decree) 1620/2007 [3]. It defines the concept of reuse, introduces the regenerated water designation, determines the qualifications necessary to carry out the activity of water reuse, procedures to obtain the demanded concession required by this law, and also defines permitted applications of reclaimed water and quality requirements in each case. It also defines compulsory minimum quality criteria for the use of the regenerated waters according to uses.

In 2010, the Catalan Water Agency released a Water Management Plan according to the Water Framework Directive (2000/60/EC) requirements. This new water plan introduces a provision for additional water sources for upcoming years, such as reclaimed water for several uses. However, this plan requires additional plans for water demand management based on the type of water usage, one of which is the Drought Management Plan (DMP). This plan must set up a policy of risk management through prevention and mitigation of drought effects. DMP is currently being developed and provides scenarios in which measures are applied to reduce demands progressively and activate alternative water sources. Moreover, the Catalan Water Reuse Programme (CWRP) has recently been enabled by the Catalan Water Agency in order to quantify water reuse possibilities and establish normative rules in the Catalan River Basin District scenario. Reuse of treated wastewater is promoted and incorporated into integrated resource planning, known as direct or planned reuse of reclaimed water, and is viewed as an alternative source for groundwater and surface water withdrawals. Water treated at wastewater treatment plants (WWTP) and then through further additional or complementary processes (tertiary treatments) in connection with reclamation has the sanitary and physicochemical quality required for certain uses. Two types of tertiary treatments of the effluents are usually applied in Catalan WWTPs depending on the end use of the water. The basic treatment encompasses coagulation, flocculation, filtration, post (additional) disinfection, and oxygen saturation. The advanced one includes additional treatment by ultrafiltration and reverse osmosis [4]. For instance, the water reclamation station "El Prat," located in the metropolitan area of the city of Barcelona, provides water for different purposes: After the basic treatment the water can be utilized for irrigation of agricultural land as well as for golf courses in the surroundings but also for industrial use. Furthermore, part of the treated wastewater is returned to the river and near wetlands to maintain minimum flow and suitable environmental water levels. At the same time, a part of the treated wastewater is directed to a reverse osmosis plant (advanced treatment) for further treatment and the water is pumped into the underground to avoid salt water from the sea to seep into the groundwater resources.

The CWRP is expected to treat 229 hm^3/year of water corresponding to the production (excluding waste) of 204 hm^3/year of reclaimed water. The Program of Urban Wastewater Treatment enabled by the Catalan Water Agency foresees that a total of 720 hm^3/year of wastewater will be treated in Catalonia on the horizon 2015. Therefore, 31% of the total wastewater treated flow will be processed on reclamation facilities or directly reused after applying proper measures. The volume of reused water is foreseen to be achieved as a sum of three components: reuse already in service (51 hm^3/year), an ever-increasing utilization of existing reuse facilities, and, finally, entry into service of new facilities under this program [5].

2 Reclaimed Wastewater Experiment in the Llobregat River

The Llobregat River (Catalonia, NE Spain) is a good example of a Mediterranean river, suffering from low flows during normal conditions (~5 m^3/s) and extraordinary peak events (maximum recorded of 2,500 m^3/s) that periodically reset the system. In addition, the river receives the effluent discharges of 63 WWTPs. The Llobregat and Ter rivers are the most important water supply system for Barcelona and nearby cities, which requires about 525 hm^3 from reservoirs to meet all demands. Of these demands, 400 hm^3 are consumed by urban uses. However, dams located in Llobragat and Ter have a maximum storage capacity of 612 hm^3 (including flood management provisions), and they are not able to store a sufficient volume of water so as to ensure the supply in dry years. Taking into account that 80% of the time during the past 40 years reservoirs were above 70% of their maximum capacity (ACA 2009), it is questionable if the most appropriate solution to the problem would be a permanent increase in total reservoir volume, or the use of external sources, especially given the high cost and elevated environmental impacts. Instead, rapidly implementable, temporary solutions such as reclaimed water and water exchange appear to be the better choice. During the years 2007 and 2008, the Llobregat River basin experienced a severe drought which affected the water supply of drinking water facilities. The Catalan Water Agency (ACA) implemented a reuse water plan using reclaimed water from the Barcelona WWTP, which usually discharges into the sea. The objective was to maintain the river flow rate in the lower Llobregat River. Before doing so, several pilot studies were carried out. A first attempt was set up in autumn 2008 (first campaign) and presented many problems in controlling the degree of actual dilution at all times since it rained frequently during the experiment. The second exploratory test (second campaign) was set up in autumn 2009; this is the experiment described in this chapter where the effects of the tertiary wastewater on the ecology and water quality of the river and consequently to the drinking water supplies were evaluated [5–7]. This study was carried out in the lower part of the Llobregat River between the cities of Molins de Rei (site R0) and Sant Joan Despí (site R2) [6]; the latter being particularly important because the intake of an important waterworks supplying drinking water to the city of Barcelona is located there. Water from the WWTP tertiary treatment of El Prat de Llobregat (site RW2) was pumped upstream ca. 16.6 km and discharged into the river at 0.2 km downstream of site R0 (Fig. 1). This test was carried out for 23 days, from October 29 to November 20 of 2009, and different dilutions between river flow and reclaimed water were tested 3:1 ($Q_{RW2}/Q_{R2} = 0.25$), 2:1 and 1:1 dilutions). At the R0 and R1 river sites, the main water-quality parameters affecting human consumption were analyzed according to the Directive 98/83/EC. In addition, other specific analyses were carried out in order to assess water supply and aquatic ecosystem damage. Three different parameters were evaluated in the river: (1) biological quality using macroinvertebrates evaluating biological indices and biomarkers responses of the caddisfly larvae *Hydropsyche exocellata* [7], (2) microbiological parameters (pathogens and

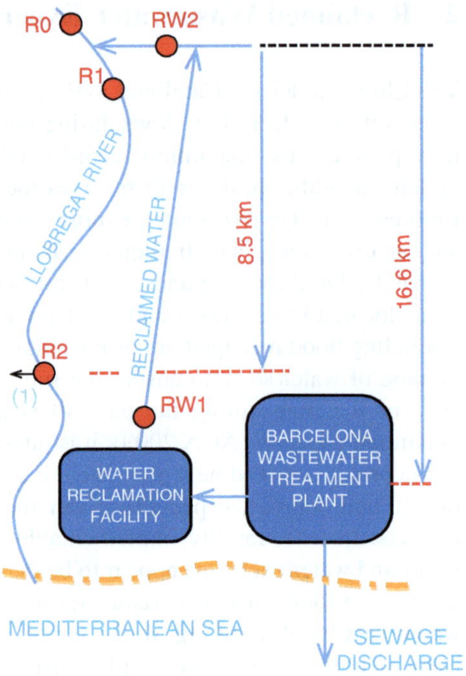

Fig. 1 Sampling points location in the lower stretch of the Llobregat River. *R0* Llobregat River at Molins de Rei; *R1* Llobregat River at Sant Joan Despí; *R2* WWTP El Prat de Llobregat, RW1 and RW2 are located in the pipe which transports reclaimed water

indicators [8]), and (3) occurrence of priority substances and emerging pollutants (pharmaceuticals, illicit drugs and estrogens, [9]). Several river points were sampled for the evaluation of the three parameters: upstream (R0) and downstream (R1 and R2) of the reclaimed water discharge site (Fig. 1) and also RW1 and RW2 sites for microbiological analysis, in order to detect the effect of reclaimed water transport on the concentration of microbiological indicators and pathogens. Microorganisms tested were: *Escherichia coli*, spores of sulfite-reducing clostridia, somatic coliphages, cytopathogenic enteroviruses, as well as total and infectious *Cryptosporidium* oocysts.

3 Chemical Analysis

In the last century, many contaminants have been detected in the aquatic environment and nowadays are considered classic pollutants like PCBs, PAH, some pesticides, etc. and included in the environmental regulatory lists. Recently, new contaminants have emerged as new contaminants, but most could have been in the environment for ages, and have remained undetected until recently. These are the emerging contaminants which are defined as compounds that are currently not covered by existing regulations of water quality (neither included in routine monitoring programs) and possible candidates for future regulation depending on research on their (eco)toxicity, potential health effects, public perception, and on monitoring data regarding their occurrence in the various environmental compartments [10, 11]. They include a diverse group of compounds including pharmaceuticals, illicit drugs, personal care products, steroids and hormones, surfactants, perfluorinated compounds, flame-retardants, industrial additives, and agents and gasoline additives as well as their transformation products [12]. Among these significant environmental trace pollutants, pharmaceuticals emerged as particularly relevant, due to several facts: (1) worldwide and continuous increase on their consumption and on their subsequent input into the environment; (2) design to be bioactive compounds which are able to cause potential effects on living organisms; (3) the possibility of antagonistic-synergistic interactions, for instance, growth inhibition of algae due to the simultaneous occurrence of a cocktail of pharmaceuticals [13].

Estrogens, both natural (e.g., estradiol, estrone, estriol, and their sulfate and glucuronide conjugates) and synthetic (e.g., ethynyl estradiol, used mainly as contraceptive), stand up for being aquatic contaminants present at very low concentrations (usually low ng/L or pg/L) but often sufficient to exert estrogenic effects, such as feminization and hermaphroditism, in aquatic organisms due to their very high estrogenic potency [14].

Illicit drugs (e.g., cocaine, heroin, cannabis, and their metabolites) represent a class of compounds recently discovered as emerging contaminants whose effects in the aquatic environment are still unknown [15]. All three compound classes were analyzed by online solid phase extraction–liquid chromatography–electrospray–tandem mass spectrometry (SPE–LC–ESI–MS/MS) following previously described methods [16–18].

3.1 *Priority Contaminants and Chemical Parameters*

The (physico-)chemical parameters ammonium concentration, conductivity, and TOC were monitored in the Llobregat River [6]. A slight increase in the river after the reclaimed water discharge was observed for all three parameters. Values for conductivity of 1,364 ± 262 μS/cm were measured in the site R0 while conductivity ranged from 2,231 to 2,765 μS/cm in R1 and R2 sites. In the same way, after the reclaimed water discharge, TOC concentrations also increased to some extent. TOC concentration varied from 4.23 mg/L in the site R0 to 8.5 mg/L in the R2 site. This is consistent with the organic load discharged by the water reclamation facility which doubled to the TOC concentration in the river [6]. Also, the concentrations of the organic target analytes increased to some degree. For instance, the concentrations of some pesticides such as terbutylazine (0.03 μg/L) and terbutryn (0.03 μg/L) were above the detection thresholds, but remained below the drinking water quality standards (Directive 98/83/EC) (<0.1 μg/L) [6]. However, the concentration of diazinon (maximum values of 0.24–0.39 μg/L) in surface waters exceeds the quality threshold for human water supply. It is worth mentioning that the water of the Llobregat River has to be treated in the drinking water facility prior becoming drinking water.

3.2 *Occurrence and Evaluation of the Contamination Loads for Emerging Pollutants*

Three classes of emerging contaminants were monitored in the Llobregat River water and in treated wastewater [9]. On the other hand, the conventional physical–chemical and microbiological contaminants, were covered by ACA. In the Llobregat River at some points, especially at drought periods, it is worth noting that WWTP effluents may represent almost 100% of the total flow of the river. Thus, relevant concentrations of organic pollutants are commonly found along the river, usually showing growing levels when moving downstream, due to the corresponding increase of WWTPs discharges and population density [19]. Moreover, the loads of sewage-borne contaminants could have been increased with the use of reclaimed waters for increasing drinking water supply. Although the reclaimed water originates from effluent wastewaters treated under efficient tertiary treatments, some compounds can be still present in the end product. Therefore, in this part of the work, we analyzed 103 emerging pollutants, including pharmaceuticals (74), illicit drugs (17), and estrogens (12), in the surface water and in the effluents pumped and discharged into the river from the WWTP el Prat de Llobregat after the tertiary treatment. Regarding pharmaceuticals in different samples, 58 compounds out of 74 target analytes were detected at least in one sample. In the treated effluent site RW2 (see map for location of the site in Fig. 1), levels of pharmaceuticals were higher than those detected in the surface waters, but still below 1,000 ng/L for the majority of the compounds with the exception of three antibiotics (azithromycin, erythromycin, and sulfamethazine) and the diuretic drug furosemide. Other compounds such as atenolol, bezafibrate, codeine, enrofloxacine, fenofibrate, metronidazole, and ofloxacine were detected occasionally at levels from 100 to 900 ng/L. At the site R0, located in the surface water upstream of the discharge of the tertiary effluent, three analgesics and anti-inflammatory drugs (acetaminophen, ibuprofen, and salicylic acid) were detected at levels higher than 100 ng/L, rarely exceeding 300 ng/L. Also other compounds were detected frequently, such as atenolol, carbamazepine, clarithromycin, erythromycin, metropolol, lorazepam, and sulfamethoxazole. Moreover, 39 compounds were detected at low ng/L levels. In general, at the site R2 the increase of concentrations was not significant for most of the compounds detected, with the exception of acetaminophen, diclofenac, erythromycin, and sulfamethazine, which were detected at levels higher than 100 ng/L or even higher for salicylic acid (500 ng/L) [9]. In order to have a better approaching on the relative influence of both the river upstream and the effluent discharge on the overall amount of pollutants detected downstream in the river, loads expressed as mass flows in mass/time units were calculated. In contrast to the direct comparison of the concentrations, the calculation of the mass loads provides a more convenient method to quantify and compare the relative contribution of each stream (river and effluent). The mass loads were calculated as the product of concentrations of emerging pollutants per flow in the three sampled sites for emerging pollutants analysis (R0, R2, and RW2). The loads for the site R1 were calculated by the sum of the load river upstream (site R0) and the load from the treated effluent (RW2) since R0 and RW2 were really close (ca. 0.2 km) so that in such short part of the river other contributions were deemed negligible. Regarding pharmaceutical pollution, the calculated loads of the effluent contribution (RW2) seem to predominate over that of the river upstream (R0) accounting for 40–65% of the total load (see Fig. 2). Comparing the results between the first and the second campaign (see section reclaimed wastewater experiment in the Llobregat River for further details), the contribution of the pharmaceuticals from the latter campaign was slightly higher than that reported in 2008 [20] possibly due to lower effluent dilution ratios. The highest effluent contributions were observed with the lowest effluent dilution (1:1) which corresponds to the last sampling date (Fig. 2).

In contrast, the relative contribution of the effluent to the total load of illicit drugs in the river downstream of the discharge point was lower than that of the river upstream (between around 30% and 50%), but, as for pharmaceuticals, this contribution was observed to increase to approximately 60–70% in the second campaign (see Fig. 3), which again could be explained by the comparatively lower river discharges. This important effluent contribution in the 2009 campaign led to increasing levels of illicit drugs downstream the river, i.e., to higher levels in R2 than in R0 (except in two sampling dates where overall illicit drug concentrations were practically the same). As compared to the previous campaign, both the number of illicit drugs detected and the levels measured in the 2009 campaign were lower. Of the five classes of illicit drugs measured, cocainics followed by amphetamine-like compounds and opiates were the most ubiquitous and abundant compounds in river water where cannabinoids and LSD and its metabolites were not detected at all. In treated wastewater the profile found was somewhat different, with a predominance of amphetamine-like compounds over opiates and cocainics (the latter are the drugs best removed in WWTPs). Of the 17 drugs of abuse analyzed, only 7 were detected namely amphetamine, ephedrine, ecstasy, methamphetamine, cocaine and its metabolite benzoylecgonine, and the opiate morphine. Individual drug concentrations did not surpass 40 ng/L in river water (the maximum concentration of benzoylecgonine was 39.6 ng/L in R0) and 80 ng/L in treated wastewater (maximum concentration for ephedrine was 79.5 ng/L).

In the case of estrogens (Fig. 3) only two free compounds, namely, estrone and diethylstilbestrol, were detected. Unlike pharmaceuticals and illicit drugs, estrogens were observed to decrease downstream the river, with levels in R0 (up to 15 ng/L) considerably higher than in R2 (below 2 ng/L and in two cases below method detection limits). In treated wastewater, only one out of the six samples analyzed was positive for estrogens (2.30 ng/L estrone in the last sampling date), thus, except in this case, contribution of the WWTP effluent to the total load of estrogens in the river downstream the discharge point was zero. Estradiol, estriol, ethynyl estradiol, and all conjugated estrogens (glucuronides and sulfates of estradiol, estriol, and estrone) were not detected in any sample. As in the case of illicit drugs, both the number of estrogens detected and the levels measured were lower in 2009 than in 2008.

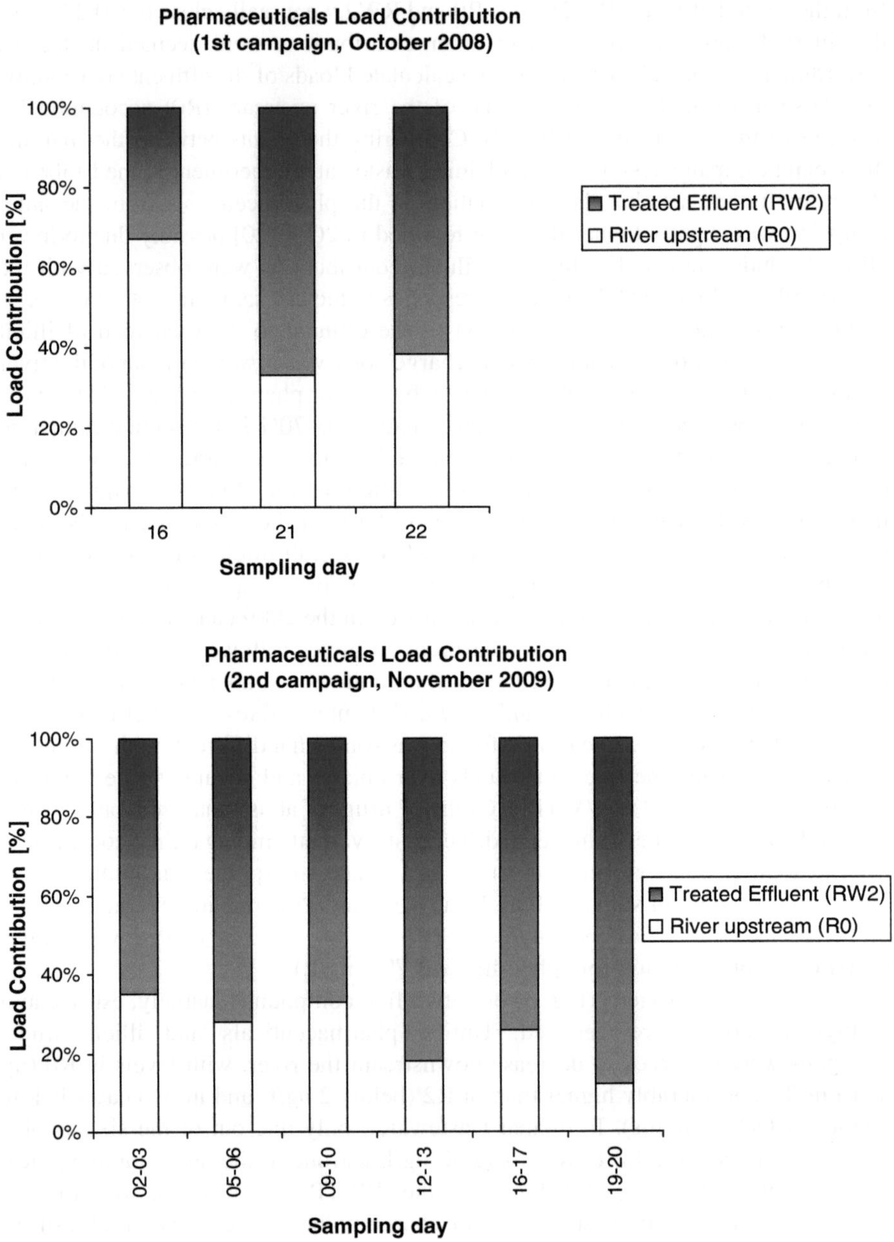

Fig. 2 Relative contribution (in percentage) of the river (R0) and the treated effluent (RW2) to the total mass load of pharmaceuticals in site R2 at different sampling dates of the two sampling campaigns (2008 and 2009)

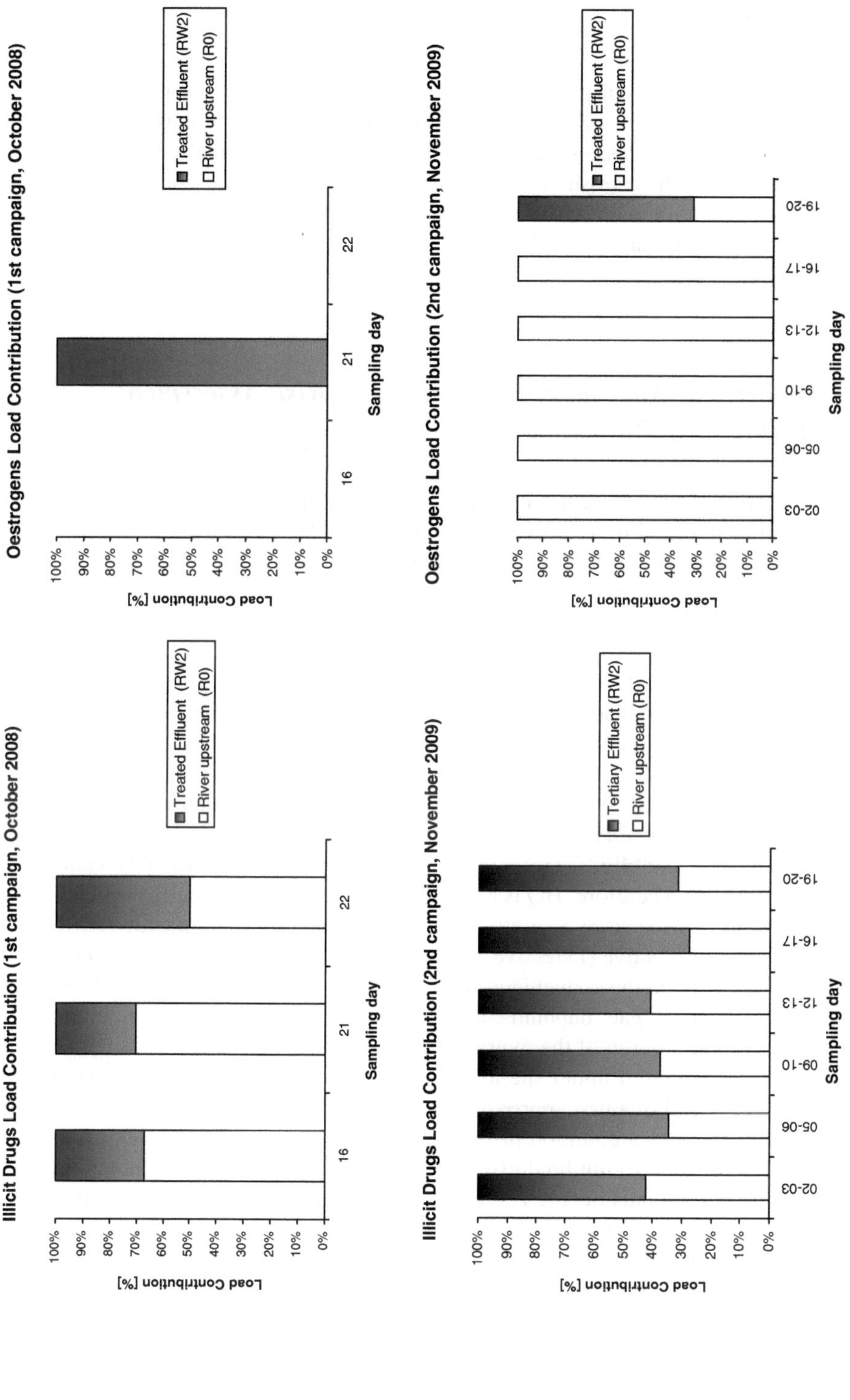

Fig. 3 Relative contribution (in percentage) of the river (R0) and the treated effluent (RW2) to the total mass load of illicit drugs and oestrogens in site R1 at different sampling dates of the two sampling campaigns (2008 and 2009)

4 Evaluation of Effects of the Reclaimed Water on the River Water

In this section, the effects of the reclaimed water discharge on the Llobregat River for the second experiment were assessed using both theoretical and experimental approaches. First, we assessed the risk of the emerging pollutants by computing hazard quotients for three freshwater organisms based on their presence and acute toxicity data. Second, the effects of the introduction of the reclaimed water on the microbial community and the biological status of the Llobregat River were evaluated.

4.1 Theoretical Approach: Environmental Risk Assessment of Emerging Pollutants

The main issue of the presence of drugs in the environment is that they are inherently bioactive and they can cause adverse toxicological effects to the environment and to humans through drinking water or food intake. While it is impracticable to evaluate the effects of exposure directly to humans and to a lesser extent to aquatic organisms, other nontarget organisms (bearing similar receptors or biomolecules to humans) are used for such purposes and generate ecotoxicological data for single compounds or mixtures to a nontarget organism. Using the ecotoxicological data the ecological risk of a target analyte can be estimated calculating the so-called hazard quotient. The hazard quotient (HQ) is basically the ratio of the exposure estimate (predicted environmental concentrations, PEC) to a "no adverse effects level" considered to reflect a "safe" environmental concentration or dose (predicted no-effect concentration, PNEC). Instead of PECs the measured environmental concentrations (MECs) are commonly used, and the acute toxicity values are used for PNECs. Therefore, HQ is MEC/PNEC. PNECs are typically calculated from EC_{50} values corrected by a safety factor of 1,000 as it is recommended by the Water Framework Directive (Directive 2000/60/EC). The Water Framework Directive [21] acknowledged the convenience of assessing the toxicity using three levels of the trophic chain. Algae, daphnids, and fish are usually employed as reference organisms. The calculation of the overall HQ for a mixture of compounds ($HQ = \sum HQ_i$) can be calculated under the assumption of concentration addition (CA) mode of action [22]. For a first approximation to estimate the toxicity of a mixture CA can be accepted as a common principle model. Thus, this entails that all the components contribute to the final effect and that neither synergic nor antagonic effects among compounds occur. Therefore, it is typically accepted that if HQ is equal or higher than 1, potential effects to the aquatic ecosystems are probable to take place. In order to take into account that chronic toxicity at lower concentrations of pharmaceuticals and illicit drugs can take place, a safety factor of 1,000 was applied in the calculation of the PNECs. EC_{50} values were extracted from the literature or they were estimated with USEPA's ECOSAR (Ecological Structure Activity Relationships, ECOSAR v1.00) model. For some compounds, more than one value was found; therefore, the lowest one was taken into account.

4.1.1 Environmental Risk Assessment of Emerging Pollutants

Table 1 shows the overall hazard quotients calculated for pharmaceuticals and illicit drugs in R0 and R2 river sites. It is worth to mention that only two compounds (ciprofloxacin and sulfamethoxazole) individually showed HQ above 1 (see Fig. 4). The overall HQs were in general low for both pharmaceuticals and illicit drugs, although in the case of algae the pharmaceuticals detected might represent a risk (overall HQ always above 1 in both R0 and R2). In general the HQ values were slightly higher for pharmaceuticals than for illicit drugs, in site R2 than in site R0, and in algae than in daphnids and in fish.

In the case of estrogens, the environmental concern focuses on their estrogenic activity, which can result in effects, such as feminization, in exposed organisms. For assessment of this risk estradiol equivalent concentrations (EEQ) were calculated for each river sample (R0 and R2) by multiplying the concentration measured of each detected compound by its corresponding estradiol equivalent factor (EEF, defined as the EC_{50} of the compound relative to the EC_{50} of 17β-estradiol) and applying the simple additive approach. In spite of the low levels of estrogens detected, EEQs above 1 were obtained for some samples from both R0 and R2, and reached 14 ng/L in the sample containing 12.2 ng/L of diethylstilbestrol (one of the most potent estrogens known with an EEF of 1.1), which alert about potential estrogenic effects in the aquatic ecosystems.

4.2 *Experimental Approach*

Introducing reclaimed water with chemical and biological contamination in the river can cause microbiological and ecological damage. Therefore, in this section the effects of the reclaimed water on the microbiological and ecological status are reviewed.

4.2.1 Effects of the Reclaimed Water on the Microbiological Quality of the River Water

Pathogens constitute recognized health hazards associated to municipal and animal related industries water reuse. However, present day reclamation technologies are able to control and reduce the risk posed by these hazards down to acceptable levels.

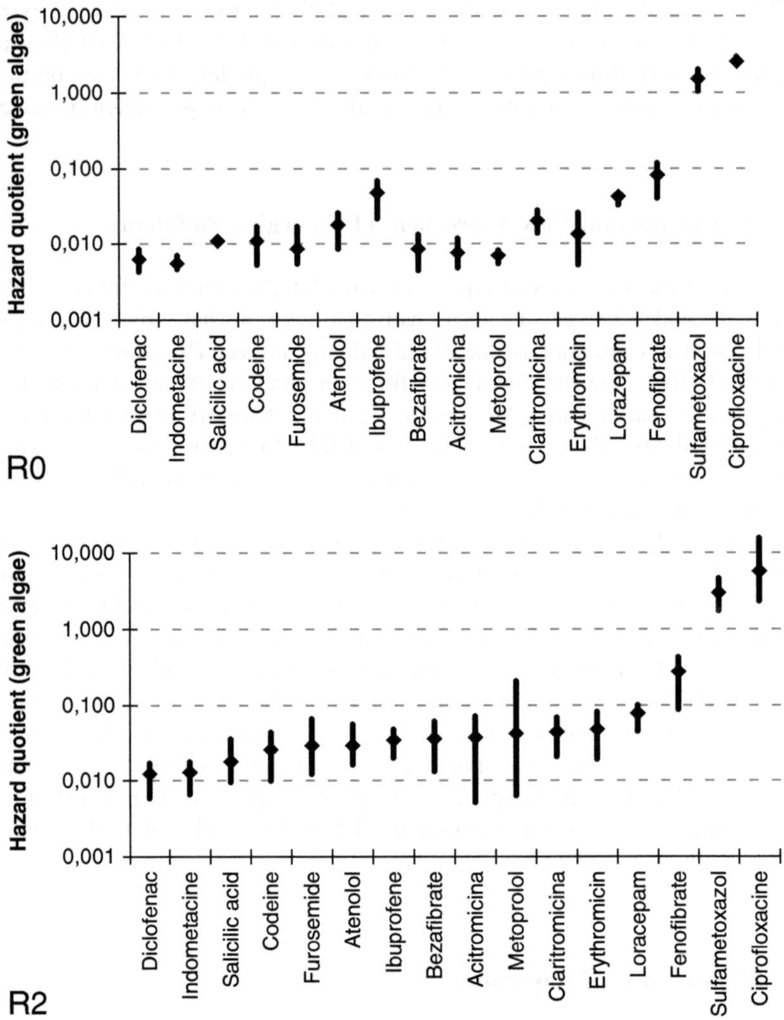

Fig. 4 Hazard quotients calculated with the detected concentration and the calculated toxicity of green algae for selected pharmaceuticals in R0 and R2 sites

Regarding the possibility of augmenting drinking water supplies with reclaimed water, a pair of reports [23, 25] recognize the feasibility of using reclaimed water to supplement drinking-water sources, but recommend adding viruses and protozoa to traditional, bacterial indicators based, microbiological water quality controls in order to better estimate the health risks linked to water-borne infectious diseases.

Pathogens and Fecal Indicators

E. coli, which is the traditional bacterial indicator, spores of sulfite reducing clostridia as potential indicators of oocysts of protozoa, somatic coliphages as potential indicators of viruses, cytopathogenic enteroviruses, and total and infectious *Cryptosporidium* oocysts were determined in sampling points RW1, which is reclaimed water delivered by the treatment plant; RW2 that is reclaimed water discharged in the river after transport upstream through a 15.6 km long pipe, R0, R1 and R2.

Methods used were either the standardized ones or the more sensitive methods available to detect "still alive" indicators and pathogens. Two different tertiary treatments were applied to the secondary effluent of a WWTP (biological-activated sludge) during the follow up of the effect of discharge of reclaimed water in the river. First one included physic-chemical (flocculation), microfiltration (10-μm pore size membranes), UV irradiation (medium pressure lamps) and chlorination; whereas the second one omitted chlorination.

The concentrations of indicators in reclaimed water were significantly different depending on whether the water was chlorinated or not. After chlorination only spores of clostridia were detected. But even in the non-chlorinated reclaimed water the values of indicators were quite low, with the following geometric means per 100 mL: *E. coli* 15 colony forming units (CFU), spores of sulfite reducing clostridia 50 CFU and somatic coliphages 40 plaque forming units (PFU). The cytopathogenic enteroviruses and infectious *Cryptosporidium* numbers were below the detection limit (<0.01 PFU per liter for enterovirus and <0.02 fluorescent foci per liter for *Cryptosporidium*) despite the treatment applied (Fig. 5). *Cryptosporidium* oocysts were still found in some samples regardless of the treatment, but the highest values never reached 1 oocyst per 10 L that is the value required for drinking water by the only regulation that includes [26]. The transport upstream through the pipe did not change significantly the numbers of the three indicators detected in the reclaimed water. The concentrations of indicators and pathogens in the river, upstream the discharge point (R0), were significantly higher than the concentrations in reclaimed water. Geometric means per 100 mL or river water were: *E. coli*, 700 CFU; spores of sulfite reducing clostridia, 1,300 CFU; and somatic coliphages, 7,000 PFU, with the lowest value for each parameter being higher than the higher values in reclaimed water only treated with UV. Low, ranging from 1 to 5 per liter, but measurable values of both studied pathogens were found in the great majority of the samples (Fig. 5). Numbers of indicators and pathogens determined in river water sampled downstream the discharge point (R1) were either significantly lower, *E. coli* and coliphages, or nonsignificantly different, spores of clostridia, enteroviruses and *Cryptosporidium*, than those upstream the discharge point. No significant differences related to the different dilutions between river flow and reclaimed water (3:1 ($Q_{RW2}/Q_{R2} = 0.25$), 2:1 and 1:1 dilutions) were detected. Thus, it can be concluded that the release of reclaimed water in the river does not impair the microbiological quality of the river water used as source water by the Sant Joan Despi Drinking Water Facility.

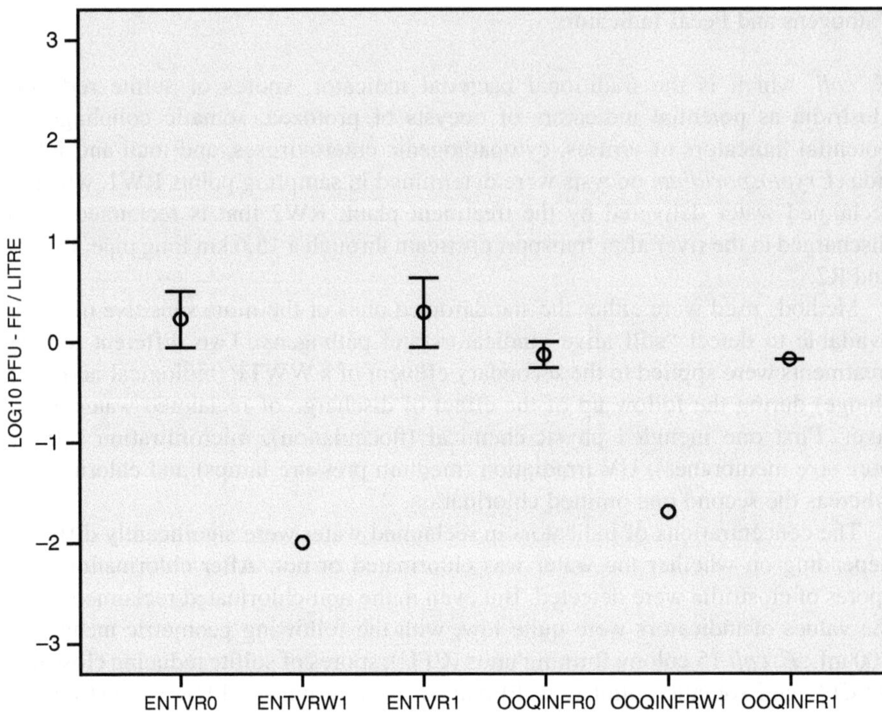

Fig. 5 Log10 values of the concentrations of infectious pathogens (mean values and 95% confidence levels) in reclaimed water (R2 samples) and river water samples (R0 and R1). Values of enteroviruses are expressed as plaque forming units (PFU) (ENTVIR) and infectious *Cryptosporidium* (INFOOQ) as fluorescent foci (FF) per liter

These results indicate that discharging reclaimed water into the river does not increment the microbiological health risks associated to the use of Llobregat river water as source water. Then, in emergency situations, at least from the infectious diseases point of view, the risks of augmenting drinking water supplies with reclaimed water can be satisfactorily and safely managed.

4.2.2 Effects of the Reclaimed Water on the Ecological Status: Macroinvertebrate Community Assemblages and Biomarkers

The reclaimed water discharged into the lower part of the Llobregat River did not produce significant changes in the structure and composition of macroinvertebrate communities [27]. The number of taxa is similar in both sites, before and after the reclaimed water discharge, with some reduction of densities downstream. Some of the differences for biological quality indices analyzed (IASPT, ICM-Star, and IMMi-T) between R0 and R1 were found already before the reclaimed water dumped into the river. The lower Llobregat River already has low water quality and low abundant species even before the discharge of reclaimed water. Biomarkers used in the analysis also indicate a low risk of sublethal contamination due to reclaimed water discharge [27]. Three out of the five biomarkers measured in field collected caddisfly larvae (*H. exocellata*) were affected by the reclaimed water input. However, this seems not to produce any changes in the macroinvertebrate community structure found in the lower Llobregat River, at least in short time.

5 Conclusions

Due to the water scarcity in the catchments' area of the Llobregat River, a Mediterranean river in Catalonia, the Catalan Water Agency designed performed several experiments for the artificial recharge of this River with reclaimed water. In this book chapter, the impact of the reclaimed water on ecology, microbial communities, and chemical contamination [24] during the campaign carried out in fall 2009 is summarized. The concentrations of emerging pollutants detected in this experiment were compared with those obtained in the previous campaign which took place in fall 2008. The water quality was evaluated for priority pollutants, three (physico-)chemical parameters (ammonium concentration, conductivity, and TOC), pharmaceuticals, illicit drugs, and estrogens. In general a null or slight increase of the water quality parameters was observed. Regarding the analyzed pesticides, only one was detected at concentrations higher than the quality threshold for human water supply, although this parameter was reduced once water was treated for urban purposes. Comparing concentrations of pharmaceuticals in river water upstream and downstream of the discharge point of the reclaimed tertiary effluent, a slight increase of concentrations was observed but it was not significant. Concerning illicit drugs and their metabolites, the impact of the discharge into the river was even lower than the one observed for pharmaceuticals. In contrast to the previous campaign (fall 2008), estrogens levels were clearly higher river upstream than downstream. In order to get a better assessment of the relative contribution of the river basin upstream and discharged sewage to the burden of the different micro-contaminants, their respective concentrations and flows should be handled together, especially if one considers the low proportion of the receiving river flow relative to the discharged tertiary effluent. Comparing the calculated loads of emerging pollutants downstream at R0 site to the R2 site (upstream), it was feasible to identify the origin of the overall pollution. While pharmaceuticals and illicit drugs loads were mostly allocated to the effluent, being the relative contribution dependent on the effluent to receiving river dilution ratio, the origin of estrogens was mainly the upstream river (except at the effluent's lowest dilution).

Finally, an evaluation of effects of the reclaimed water on the river water was done. With a theoretical approach, the ecotoxicity risk for pharmaceuticals and illicit drugs was assessed. First, hazard quotients were calculated based on the concentration ranges found and the ecotoxicological data reported in the literature for three reference organisms belonging to different trophic level (fish, dapnids, and algae), as it is recommended by the WFD. Comparing the hazard quotients for the river water upstream and downstream of the discharge point, differences were appreciable but not significant. In general, no relevant risks were identified, with the only exception of pharmaceuticals vs. algae that were likely to occur both in river upstream and downstream. In a similar fashion, estrogenicity was evaluated using EEQs for the estrogenic compounds identified. In the present case, river upstream showed higher estrogenicity than river downstream, attributable to estrone and diethylstilbestrol. An experimental approach, in turn, was used for the evaluation of the effects of the reclaimed water on the microbiological and on the ecological status. Several organisms and biomarkers were used to detect the effects of the wastewater, concluding that the risks of supplementing drinking water supplies with reclaimed water can be satisfactorily and safely managed.

Acknowledgments This work has been funded by the Spanish Ministry of Economy and Competitiveness through the projects SCARCE (Consolider-Ingenio 2010 CSD2009-00065) and CEMAGUA (CGL2007-64551/HID). It reflects only the author's views and the European Community is not liable for any use that may be made of the information contained therein. SP acknowledges the contract from the Ramón y Cajal Program of the Spanish Ministry of Economy and Competitiveness. The Catalan Water Agency provided useful information on water management in the Catalan River Basin District and its water reuse programme. Full cost recovery of all analysis carried out in the lower Llobregat River and the reuse facility implementation were supported by the Catalan Water Agency.

References

1. Christensen JHB, Busuioc A, Chen A, Gao X, Held I, Jones R, Kolli R, Kwon W, Laprise R, Magaña Rueda V, Mearns L, Menéndez C, Raisären J, Rinke A, Sarr A, Whetton P (2007) Regional climate change projections. In: Solomon S, Qin D, Manning M et al (eds) The physical science basis contribution of working group to the fourth assessment report of the intergovernmental panel on climate change. Cambridge University Press, Cambridge/New York
2. Aggeli K, Kalavrouziotis IK, Bezergianni S (2009) A proposal of a treated wastewater reuse design system in urban areas. Fresen Environ Bull 18(7B):1295–1301
3. Real D (1620/2007) De 7 de diciembre, por el que se establece el regimen juridico de la reutilizacion de las aguas depuradas. http://www.Boe.Es/boe/dias/2007/12/08/pdfs/a50639-50661.Pdf (in Spanish)
4. Cazurra T (2008) Water reuse of south Barcelona's wastewater reclamation plant. Desalination 218(1–3):43–51
5. Agency A-CW (2009) Programa de reutilització de l'aigua a catalunya. Agència catalana de l'aigua. Deparament de territori i sostenibilitat. http://aca-web.Gencat.Cat/aca/documents/ca/planificacio/reutilitzacio/prac_v_3_1.Pdf, 75 pp
6. Munné APJ, Tirapu L, Solà C, Manzano A, Matia L, Martín-Alonso J, Ginebreda A, Jofre J, Lucena F, Prat N (2012) Reclaimed water discharge into lower Llobregat river. Bringing solutions to manage drought periods in a Mediterranean urban area. Environ Sci Poll Res (in press)
7. Prat N, Rieradevall M, Barata C, Munné A (2012) Assessing the effects of tertiary treated wastewater reuse on a Mediterranean river (Llobregat, NE Spain): macroinvertebrate community assemblages, biomarkers and ecological status. Ecological indicators (submitted)

8. Rubiano M-E, Augello-Barcello M, Casas-Mangas R, Jofre J, Lucena F (2011) Assessing the effects of tertiary treated wastewater reuse on a Mediterranean river (Llobregat, NE Spain), Part iii: pathogens and indicators. Environ Sci Pollut Res. doi:10.1007/s11356-011-0562-9
9. López-Serna R, Cristina P, Blanco J, Pérez S, Ginebreda A, López de Alda M, Petrović M, Munné A, Barceló D (2011) Assessing the effects of tertiary treated wastewater reuse on the presence emerging contaminants in a Mediterranean river (Llobregat, NE Spain). Environ Sci Pollut Res. doi:10.1007/s11356-011-0596-z
10. Petrovic M, Cristina P, López de Alda M, Ginebreda A, Gros M, Radjenovic J, Barceló D (2010) Water scarcity in the Mediterranean: perspectives under global change. In: Sergi S, Barceló D (eds) The handbook of environmental chemistry, Springer-Verlag Berlin Heidelberg, vol 8. pp 1978–1228
11. Farré M, Pérez S, Kantiani L, Barceló D (2008) Fate and toxicity of emerging pollutants, their metabolites and transformation products in the aquatic environment. Trends Anal Chem 27 (11):991–1007
12. NORMAN (2009) 6th eu framework program project (network of reference laboratories and related organizations for monitoring and bio-monitoring of emerging environmental pollutants). http://www.norman-network.net
13. Santos AA, Fachini A, Pena A, Deleure-Matos C, Montenegro MCBSM (2010) Ecotoxicological aspects related to the presence of pharmaceuticals in the aquatic environment. J Hazard Mater 175:45–95
14. Kuster M, Lopez de Alda MJ, Rodriguez-Mozaz S, Barceló D (2007) Analysis of steroid estrogens in the environment. In: Petrović M, Barceló D (eds) Analysis, fate and removal of pharmaceuticals in the water cycle, vol 50, Comprehensive analytical chemistry series. Elsevier, Amsterdam
15. Postigo C, de Alda López M, Barcelò D (2011) Occurrence of illicit drugs in wastewater in Spain. In: Castiglioni S, Zuccato E, Fanelli R (eds) Illicit drugs in the environment: occurrence, analysis, and fate using mass spectrometry. Wiley, Hoboken
16. López-Serna R, Pérez S, Ginebreda A, Petrović M, Barceló D (2010) Fully automated determination of 74 pharmaceuticals in environmental and waste waters by online solid phase extraction-liquid chromatography-electrospray-tandem mass spectrometry. Talanta 83(2):410–424
17. Postigo C, Lopez De Alda MJ, Barceló D (2008) Fully automated determination in the low nanogram per liter level of different classes of drugs of abuse in sewage water by on-line solid-phase extraction-liquid chromatography-electrospray-tandem mass spectrometry. Anal Chem 80(9):3123–3134
18. Rodriguez-Mozaz S, Lopez De Alda MJ, Barceló D (2005) Picogram per liter level determination of estrogens in natural waters and waterworks by a fully automated on-line solid-phase extraction-liquid chromatography-electrospray tandem mass spectrometry method. Anal Chem 76(23):6998–7006
19. Huerta-Fontela M, Galceran MT, Martin-Alonso J, Ventura F (2008) Occurrence of psychoactive stimulatory drugs in wastewaters in north-eastern Spain. Sci Total Environ 397 (1–3):31–40
20. Köck-Schulmeyer M, Ginebreda A, Postigo C, López-Serna R, Pérez S, Brix R, Llorca M, Alda MLD, Petrović M, Munné A, Tirapu L, Barceló D (2010) Wastewater reuse in mediterranean semi-arid areas: the impact of discharges of tertiary treated sewage on the load of polar micro pollutants in the Llobregat river (NE Spain). Chemosphere 82(5):670–678
21. Sanderson H, Johnson DJ, Wilson CJ, Brain RA, Solomon KR (2003) Probabilistic hazard assessment of environmentally occurring pharmaceuticals toxicity to fish, daphnids and algae by ecosar screening. Toxicol Lett 144(3):383–395
22. Bliss CI (1939) The toxicity of poisons applied jointly. Ann Appl Biol 26(3):585–615
23. Council USNR (1998) Issues in potable reuse: the viability of augmenting drinking water supplies with reclaimed water. National Research Council, National Academy Press, Washington

24. Perez S, Kock M, Tong L, Ginebreda A, Lopez-Serna R, Postigo C, Brix R, de Alda ML, Petrovic M, Wang Y, Barcelo B (2012) Waste Water Treatment and reuse in the Mediterranean Region. In: Barceló D, Petrovic M (eds). The Handbook of Environmental Chemistry, Vol 11. pp 249–294
25. Commission NW (2009) Using recycled water for drinking: National water commission position. Canberra: Australian Government. http://www.Nwc.Gov.Au
26. (DWI) Dwi (1999) Water supply (water quality) (amendments) regulations. Laboratory and Analytical Procedures. Revision 2 October 1999. Revision 3 November 2000. London
27. Commission Decision (2002/657/EC) of 12 August 2002 Implementing Council Directive 96/23/EC concerning the performance of analytical methods and the interpretation of results OJot Commission decision (2002/657/ec) of 12 August 2002 implementing council directive 96/23/ec concerning the performance of analytical methods and the interpretation of results. Off J Eur Commun (Brussels, Belgium) L221:8–36

Ecosystem Services in an Impacted Watershed

The Relevance of Hydrology, Human Influence, and Global Change

Marta Terrado, Jordi Honey-Rosés, Vicenç Acuña, and Sergi Sabater

Abstract Climate change previsions in the Mediterranean regions are associated with more frequent extreme climatic conditions, which could alter water availability and impact the delivery of ecosystem services. In this chapter, the vulnerability of hydrological ecosystem services to recently observed climatic extremes in the Llobregat River basin is assessed. Provisioning (water) and regulating services (erosion control and water purification) were quantified under mean climatic conditions and subsequently compared to their provision under wet and dry conditions. Results stress that in semiarid basins submitted to chronic human pressure, hydrological services are very sensitive to climatic extremes. Provisioning services are important in mean and wet climatic conditions and are the most threatened in dry conditions when their reduction can approach 100%. Conversely, water purification (regulating service) has its largest contribution to human well-being in dry conditions. These results constitute a lower boundary of the possible benefits provided by ecosystem services in the basin.

Keywords Climate change, Hydrological ecosystem services, Mediterranean basin, Water scarcity

---

M. Terrado (✉), J. Honey-Rosés, and V. Acuña
Institut Català de Recerca de l'Aigua (ICRA), Edifici H2O, Parc Científic i Tecnològic UdG, Emili Grahit 101, 17003 Girona, Spain
e-mail: mterrado@icra.cat

S. Sabater
Institut Català de Recerca de l'Aigua (ICRA), Edifici H2O, Parc Científic i Tecnològic UdG, Emili Grahit 101, 17003 Girona, Spain

Institut d'Ecologia Aquàtica, Universitat de Girona, Campus de Montilivi, 17071 Girona, Spain

S. Sabater et al. (eds.), *The Llobregat: The Story of a Polluted Mediterranean River*, Hdb Env Chem (2012) 21:347–368, DOI 10.1007/698_2012_150,

Contents

1 Ecosystem Services: Measuring the Value of Nature

Humans receive various benefits from the natural environment in the form of goods and services, such as food, wood, clean water, energy, protection from floods and soil erosion, among others. These benefits that humans obtain from ecosystems had been broadly referred to as "ecosystem services" [1]. The concept of ecosystem services dates back at least to the 1970s but gained importance in the scientific literature in the 1990s [2–4]. It was mainstreamed by the Millennium Ecosystem Assessment (MA) [1], which emphasized the need to understand and quantify the services provided by ecosystems in order to avoid their loss. Among the different classifications for ecosystem services, de Groot et al. [5] established four main types based on the groups previously defined by the MA (Table 1). The main difference in this categorization is that supporting (or habitat) services are limited to the nursery and genepool function and biodiversity is not recognized as a separate service.

There are important connections among nature, its preservation and destruction, human welfare, and the economic value. The *Economics of Ecosystems and Biodiversity* (TEEB) project [6] provides a framework for linking ecosystems to human well-being. Ecosystem services are generated by ecosystem functions, which in turn are underpinned by biophysical processes or structures. Ecosystem functions are thus intermediate between ecosystem processes and services and can be defined as "the capacity of ecosystems to provide goods and services that satisfy human needs, directly or indirectly" [4]. The actual use of a service is what provides benefits, which in turn can be valued in monetary terms.

Ecosystems play an important role in regulating climate since they absorb ca. half of human-made carbon emissions. Biodiversity and ecosystem services help in the adaptation and mitigation of the effects of climate change, but climate change is simultaneously altering natural systems. Thus, the continuing loss of biodiversity and the degradation of ecosystems weaken their ability to provide essential services.

Table 1 Classification of ecosystem services (after [1, 5])

Category	Description	Services
Provisioning	Services that describe the material outputs from ecosystems	– Food – Water – Fiber, fuel and other raw materials – Genetic materials: genes for resistance to plant pathogens – Biochemical products and medicinal resources – Ornamental species and/or resources
Regulating	Services that ecosystems provide by acting as regulators	– Air quality regulation (i.e., capturing dust particles) – Climate regulation – Natural hazard mitigation – Water regulation – Wastewater treatment – Erosion protection/control – Soil formation and regeneration – Pollination – Biological regulation
Habitat or supporting	Services that ecosystems provide as living spaces for animals and plants. Maintain diversity of different breeds of plants and animals	– Nursery habitat – Genepool protection
Cultural	Non-material benefits that people obtain from the contact with ecosystems	– Aesthetic: appreciation of natural scenery (other than through other deliberate recreational activities) – Recreational: opportunities for tourism and recreational activities – Inspiration for culture, art, and design – Cultural heritage and identity: sense of place and belonging – Spiritual and religious inspiration – Education and science: opportunities for formal and informal education and training

In fact, the levels of many of the benefits that human populations derive from the environment have plunged over the past 50 years, with human impacts being the root cause [1]. Further declines are projected over the coming decades because of factors such as population growth, changing land use, economic expansion and global climate change. The impacts of cumulative pressures on ecosystems may not be perceived for many years, until some tipping point is reached which catalyze rapid or nonlinear changes [7]. In order to more effectively manage ecosystem services, there is a need to understand how incremental impacts on biodiversity and ecosystems affect the goods and services they provide. For this reason, the development of methods for measuring the biophysical and economic value of natural ecosystems has been identified as a key element for their preservation.

Such methods can help integrate nature's value, often regarded exclusively as an environmental issue, into management decisions. While valuation exercises have generated criticism [8], without a mechanism for comparing management alternatives, decision makers tend to ignore the contribution of ecosystems and by default their value becomes zero.

Costanza et al. [2] estimated the value of ecosystem services at $\$33 \times 10^{12}$ per year globally. However, this estimation has been criticized because it attempted to measure the total value of the globe's ecosystems rather than the marginal value of particular sites. Nevertheless, this seminal work did bring the value of ecosystem services into focus. But yet the process for valuing ecosystems remains complex, since it requires integration of ecology and economics in an interdisciplinary framework. This is necessary so that natural resource managers can more easily incorporate nature's value into decision-making. Despite the existing attempts to place monetary values on ecosystem services, many challenges still remain, including [6]:

- Measuring the quantity and quality of services provided by ecosystems in various possible states to avoid generalization.
- Catastrophic changes in ecosystem services involve nonlinear or abrupt shifts. The location of these thresholds remains difficult to predict, as well as if these changes are reversible. The risks associated with catastrophic changes in ecosystems remain poorly quantified.
- The links between ecosystem processes and the benefits they provide vary in complexity and directness. When discussing the benefits humans derive from ecosystems, it is useful to distinguish between final services (e.g., crop provision and clean water provision) that provide benefits directly relevant for human welfare and intermediate services that serve as inputs for the production of other services (e.g., pollination and water regulation). The economic value of pollination, for example, cannot be assessed separately from that of the provision of crops. Focusing on the contribution of services to final benefits to people will avoid double counting, but can put shade on the intrinsic and nonuse value of the other benefits.

2 Vulnerability of Hydrological Ecosystem Services to Climatic Extremes in the Llobregat River Basin

Ecosystems are continuously threatened by human activities and stand to be further affected by climate change. This will impact the supply of ecosystem services that are essential for human well-being. Resulting changes can have positive consequences (like the increase in forest area and productivity), or can increase vulnerability as a result of decreasing the supply of ecosystem services (e.g., declining soil fertility or water availability, or increasing the risk or forest fires), especially in semiarid regions.

The Mediterranean region appears as one of the most vulnerable to global change [9]. The IPCC predicts that in 2050 annual average runoff will have decreased by 10–30% over dry regions at mid-arid and semiarid low latitudes, some of which are presently already water-stressed areas. Scientists generally agree that the global hydrological cycle will intensify as a result of a global increase in temperatures, rainfall concentration in shorter periods of the year, and more extended droughts, suggesting that extreme conditions will become more common [10]. Drought is considered to be a recurrent climatic phenomenon applicable to any climatic region. It can be defined as a "sustained and regionally extensive occurrence of below average water availability" [11], and when it has a structural character it becomes water scarcity [12]. Changes in water provisioning due to water scarcity or larger periods of drought have wide-range impacts through effects on other ecosystem services, including disruption of public water supply, reductions in power generation, industrial, agricultural, and forestry production, and degradation of the environment with reduced habitat for flora and fauna, among others. In financial terms the cost of drought is also substantial: in Europe, for example, droughts are estimated to cost over €13 $\times$ 10^9 in 2003 [13], and at least €100 $\times$ 10^9 over the last 30 years [14].

This chapter estimates different ecosystem services provided by the Llobregat River basin, assessing the effects that recently observed climatic extremes have on their supply. A spatially explicit modeling tool is applied to evaluate the delivery of three hydrological services – one provisioning (water) and two regulating (erosion control and water purification or nutrient retention) [5]. The Llobregat basin constitutes an example of highly populated, impacted, and severely exploited area in the Mediterranean region. This human modification of the landscape is especially evident at the lower part of the basin. The river is the main water supply for Barcelona and its metropolitan area (more than three million people). Climate in the basin is Mediterranean with a strong seasonal fluctuation in temperature and rainfall, which mainly occurs in spring (March to June) and autumn (September to December). There are important geographic differences between the upper and lower parts, as annual rainfall at the Pyrenees exceeds 1,000 mm but is less than 600 mm near the coast (see the chapter about the hydrology of the Llobregat basin from Marcé et al. in this book [43]). Three reservoirs are located in the upper part of the basin: *La Baells* (115 hm^3), *La Llosa del Cavall* (80 hm^3), and *Sant Ponç* (24 hm^3).

2.1 Description of the Selected Hydrological Ecosystem Services

This chapter describes services considered crucial in the study area, but also in many Mediterranean watersheds: water provisioning, water purification, and erosion control. In semiarid areas such as the Llobregat basin, water scarcity can constrain economic activities since water constitutes a key element in ensuring industrial development, population growth, and improvement of living standards. Because of the high pressures coming from the mentioned activities, water quality

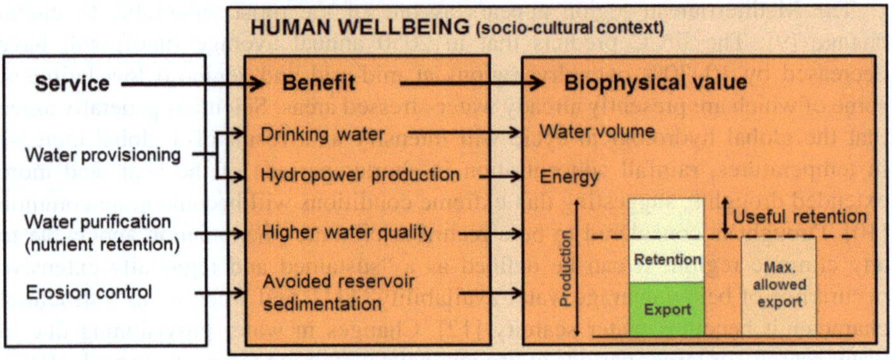

Fig. 1 Conceptual approach followed to link ecosystem services and human well-being

has become a major issue. Furthermore, erosion is a major concern in many semiarid areas worldwide and is expected to further increase owing to land use changes and increasing drought intensity and frequency. The mentioned services were estimated after the conceptual framework of Haines-Young and Potschin [15], which describes the pathway from ecosystem structures and processes to human well-being (Fig. 1). In the mentioned approach, "service" represents the supply of all possible benefits, while "benefit" involves the use of a service by humans, even though no change in actual well-being is projected. Only certain benefits were assessed for each service, in the terms defined in the "biophysical value" (Fig. 1).

Hydropower production and available water for drinking were the benefits assessed for the water provisioning service. Drinking water constitutes the most important annual consumptive demand of water resources in the Llobregat basin (65%), followed by industry (25%), agriculture (8%), and livestock (2%) [16]. Water purification is a regulating service provided by ecosystems through the retention of pollutants, ultimately preventing them from reaching the water course. The assessed benefit provided by this service is a higher water quality, defined in terms of total nitrogen (TN) and total phosphorus (TP) concentration. Erosion control is a regulating service provided by ecosystems through sediment and soil retention. The hydrological-based benefit assessed for erosion control was the avoided sedimentation in reservoirs, which can affect their water capacity and functioning for hydropower generation. We are aware that the selected services represent only a part of all the delivered hydrological services in the basin. Therefore, we are not aiming to quantify the total value of services in the basin, but only the effects of climate on the three considered ones.

The mentioned services were calculated at the annual scale for average and extreme climatic conditions. Average conditions of rainfall and evapotranspiration were obtained from Ninyerola et al. [17] for the period 1951–2000, and extreme (dry and wet) climatic conditions were obtained from Llebot [18] in a study covering a 30-year period (1971–2000). Dry and wet conditions were calculated by averaging the rainfall and evapotranspiration data of the five driest and wettest years of these series. The impact of climatic extremes on the provisioning of ecosystem services was then assessed by comparing their provision under average conditions against those under extreme conditions.

2.2 Integrated Valuation of Ecosystem Services and Tradeoffs

InVEST, or *Integrated Valuation of Ecosystem Services and Tradeoffs* [19], was used to assess the impact of climatic extremes on the selected hydrological services in the Llobregat basin. InVEST is a spatially explicit tool consisting of a suite of models that use land use and land cover patterns to estimate levels and economic values of ecosystem services. The model runs in a gridded map at an annual average time step, and results can be reported in either biophysical or monetary terms, depending on the needs and the availability of data. Information requirements and outputs for the modeling of the selected services are given in Table 2 and described below.

Water provisioning or the relative contribution of the different parts of the landscape to water supply was based on the land use patterns affecting annual surface water available for drinking and hydropower production. The amount of water provisioned from each cell in the landscape or water yield (Y_{jx}) was calculated as the annual amount of rainfall that did not evapotranspire (*Service*; Table 2). The evapotranspiration term represented both evaporation due to climatic demand and transpiration determined by the cell vegetation characteristics [20, 21]. In this way, land use and land cover played a crucial role in determining the amount of water arriving downstream. No surface–groundwater interactions or temporal dimension of water supply was considered in the model. Only the fraction of water from every cell available for drinking purposes was considered, obtained as the remaining water fraction after removal of demand for other consumptive uses [16] and the volume corresponding to the environmental flow required at the outlet of the basin [22]. Water from every cell nonavailable for drinking was called u_x. The contribution of each cell to the total water yield available for drinking (c_x) was then determined by (1):

$$c_x = \frac{Y_{jx} - u_x}{V_{\text{out}}} \tag{1}$$

where V_{out} was the total amount of water reaching the outlet of the watershed (or point of interest). Available water was then compared to the drinking water demand, since only the actual use of water was considered a benefit (*Benefit*; Table 2)

The same calculation was performed for hydropower production, previously subtracting the amount of water devoted to any consumptive use from every cell (u_z), since it was considered nonavailable for energy generation (*Benefit*; Table 2).

Table 2 Data requirements and outputs for the selected ecosystem services

ES	Step	Data requirements	Process	Output
Water provisioning – drinking water	Service	DEM (m) Land use/land cover (LULC) Effective soil depth (mm) Average annual rainfall (mm) Average annual reference evapotranspiration (mm) Plant available water content ([0, 1]) Maximum root depth (mm) Evapotranspiration coefficient Zhang coefficient ([0, 10])	Calculates cell level yield as difference between rainfall and evapotranspiration	Annual average water yield (mm year)
	Benefit	Consumptive use by LULC (m^3 year) Watershed above the point of interest	Subtracts water consumed for other uses and identifies points of extraction	Annual average water yield available for drinking purposes (mm year)
Water provisioning – hydropower production	Service	Same as for drinking water	As for drinking water	As for drinking water
	Benefit	Calibration coefficient Turbine efficiency (%) Reservoir fraction for hydropower (m^3 year) Average annual head (m)	Estimates power generated by water available for hydropower	Energy production (kWh year)
Water purification – NP retention	Service	DEM (m) LULC Soil depth (mm)	Calculates nutrient export and retention	Nutrient export (kg year) Nutrient retention (kg year)

		Water yield (mm, output from water provisioning) Export coefficient (g ha^{-1} year) Nutrient filtration efficiency (%)		
	Benefit	Allowed level of nutrient pollution (kg year) Watershed above point of interest	Subtracts retention equal to amount of maximum allowed level and identifies treatment facility locations	Nutrient retention for water quality (kg year)
Erosion control	Service	DEM (m) LULC Rainfall erosivity (*R*) (MJ mm ha^{-1} h^{-1} year) Soil erodibility (*K*) (Mg ha h ha^{-1} MJ^{-1} mm^{-1}) Crop factor (*C*) Management practice factor (*P*) Sediment retention efficiency (%) Slope threshold	Calculates sediment export and retention at each cell using USLE and routing	Annual average erosion (kg year) Annual average sediment retention (kg year)
	Benefit	Reservoir dead volume (m^3) Watershed above point of interest	Subtracts sediment equal to dead volume	Annual average sediment retention to reservoirs (kg year)

The contribution of each cell to the total water yield available for hydropower production (c_z) was determined by (2):

$$c_z = \frac{Y_{jz} - u_z}{V_{\text{in}}} \tag{2}$$

V_{in} corresponding to the inflow water into the dam used to generate energy. The amount of energy that could potentially be generated at each cell according to its water yield for hydropower (hp_z) was determined by (3):

$$\text{hp}_z = c_z \cdot 0.00272 \cdot \beta \cdot h_{\text{d}} \cdot V_{\text{in}} \tag{3}$$

where β corresponded to the power station turbine efficiency and h_{d} was the water height behind the dam at the turbine (obtained from [23]). The 100% of the water entering the reservoir was considered to be used for hydropower production. The total amount of hydropower produced was calculated as the sum of the hydropower from all the cells located within the basin's territory upstream reservoirs. Although more than 100 small hydraulic plants exist in the Llobregat basin, the lack of information about diversion concessions constrained the study to use only power stations located in reservoir systems. Seven gauge stations located along the main stem of the Llobregat River were used for calibration of the water provisioning model.

The *water purification* service, i.e., the relative contribution of the different parts of the landscape to nutrient retention, was estimated in terms of TN and TP. Although these single measures ignore many other sources, they provide a proxy for water pollution and their removal is indeed representative of the water purification service. The ecosystem function supporting this service (nutrient retention) was calculated as the difference from the received (either from upstream cells or generated within the same cell) and exported nutrients. The nutrient load produced in each cell was estimated from export coefficients, corresponding to annual averages of nutrient fluxes. Nutrient load retained by each cell was a function of the retention coefficients associated with vegetation covers [24] (*Service*; Table 2). Nutrient load production (P) was calculated as the sum of total nutrient exported and retained loads (Fig. 1). The service water purification was calculated at the point of interest (drinking water treatment plant near the outlet of the basin), where P was compared to the drinking water quality standards (drinking WQS) established by the European Drinking Water Directive 80/778/EEC [25] (Fig. 1). This was to calculate the nutrient load exceeding the quality standards that would potentially reach the treatment plant if nutrient retention was nonexistent (*Benefit*; Table 2). The drinking WQS was initially given as a concentration and transformed into loads, adopting therefore a different value in each climatic condition. Watershed retention of nutrients leading to river concentrations below the drinking WQS was not accounted as a final benefit given that no water treatment is mandatory below this limit. The amount of water purification service provided by each cell based on the retention that natural vegetation and soil provide, wp_x, was calculated as defined in (4):

$$\mathrm{wp}_x = (P - \mathrm{WQS}) \cdot c_x \tag{4}$$

where c_x indicated the contribution of each cell to total retention in the basin ($\mathrm{retention}_x$/total retention). In the case that $P < \mathrm{WQS}$ the value of the water purification service equaled zero. Additionally to the drinking WQS, ecological WQS for the so-called good environmental status (European Water Framework Directive, [26]) were also considered. TN and TP dynamics were calibrated with the estimates obtained using the FLUX software (US Army Corps of Engineers), which was in turn fed by nutrient data from a water quality station of the Catalan Water Agency at the outlet of the basin. Nutrients were initially obtained in terms of nitrate and phosphate concentrations and were subsequently transformed to TN and TP using the ratios: $\mathrm{N\text{-}NO_3/TN} = 0.75$ and $\mathrm{P\text{-}PO_4/TP} = 0.48$ [27] to make them comparable to model predictions.

The service *erosion control*, i.e., the relative contribution of the different parts of the landscape to sediment retention, was estimated considering the land use patterns affecting sedimentation in downstream reservoirs. The function sediment retention supporting this service was calculated as the difference from received (from upstream cells) and exported sediment. The amount of soil eroded from each cell was estimated using the universal soil loss equation (USLE) [28], while the retained amount of sediment by each cell was a function of the retention coefficients associated with vegetation covers (*Service*; Table 2). Only sheet-wash erosion was included in the model (no rill-inter-rill, gullies, or stream bank erosion considered). The landscape upstream the reservoir was assumed to have a maximum soil export allowance equal to the reservoir dead volume, res_deadvol, and corresponding to the volume that, when filled with sediment, impacts the reservoir function [23]. Erosion control was calculated at the point of interest (reservoirs), where total sediment production was compared to the maximum soil export allowance to obtain the sediment load exceeding the reservoir dead volume that would reach the reservoir if sediment retention was nonexistent (*Benefit*; Table 2). Only the sediment excess, calculated as the difference between sediment production and maximum allowed export, constituted the erosion control service (Fig. 1). The amount of erosion control service provided by each cell based on the sediment retention that vegetation provides, ec_x, was calculated as defined in (5):

$$\mathrm{ec}_x = \mathrm{ret}_x - \frac{\mathrm{res\_deadvol}}{n\_\mathrm{cells}} \tag{5}$$

where ret_x was the total sediment retained by each cell and *n*_cells the number of cells in the landscape upstream the reservoir. One of the parameters used in this model to calculate the erosion control service was the rainfall erosivity factor, which was obtained from the rainfall series of the Catalan Meteorological Service [29]. The erosivity factor is the only factor in the model susceptible to experience changes under different climatic conditions. Therefore, it was used for simulation of dry and wet extremes. Calibration of erosion control was done at *La Baells* headwater reservoir. Temporal changes in the bathymetry in this reservoir [30] were used. Model results were then validated with data reported by Liquete et al. [31] at the outlet of the basin.

2.3 Delivery of Ecosystem Services and Quantification of Their Contribution to Human Well-being

The approach followed for mapping and quantifying the selected hydrological ecosystem services in the Llobregat basin comprises of three parts: the calibration of the applied models (Sect. 2.3.1), the analysis of the impact of recently observed climatic extremes (dry and wet) on the delivery of ecosystem services in the basin (Sect. 2.3.2), and the analysis of the impact of climatic extremes on some of the benefits received by humans (Sect. 2.3.3).

2.3.1 Model Calibration

Model fits for the three selected hydrological services were good, especially for water provisioning (Table 3). In the initial calibration of the water supply (black dots in Fig. 2), water demands in the model were considered 100% consumptive, but this estimate under-predicted by 20% the amount of water at the river mouth. The model fit improved when return flows from waste water treatment plants (WWTPs) and industry were added to the whole calculation (white dots in Fig. 2). With this adjustment, prediction errors were always lower than 8% when considering water returns from non-fully consumptive uses. The model fit for exported TN and TP at the outlet of the basin was also good (Table 3). However,

Table 3 Observed values used for model calibration and predicted values for mean, dry, and wet conditions

Parameter	Observed	Predicted		
		Mean	Dry	Wet
Water supply ($\times 10^6$ m^3 year)	606	606	121	1,586
TN export (Mg year)	6000[a]	5,998	5,312	6,384
TP export (Mg year)	420	422	385	448
Sediment export-1 (Gg year)	200[b]	150	66	246
Sediment export-2 (Gg year)	602–1,418[c]	1,535	719	2,623

Data sources are
[a]Ludwig et al. [27]
[b]Catari et al. [30]
[c]Liquete et al. [31]

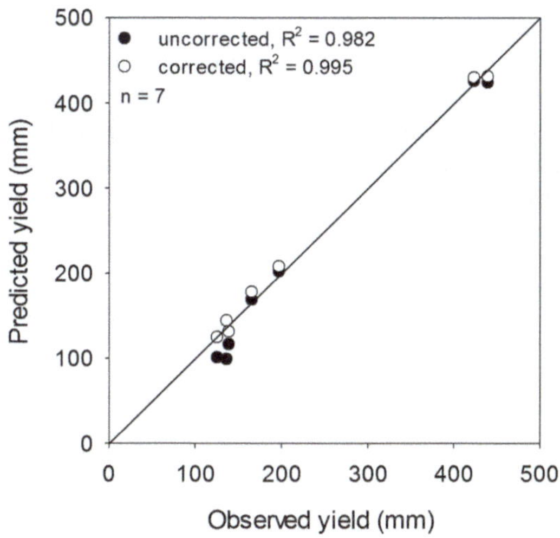

Fig. 2 Observed versus predicted annual water yield at seven different gauging stations along the mainstem of the Llobregat River, and effect of correcting the predicted discharge values by the returns from nonconsumptive uses

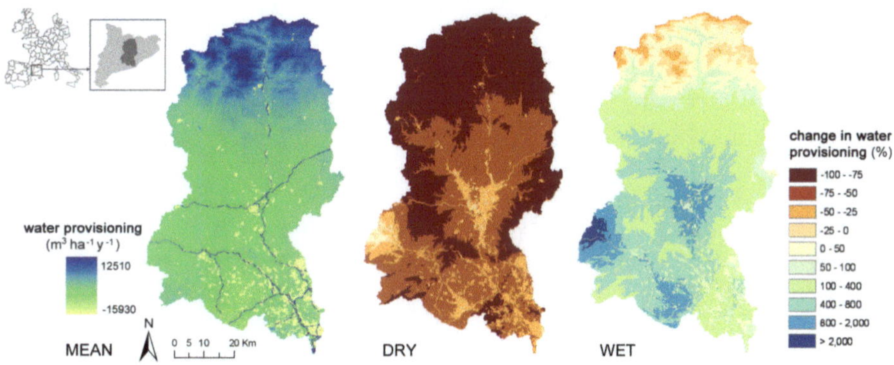

Fig. 3 Mean annual water provisioning in the Llobregat basin, and changes under dry and wet climate conditions

the model under-predicted exported loads of sediment to *La Baells* reservoir (1; Table 3) and over-predicted exported loads at the outlet (2; Table 3).

2.3.2 Ecosystem Services and Impact of Climatic Extremes

Total annual water volume at the outlet of the basin experienced 80% decrease in dry conditions and 160% increase in wet conditions (Table 3). Higher values of water provisioning occurred in high-altitude areas, characterized by important water yield production and low demand (Fig. 3). These locations were the ones experiencing the highest decrease during dry periods (from 75% to 100%, Fig. 3). However, the central and southern areas of the basin experienced lower changes during drought conditions and in some parts were estimated to exceed 800% increase in provisioning during wet years (Fig. 3). Despite the differential change between uplands and lowlands, the former continued being the water source both in wet and dry conditions.

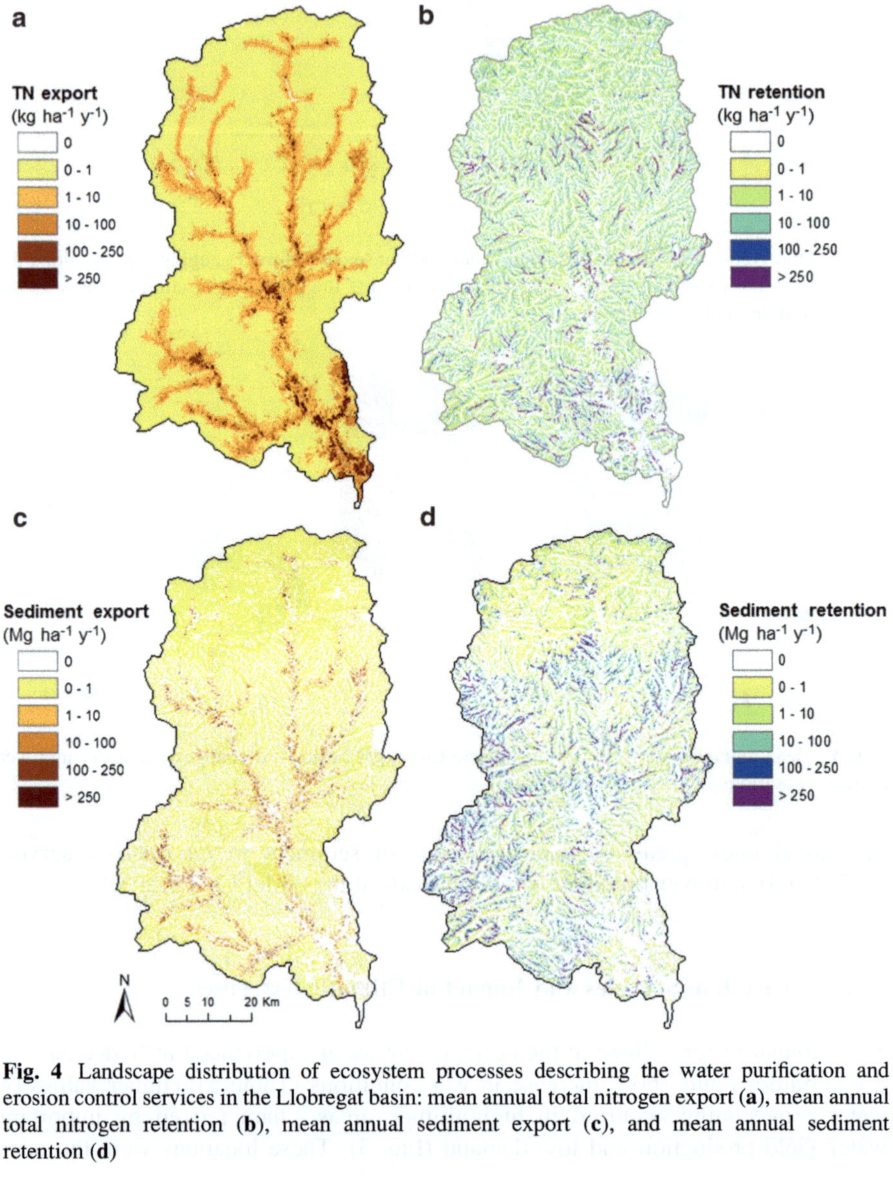

Fig. 4 Landscape distribution of ecosystem processes describing the water purification and erosion control services in the Llobregat basin: mean annual total nitrogen export (**a**), mean annual total nitrogen retention (**b**), mean annual sediment export (**c**), and mean annual sediment retention (**d**)

Water purification results need to be interpreted considering both nutrient retention and nutrient export. The sum of exported and retained loads equals to the nutrient production in the basin. The retention magnitude depends not only on the characteristics of a particular landscape unit but also on the loads entering from upstream areas to that landscape unit. The export to the stream or point of interest depends on the retention capacity of vegetation downslope. As such, landscape areas located closer to the river network presented the highest values of nutrient export to the stream (Fig. 4a). This was especially the case at the lower river course, where export from parcels closer to the river had fewer opportunities to be retained before reaching the point of interest. Nutrient retention did not follow the same pattern as that observed for the export. It was higher in flat areas contiguous to steep slopes and presented moderate values close to the river network (Fig. 4b). Patterns were similar for both TN and TP. The spatial patterns identified for sediment export and retention followed a similar behavior as that observed for nutrients. The areas neighboring the river channel produced the highest sediment export values (Fig. 4c). Sediment retention was lower at the upper part of the basin where flowpaths were shorter and there was less opportunity for a pixel to accumulate sediment from upstream sources (Fig. 4d).

Effects of climatic extremes on annual exported nutrient loads were low and closely related to changes in the water yield. TN and TP loads reaching the outlet were estimated to be 11% and 9% lower, respectively, in dry conditions, and around 6% higher in wet conditions (Table 3). Nutrient exports considerably increased in wet periods (from 50% to 200%) in the central and southern areas of the basin. The northern areas experienced an important decrease in nutrient export during dry conditions. Annual sediment export was reduced 53% in dry conditions and increased 70% in wet conditions (Table 3). The effects of climatic extremes on soil erosion were not homogeneous throughout the basin. The southwestern areas were identified as the most vulnerable to Hortonian runoff (described by the rainfall erosivity factor), where erosion increased from 75% to 100% in wet conditions. As a result of the higher erosion, the value of sediment retention also increased during the wet period.

Only processes occurring on the land phase were considered in the water purification and erosion control models. The river channel was modeled as a pipe where nutrients and sediment reaching the fluvial network were directly exported to the point of interest. However, in-stream processes are essential for the correct modeling of nutrient and sediment dynamics, since the river channel can perform as sink and/or source of nutrients [32, 33]. In the case of the Llobregat basin, around 23% of nitrates and 40% of phosphates entering the watercourse are estimated to be retained or self-purified by the Llobregat River and its tributaries (see chapter from Aguilera et al. in this book [44]). Apart from the diffuse sources of nutrients coming from the land phase, the Llobregat River is also characterized by important point discharges from WWTPs [34]. As point sources are directly discharged into the river, they are not considered in the model either, but were added a posteriori in order to avoid imbalances. Like in the case of nutrients, various studies point out the relatively large contribution of channel erosion and deposition processes to the total sediment yield [35, 36], which were not included in the applied model either.

2.3.3 Benefits of Ecosystem Services and Impact of Climatic Extremes

Drinking water provisioning in the Llobregat basin accounted for 603 $\times$ 10^6 m^3 year. This was the potential water quantity available for drinking use, while the real water demand in the basin is around 300 $\times$ 10^6 m^3 year. As supply exceeds demand, benefits correspond to the volume of demanded water. Drinking water demand remains approximately constant in the basin unless exceptional water restrictions are enforced during prolonged droughts. The highest drinking water provision was in areas of the northern part of the basin and upstream from urban areas (Fig. 5a). Northern areas function as regional "water towers" [37]. Water provisioning for hydropower production in the basin was quantified at 304 $\times$ 10^6 m^3 year, allowing the production of up to 65,270 MWh year of energy. This value corresponds to the potential energy generation in the case all available water was used for hydropower generation, but power stations are not working continuously and the actual amount of produced electricity would be lower. Areas located at higher altitudes also contributed the most to hydropower generation (>1,000 kWh ha^{-1} year) (Fig. 5b). Areas upstream of *La Llosa del Cavall* reservoir received larger values for hydropower production per landscape unit than those upstream of *La Baells*. These results reflected the double use of water generated within the territory upstream *La Llosa* for energy production, which is afterward directed to *Sant Ponç* reservoir. Water purification was quantified as 6,168 Mg of nutrients. In average conditions, the benefit obtained from this service was exclusively derived from the TN retention, since the established drinking WQS was only exceeded by nitrogen loads but not by phosphorus loads reaching the river (Fig. 5c).The annual retention amounted to 15,475 Mg of nutrients using the ecological WQS instead of the drinking WQS. In this case, both TN and TP retention contributed to the benefit obtained by the water purification service, since ecological standards were exceeded by both nutrients. Quantification of erosion control gave an annual retention of 980 Gg of sediments, which contributed to avoid reservoir sedimentation. Only the landscape upstream of the reservoirs was assessed for benefits in terms of avoided reservoir sedimentation. Again, the landscape upstream *La Llosa del Cavall* reservoir contributed to avoid sedimentation in two adjacent reservoirs (Fig. 5d).

Water provisioning for drinking water decreased to 2.4 $\times$ 10^6 m^3 year (nearly 100%) in dry conditions, while the increase in wet conditions amounted ca. 150% and reached 1,488 $\times$ 10^6 m^3 year. In dry conditions demand was not fulfilled and benefits equaled the total water provision (Fig. 6). In spite of the increase in water provisioning during wet conditions, benefits continued being 300 $\times$ 10^6 m^3 year (like in average conditions), corresponding to the demand of drinking water in the basin. Hydropower production decreased (around 98%) in dry conditions, when water provisioning allowed to generate 988 MWh year of energy, and slightly increased (around 7%) in wet conditions, when 70,133 MWh year were produced (Fig. 6). Climate extremes also affected water purification. In dry conditions the water purification service increased around 127%, retaining an annual amount of 13,996 Mg of nutrient. Retention of both TN and TP was more beneficial in dry

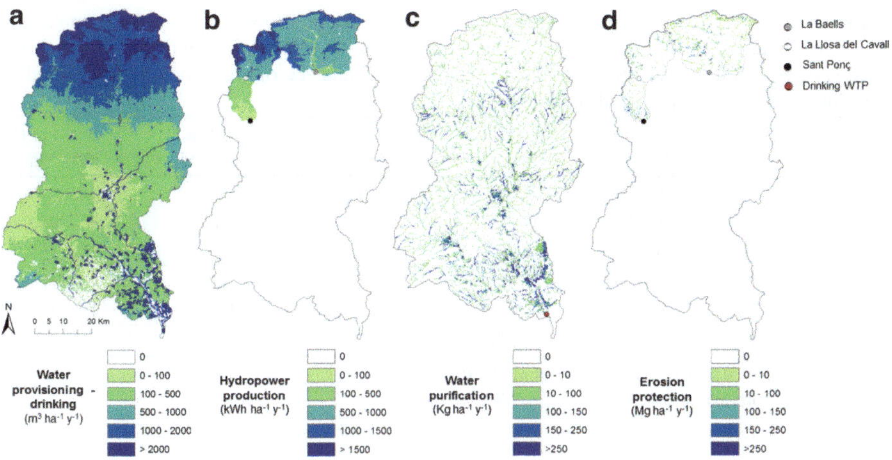

Fig. 5 Landscape allocation of annual hydrological ecosystem services provided by the Llobregat basin for: water provisioning for drinking use (**a**), water provisioning for hydropower production (**b**), water purification for drinking water (**c**), and erosion control for reservoir maintenance (**d**)

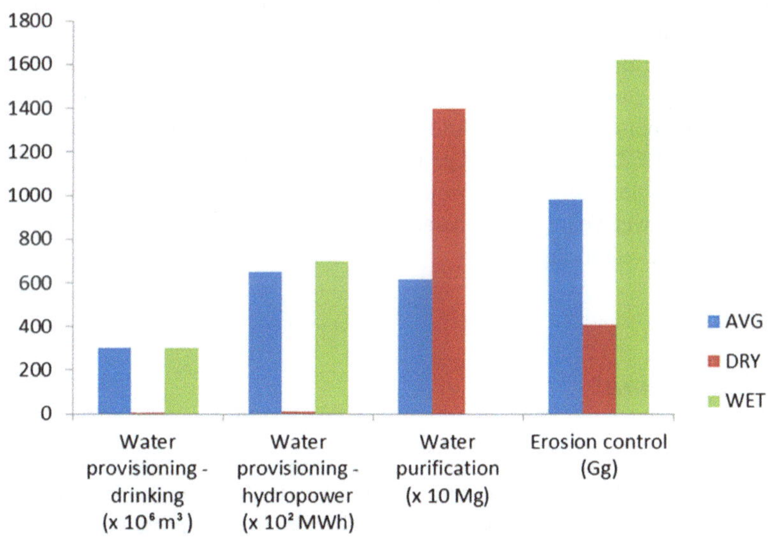

Fig. 6 Annual change in the provision of hydrological ecosystem services in the Llobregat basin under mean and extreme (dry and wet) climatic conditions

conditions because nutrient production exceeded the drinking WQS (Table 4). However, the drinking WQS thresholds were not exceeded for either nutrient in wet conditions, and therefore, the benefit was zero (Fig. 6). Despite not being problematic in drinking terms, environmental concentrations of nitrogen and phosphorus were

Table 4 Quantification of the water purification service in the Llobregat basin under mean and extreme climatic conditions according to nutrients production (export + retention) and maximum load allowed fulfilling drinking water quality standards (a) and ecological standards established by the European Water Framework Directive (b)

Climate conditions	Nutrient	Production (Mg year)	Drinking WQS (Mg year)	Service (a) (Mg year)	Ecological WQS (Mg year)	Service (b) (Mg year)
Mean	TN	16,810[a]	10,642	6,168	2,128	14,682
	TP	1,115[a]	3,215	0	321	793
Dry	TN	15,054[b]	1,596	13,458	319	14,735
	TP	1,020[a]	482	538	48	972
Wet	TN	18,943[a]	23,957	0	4,791	14,151
	TP	1,232	7,238	0	724	508

[a]Nutrient export exceeded the ecological WQS
[b]Nutrient export exceeded the drinking WQS

important since they could be sufficient to trigger eutrophication, biodiversity loss, and ultimately water quality problems [38, 39]. Ecological WQS were more restrictive than drinking WQS and, as a result, both TN and TP loads exported to the water treatment plant already exceeded ecological standards in mean climatic conditions. This meant that all the retention of TN and TP in the basin, corresponding to 15,475 Mg year of nutrient, was producing a benefit in terms of ecological water quality (Table 4). Benefits from erosion control for reservoir maintenance decreased in dry conditions (around 58%), when sediment retention amounted to 410 Gg year. The benefits increased in wet conditions (around 65%), when the amount of retained sediment was 1,622 Gg year (Fig. 6). In the Llobregat basin, dry years were characterized by less aggressive rains for which erosion was lower, and so was retention, while wet years were characterized by more intense rains that increased both erosion and retention.

2.4 *Conclusions*

The vulnerability of ecosystem services to climatic extremes was assessed with the aim of identifying the magnitude of change of key hydrological ecosystem services in extreme dry and wet conditions. Not surprisingly, results indicate that provisioning services are more relevant in mean and wet climatic conditions, while the regulating water purification service had its largest contribution to human well-being in dry conditions.

Results indicate that urgent measures are needed for areas like the Llobregat basin, which show chronic human impact and intensive water use. Either proactive or reactive management of basins should be envisaged for adapting to climate change [40]. The former includes management actions such as restoration, land purchase, and measures suitable for increasing or maintaining resilience in rivers. Reactive measures involve responding to problems as they arise by repairing damage or mitigating ongoing impacts. The reactive approach is not the most desirable strategy for anticipating and adapting to the changing circumstances of climate change. Protection of water towers in semiarid regions and regions expected to experience dramatic changes will help ensure water provisioning at the basin scale in dry conditions, that is, when the service is more at risk. Is also obvious that protection of water resources is not sufficient in the case consumption rates continue or increase in the future. Interventions to enhance the provision of regulating services should focus on areas neighboring the river network, where obtained benefits per surface area are estimated to be the highest. The aquatic phase also plays an important role in the provision of regulating services. This suggests the inclusion of in-stream processes in order to gain modeling accuracy. Overall, this study in the Llobregat basin stresses that in semiarid basins under chronic human impact hydrological ecosystem services are very sensitive to climatic extremes, and that usually service supply and demand areas show a spatial and temporal decoupling. Both aspects are relevant and need to be considered in basin management in semiarid regions.

3 Limitations and Next Steps in Ecosystem Services Research in the Llobregat Basin

The modeling approach followed here is straightforward and requires relatively few inputs. The models applied are well suited for identifying patterns in the provision of ecosystem services. Areas of high or low service production across the landscape have been identified, allowing estimation of where tradeoffs and synergies among services can occur under current conditions. However, various limitations were faced during the process of evaluation of ecosystem services in the basin. Some of them are described below.

3.1 Conceptual Approach

The conceptual approach adopted in this chapter (Fig. 1) constitutes one of the possible ways of assessing benefits provided by the selected services; however, it is not the only one. Water provisioning was evaluated according to its use for generating hydroelectricity as well as for drinking water (main consumptive use in the Llobregat basin). But other benefits are obtained in the basin from the water provisioning service, such as water for industry, agriculture, or livestock. In the case of water purification, the benefit of attaining a higher water quality was assessed, but only nutrients (nitrogen and phosphorus) were considered, while other pollutants coming from the different activities existing in the basin territory are also responsible of water quality. A complex mixture of pollutants coming from different sources has been identified in the surface waters of the Llobregat

River: pesticides, pharmaceutical products, alkylphenols, hydrocarbons, heavy metals, and chlorides from the salt mining among others [38, 41]. These products are responsible for eutrophication, acidification, salinization, and toxicity problems in the river. The benefits of avoided sedimentation in reservoirs were assessed for the erosion control service. However, sediment retention by the landscape can report other benefits to populations, such as the prevention of losses in soil fertility (since more fertilizers need to be applied in agricultural lands with high erosion), or the prevention of the increase in the cost of water treatment. In addition, services related to freshwater were exclusively focused. Others like flood mitigation or genepool protection [5] were not accounted. Therefore, estimates presented in this chapter constitute a lower boundary of the possible benefits generated by the basin.

3.2 Interactions Among Ecosystem Services

The presented results correspond to the provision of the different ecosystem services individually, while in fact services interact [42]. Often changes in one service alter the provision of another, or multiple services may respond to the same driver of change and amplify interactions. Tradeoffs arise when the provision of one service is enhanced at the cost of reducing the provision of another, and synergies appear when multiple services are enhanced simultaneously. A more realistic valuation of the ecosystem services selected in this chapter would have been obtained if interactions among water provisioning, erosion control, and water purification had been considered. However, there is still little information about when to expect tradeoffs or synergies, and the mechanisms that cause them.

3.3 Inclusion of In-Stream Processes

In-stream processes are identified to have an important contribution in the quantification of regulating services in the Llobregat basin. In order to improve obtained results, work is being currently developed to include the aquatic phase in the applied model.

References

1. Millennium Ecosystem Assessment (2005) Ecosystems and human well-being: current state and trends. Millennium Ecosystem Assessment, Washington
2. Costanza R, d'Arge R, de Groot RS, Farber S, Grasso M, Hannon B, Limburg K, Naeem S, O'Neill RV, Paruelo J, Raskin RG, Sutton P, van den Belt M (1997) The value of the world's ecosystem services and natural capital. Nature 387:253–260

3. Daily GC (1997) Nature's services: societal dependence on natural ecosystems. Island Press, Washington
4. de Groot RS (1992) Functions of nature: evaluation of nature in environmental planning, management and decision-making. Wolters Noordhoff BV, Groningen
5. de Groot RS, Alkemade R, Braat L, Hein L, Willemen L (2010) Challenges in integrating the concept of ecosystem services and values in landscape planning, management and decision making. Ecol Complex 7:260–272
6. European Communities (2008) The economics of ecosystems and biodiversity (TEEB). Interim report
7. Gordon LJ, Peterson GD, Bennett EM (2008) Agricultural modifications of hydrological flows creates ecological surprises. Trends Ecol Evol 23:211–219
8. McCauley DJ (2006) Selling out on nature. Nature 443(7107):27–28
9. Schröter D et al (2005) Ecosystem service supply and vulnerability to global change in Europe. Science 310:1333–1337
10. Hisdal H, Stahl K, Tallaksen LM, Demuth S (2001) Have streamflow droughts in Europe become more severe or frequent? Int J Climatol 21:317–333
11. Tallaksen L, van Lanen H (2004) Hydrological drought – processes and estimation methods for streamflow and groundwater. In: Developments in water sciences, vol 48. Elsevier B.V., Amsterdam
12. Sabater S, Tockner K (2010) Effects of hydrologic alterations on the ecological quality of river ecosystems. In: Sabater S, Barceló D (eds) Water scarcity in the Mediterranean. Perspectives under global change, vol 8. The handbook of environmental chemistry. Springer, Berlin, pp 15–39
13. COPA-COGECA (2003) Assessment of the impact of the heat wave and drought of the summer 2003 on agriculture and forestry. Fact sheets of the Committee of Agricultural Organizations in the European Union and the General Committee for Agricultural Cooperation in the European Union
14. Commission of the European Communities (2007) Addressing the challenge of water scarcity and droughts in the European Union. CEC, Brussels
15. Haines-Young R, Potschin M (2010) The links between biodiversity, ecosystem services and human well-being. In: Raffaelli D, Frid C (eds) Ecosystem ecology: a new synthesis. BES ecological review series. CUP, Cambridge (Chapter 6)
16. Catalan Water Agency (2002) Description and prediction study about water demand in the internal basins of Catalonia and the Catalan Ebro basins. Department of Environment, Catalan Government
17. Ninyerola M, Pons X, Roure JM (2000) A methodological approach of climatological modeling of air temperature and precipitation through GIS techniques. Int J Climatol 20: 1823–1841
18. Llebot JE (2010) Second report of climate change in Catalonia. Institute of Catalan Studies and Catalan Government, Barcelona
19. Tallis H, Polasky S (2011) Assessing multiple ecosystem services: an integrated tool for the real world. In: Kareiva P, Tallis H, Ricketts TH, Daily GC, Polasky S (eds) Natural capital. Theory and practice of mapping ecosystem services. Oxford University Press, New York
20. Canadell J, Jackson RB, Mooney H (1996) Maximum rooting depth of vegetation types at the global scale. Oecologia 108:583–595
21. Lazzara P, Rana G (2010) The crop coefficient (kc) values of the major crops grown under Mediterranean climate. Mediterranean Dialogue on Integrated Water Management, FP6 INCO-MED Funded Project
22. Catalan Government (2006) Resolution MAH/2465/2006 through which is approved the sectorial planning of environmental flows in the internal basins of Catalonia. Department of Environment and Housing, Official Journal of the Catalan Government
23. Catalan Water Agency (2010) INUNCAT, Especial emergency plan for floods in Catalonia. Catalan Government

24. Reckhow KH, Beaulac MN, Simpson JT (1980) Modeling phosphorus loading and lake response under uncertainty: a manual and compilation of export coefficients. U.S. Environmental Protection Agency, Washington
25. European Council (1980) Council Directive 80/778/EEC of 15 July 1980 relating to the quality of water intended for human consumption as amended by Council Directives 81/858/EEC and 91/692/EEC
26. European Council (2000) Water Framework Directive 2000/60/EC of the European Parliament and of the Council of 23 October 2000 establishing a framework for community action in the field of water policy.
27. Ludwig W, Dumont E, Meybeck M, Heussner S (2009) River discharges of water and nutrients to the Mediterranean and Black Sea: major drivers for ecosystem changes during past and future decades? Progr Oceanogr 80:199–217
28. Wischmeier WH, Smith DD (1978) Predicting rainfall erosion losses – a guide to conservation planning, vol 537. U.S. Department of Agriculture, Washington
29. Catari G, Gallart F (2010) Rainfall erosivity in the upper Llobregat basin, SE Pyrenees. Pirineos Rev Ecol Montaña 165:55–67
30. Catari G, Gallart F, Josa R (2009) Quantifying soil erosion with GIS-based RUSLE in La Baells reservoir (Llobregat River basin), SE Pyrenees, Spain. Geophys Res Abstr 11
31. Liquete C, Canals M, Ludwig W, Arnau P (2009) Sediment discharge of the rivers of Catalonia, NE Spain, and the influence of human impacts. J Hydrol 366:76–88
32. Mulholland PJ (1992) Regulation of nutrient concentrations in a temperate forest stream: roles of upland, riparian, and instream processes. Limnol Oceanogr 37(7):1512–1526
33. Mulholland PJ, Hill WR (1997) Seasonal patterns in streamwater nutrient and dissolved organic carbon concentrations: separating catchment flow path and in-stream effects. Water Resour Res 33:1297–1306
34. Marti E, Aumatell J, Godé L, Poch M, Sabater F (2004) Nutrient retention efficiency in streams receiving inputs from wastewater treatment plants. J Environ Qual 33:285–293
35. Poesen J, Nachtergaele J, Verstraeten G, Valentin C (2003) Gully erosion and environmental change: importance and research needs. Catena 50(2–4):91–133
36. de Vente J, Poesen J, Verstraeten G, Rompaey AV, Govers G (2008) Spatially distributed modeling of soil erosion and sediment yield at regional scales in Spain. Global Planet Change 60:393–415
37. Viviroli D, Dürr HH, Messerli B, Meybeck M, Weingartner R (2007) Mountains of the world, water towers for humanity: typology, mapping, and global significance. Water Resour Res 43. doi:10.1029/2006WR005653
38. Ricart M, Guasch H, Barceló D, Brix R, Conceiçao MH, Geiszinger A, de Alda MJL, López-Doval JC, Muñoz I, Postigo C, Romaní AM, Villagrassa M, Sabater S (2010) Primary and complex stressors in polluted Mediterranean rivers: pesticide effects on biological communities. J Hydrol 383:52–61
39. Sabater S, Sabater F, Tomas X (1987) Water quality and diatom communities in two Catalan rivers (NE Spain). Water Res 21:901–911
40. Palmer MA, Reidy CA, Nilsson C, Flörke M, Alcamo J, Lake PS, Bond N (2008) Climate change and the World's river basins: anticipating management options. Front Ecol Environ 6:81–89
41. Terrado M, Lavigne M-P, Tremblay S, Duchesne S, Villeneuve J-P, Rousseau AN, Barceló D, Tauler R (2009) Distribution and assessment of surface water contamination by application of chemometric and deterministic models. J Hydrol 369:416–426
42. Raudsepp-Hearne C, Peterson GD, Bennett EM (2010) Ecosystem service bundles for analyzing tradeoffs in diverse landscapes. P Natl Acad Sci USA 107:5242–5247
43. Marcé R, Honey-Rosés J, Manzano A, Moragas L, Catllar C, Sabater S (2012) The Llobregat River Basin: A paradigm of impaired rivers under climate change threats. Hdb Env Chem DOI 10.1007/698_2012_152
44. Aguilera R, Sabater S, Marcé R (2012) In-stream nutrient flux and retention in relation to land use in the Llobregat River Basin. Hdb Env Chem DOI 10.1007/698_2012_143

Index

S. Sabater et al. (eds.), *The Llobregat: The Story of a Polluted Mediterranean River*, Hdb Env Chem (2012) 21:369–374, DOI 10.1007/698_2012,

MIX
Papier aus verantwortungsvollen Quellen
Paper from responsible sources
FSC® C105338

If you have any concerns about our products,
you can contact us on
ProductSafety@springernature.com

In case Publisher is established outside the EU,
the EU authorized representative is:
Springer Nature Customer Service Center GmbH
Europaplatz 3, 69115 Heidelberg, Germany

Printed by Libri Plureos GmbH
in Hamburg, Germany